# THE LABYRINTH OF TECHNOLOGY

Why does modern technology succeed so brilliantly in some respects and simultaneously fail in others? While he was completing a doctoral thesis in mechanical engineering in the late 1960s and early 1970s, Willem Vanderburg became convinced that the environmental crisis and the possible limits to growth would require a fundamental change in the engineering, management, and regulation of technology.

In this volume he exposes the limitations of conventional approaches in these fields. Modern societies urgently need to rethink the intellectual division of labour in science and technology and the corresponding organization of the university, corporation, and government in order to get out of a self-destructive pattern where problems are first created by some, then dealt with by others, making it almost impossible to get to the root of anything. The result is what he calls the labyrinth of technology, a growing patchwork of compensations that merely transfer problems from one place to another. The author's diagnosis suggests the remedy: a new, preventive strategy that situates technological and economic growth in its human, societal, and biospheric contexts, and calls for a synthesis of methods in engineering, management, and public policy, and of approaches in the social sciences and humanities. He also suggests that this same synthesis can be applied in medicine, law, social work, and other professions.

*The Labyrinth of Technology* is a unique and invaluable text for students, academics, and laypersons in all disciplines, and speaks to those who are torn between the benefits that modern technology provides and the difficulties it creates in our individual and collective lives.

WILLEM H. VANDERBURG is Professor in the Department of Sociology, the Institute for Environmental Studies, and the Faculty of Applied Science and Engineering at the University of Toronto. He is the founding Director of the Centre for Technology and Social Development.

# The Labyrinth of Technology

*Willem H. Vanderburg*

UNIVERSITY OF TORONTO PRESS
Toronto Buffalo London

Toronto Buffalo London

Printed in Canada

ISBN 0-8020-4431-X (cloth)
ISBN 0-8020-8385-4 (paper)

Printed on acid-free paper

---

**Canadian Cataloguing in Publication Data**

Vanderburg, William H.
The labyrinth of technology

Includes bibliographical references and index.
ISBN 0-8020-4431-X (bound) ISBN 0-8020-8385-4 (pbk.)

1. Technology and civilization. 2. Technology – Social aspects. I. Title.

CB478.V38 2000 303.48'3 C00-931128-9

---

This book has been published with the help of a grant from the Humanities and Social Sciences Federation of Canada, using funds provided by the Social Sciences and Humanities Research Council of Canada.

University of Toronto Press acknowledges the financial assistance to its publishing program of the Canada Council for the Arts and the Ontario Arts Council.

University of Toronto Press acknowledges the financial support for its publishing activities of the Government of Canada through the Book Publishing Industry Development Program (BPIDP).

*To my senior colleagues who helped to keep the door open to this kind of research in the modern university: Ursula M. Franklin, the late James M. Ham, the late Morris Wayman*

# Contents

# Preface

It is my contention that modern economies could deliver goods and services with greater efficiency and with fewer burdens on human life, society, and the biosphere than they do at present. In making use of science and technology, economic growth employs an intellectual division of labour that first creates these burdens, and then mitigates them, if required by law to do so, and then only in ways that rarely get to the root of any problem. Consequently, modern civilization is lost in a labyrinth of technology created by its social and environmental implications. We have all been witnesses as our civilization steadily loses ground despite widespread activity intended to reduce the social and environmental burdens imposed by modern ways of life. It is time to recognize that we are trapped in the labyrinth of technology. The metaphor seems appropriate since all too often we think we have successfully dealt with an issue, only to have it reappear somewhere else in a different form. The results are falling standards of living and the ongoing decline of all living systems.

The nations of the world live in a shadow-world of economic bookkeeping that adds the costs of maintaining and expanding this labyrinth to those of the goods and services delivered and to the wealth created. The intellectual division of labour is creating a growing poverty of nations that is steadily undermining the creation of wealth. Some estimates suggest reductions in net wealth creation of up to 40 per cent during the last few decades.

This grim prognosis has spurred me to advance a proposal for turning the present situation around – namely, the creation of an approach for the engineering, management, and regulation of modern technology that proactively prevents social and environmental burdens. The evidence mounts in support of the contention that it is almost always cheaper, and certainly wiser, to prevent such burdens instead of perpetuating our current practice of first creating and then mitigating them.

It is the same shadow-world of economic bookkeeping that has failed to warn us that self-interested behaviour creates, in addition to an invisible hand, an even larger invisible elbow, particularly under the current intellectual division of labour. Adam Smith had it wrong. The last few decades, in particular, have made it abundantly clear that self-interested behaviour does what it always did: undermine communities, nations, and their ecosystems. In the long run, it threatens the ability of economies to deliver the goods and services people desire because it 'mines' the ability of communities, nations, and ecosystems to supply these economies with a healthy and capable workforce, and to provide as well as receive flows of matter and energy that can be neither created nor destroyed by economic activity. The reciprocal dependence among economies, society, and the biosphere was initially not very visible, but, as the scale of the human economy steadily grew in relation to the biosphere, the problem became recognized as the environmental crisis. However, its parallel within society still receives little attention. This book is a prescription for a set of preventive approaches to the problem of the poverty of nations.

This work is the result of a long intellectual and practical journey that began when I found myself wandering in the same labyrinth of technology while completing my doctorate in mechanical engineering. Several of us spent many hours, fuelled by even more cups of coffee, discussing the implications of the report of the Club of Rome and the growing concern over the environmental crisis. It dawned on me that my education was not preparing me for this kind of world. I did not know how to answer the simplest questions, such as: If I specify a particular kind of metal as the material from which a part should be made, as opposed to some other material such as a plastic, will I add or subtract from the burden imposed on the biosphere? It pulled the rug out from under my professional feet. How could I be an engineer with a responsibility to society, including the building of a viable technology and industry, if I could not answer these kinds of questions? I took off my iron ring and went to work.

The NATO Committee for Challenges to a Democratic Society provided me with the means to go to France in search of answers to my questions. Under the mentorship of Jacques Ellul, I studied the social sciences and humanities to discover what these disciplines knew about technology that was essential for engineering and thus should become a part of undergraduate and graduate professional education. This was a demanding task because the social sciences and humanities appear to be quite comfortable describing our modern world with minimal reference to technology, while I had come out of a 'professional culture' that had gone about its business with minimal reference to the human, societal, and biospherical contexts. No wonder civilization had difficulty finding its way out of the labyrinth of the social and environmental effects of modern technology.

I began my research with a study of professional 'cultures' in terms of what communities of specialists have in common, and how that leads them to make sense of and act on portions of the social or natural world. The writings of Polanyi and Kuhn influenced me greatly but raised many new questions. Kuhn's withdrawal from the concept of a paradigm and its replacement with that of a disciplinary matrix led, in my estimation, to a loss of understanding somewhat analogous to a dissolution of the culture of a society into its constituent elements. Deciding to make an intellectual detour, I examined how babies and young children learn to make sense of and live in the world by acquiring a culture through socialization. This led to the publication of *The Growth of Minds and Cultures* (1985), on which this work builds. Professional education may be regarded as secondary socialization into a professional 'culture,' thereby providing insight into the similarities among and differences between the ethos of a culture and that of a profession.

Next, I examined professional education, beginning 'at home,' with engineering. During their undergraduate education, what do future engineers learn about how technology influences human life, society, and the biosphere, and how can, and should, this understanding be used in negative-feedback mode to adjust design and decision making to prevent or greatly reduce harmful effects? Struggling with these questions amounts to creating windows around the perimeter of a specialty, permitting its practitioners to look out on the world and 'steer' a particular technology so as to avoid 'collisions' with its contexts. The results of my study confirmed what many of us have suspected for years: minimal use is made of context, which explains the many unwanted and undesirable social and environmental implications of modern technology despite a rhetorical dedication to keeping the public interest paramount. Next, I undertook a comprehensive comparative analysis of conventional versus state-of-the-art methods used in the engineering, management, and regulation of modern technology in the areas of materials and production, energy, work, and the built habitat to deal with social and environmental issues. It confirmed the obvious: that it is much cheaper and preferable to prevent problems than to first create and then mitigate them. An examination of other professions involved in modern technology revealed a similar minimal use of context considerations.

My research drew the attention of the former premier's Council of Ontario, which had come to similar conclusions along a different path. It appointed me to co-chair a 'round table' discussion of professional education and the long-term viability of the province. Things came to an abrupt halt when the council was disbanded by a newly elected Conservative government. I was then asked to join a team preparing to undertake a similar inquiry for reforming professional education in another country. This, too, came to a halt when one of the driving forces decided to run for high political office. However, I am convinced it will be merely a matter of time before the current economic world-view is shattered by the growing poverty of nations associated with the labyrinth of technology.

Preventive approaches operate at the level of individual technologies, and not the level of a society's technology. However, since the latter is more than the sum of its constituent technologies, I frequently refer to another analysis, to be published separately, of how technology as a whole influences human life, society, and the biosphere. The present work anticipates this analysis, which will examine the course many nations took when they began to industrialize, where this course has taken us thus far, and where it is likely to take us without a decisive intervention to ensure that our common future on this planet will be both humane and sustainable. Preventive approaches for the engineering, management, and regulation of modern technology will prove to be a necessary but not sufficient part of such an intervention. For centuries, Western civilization has sought to understand the world in terms of what we call a 'mechanistic' world-view, which amounts to conceptualizing living beings and systems in terms of dead machines. The results are obvious: we have prospered where machines excel and we have encountered serious problems where machines fail. My hope is that, as a civilization, we can get back to life.

From this brief sketch, it is obvious that this work touches on a great many disciplines and professions. As such, it differs from discipline-based research and scholarship. Therefore, there are several things I would like to draw to the attention of the reader. This work is not the result of a kind of intellectual stitching together of the findings of a great many disciplines. There is no science of the sciences that would allow for such an intellectual enterprise. Particularly with respect to the role of science and technology, the findings of the social sciences and humanities, on the one hand, and the natural and applied sciences, on the other, are not pieces from the same intellectual puzzle. Besides, no single puzzle pieces are produced by the disciplines in these 'intellectual cultures,' because each of these disciplines comprises different schools. To put it somewhat simplistically, a 'fact' has many dimensions of meaning and significance, each established in the context of a particular discipline. For each dimension, all the others are theoretical externalities with varying degrees of meaning and significance. In other words, this kind of reflective research, as distinguished from discipline-based frontier research, is essentially different from that of most scholarly works. It is a dialectical puzzling together of many findings, under the constraint that they help explain larger entities, including modern ways of life. The intellectual division of labour in science and technology creates knowledge externalities before many of them contribute to market externalities. Hence this work is an interpretive essay intended for a broad audience.

Reflective research requires a different approach to referencing. Many subjects covered in this work are treated differently than they would be from the vantage points of particular disciplines. Wherever possible, I rely on landmark works dealing with particular subjects that provide excellent entry points into the literature. In a few cases, I have relied on popular works if they provide a responsible and balanced interpretation of the literature and include entry points into it. Where this was not possible, I had little choice but to be selective since it was clearly impossible to proceed the way one does in discipline-based research. Part Four, dealing with four areas of application, is backed by a number of annotated bibliographies. The findings are expected to be developed further in separate volumes.

A second matter I wish to draw to the attention of my reader is that this kind of 'action research'–driven inquiry will inevitably undermine the conventional wisdom and beliefs that underlie current practices. For example, when I ask my colleagues or students what science cannot tell us about ourselves and our world, it is difficult to get clear answers. The same is true when I ask for a list of issues that technology cannot help us resolve. Then I begin to tease them and say: 'Does this mean that science in the domain of knowledge, and technology in the domain of action are omnipotent or, to use religious language, are all-powerful secular gods capable of doing all things for all people at all times?' We know this cannot be the case because, as human creations, they have their limitations, being useful for certain things, useless for others, and irrelevant to still others. Yet if we have a difficult time thinking about the limitations of our most powerful creations in the modern world, are we not collectively like a contractor who thinks his hammer can fix everything in a house? Would we hire such a contractor? Obviously not, and this means that this interpretive essay will have an iconoclastic dimension. Modern cultures have given a very high value to science and technology for the obvious reason that it is difficult to imagine who we would be, how we would live, and what our world would be like without them. In some sense, they have created the world of the twentieth century and the human life within it, with the result that any critical assessment is often greeted as science and technology–bashing, and therefore heretical.

I am particularly grateful for the patient and kind assistance of many people, including my French mentor, the late Jacques Ellul, who taught me the dialectical method for doing interdisciplinary research. Upon my return to Canada, I received much encouragement and support from my senior colleagues, Ursula Franklin and the late Jim Ham, who helped to create a niche within the University of Toronto where I could undertake interdisciplinary research and teaching. In coping with an overload of information, I am deeply grateful to my principal research associate, Namir Khan, whose help has been invaluable. I am indebted to Lynn Macfie and Pauline Brooks, as well as to the many student researchers who will be acknowledged in the annotated bibliographies. I owe special thanks to the mostly anonymous volunteers associated with the Ontario Audio Library and PAL Reading Services for their gift of making materials accessible on tape. Last, but not least, I owe a great debt to my wife, Rita Vanderburg, who has been my patient editor and helped me clarify my arguments. I am also indebted to the Social Sciences and Humanities Research Council of Canada for supporting this research throughout the years, and to the Humanities and Social Sciences Federation of Canada for assisting with the publication of this work. I hope this investment of public funds will help create a more viable Canada. Thanks also go to the Colorado School of Mines, which appointed me as the Hennebach Distinguished

Visiting Professor in the Humanities during 1997–8, thereby giving me a great deal of time to complete this work.

Bill Vanderburg
Toronto
1999

# PART ONE:

# PREVENTIVE APPROACHES

# 1

# Preventive Approaches as a New Technology and Economic Strategy

## 1.1 What Are Preventive Approaches?

Imagine that a group of scientists and engineers, having examined how technological development can best serve the public interest, recommends that the governments of the industrially advanced nations make healthy human lives, communities, and ecosystems their top priority. The reaction is bound to be overwhelmingly sceptical, if not downright hostile: such a recommendation would be taken as proof that these people spend too much time in the ivory towers of academe! Such people, it would be claimed, do not seem to appreciate that the results achieved in pursuit of these noble objectives would be in diametric opposition to their good intentions. Implementing such a policy would require heavy public investment, which would increase the deficit, in turn slowing down the economy and reducing wealth production, making even fewer resources available for reaching the original objectives. In other words, these people have it backwards: What society should concentrate on is making the economy grow as fast as possible in order to produce maximum wealth, which would circulate through society. This wealth would offer us new political choices because it would lead to a greater availability of resources. In the meantime, we have to be realistic in our expectations, given the difficulties being experienced by the industrially advanced nations. Anyone can understand the moral argument of groups promoting social or environmental issues, insisting that people were not made for the economy but the economy for people, and arguing that we ought to recognize our ultimate dependence on the biosphere. Of course, this is the bottom line but, in the meantime, we have to face facts and recognize our constraints. This kind of reasoning has become so mainstream that the entire political spectrum has narrowed.[1]

The economy is widely regarded as the engine that must power everything else by circulating wealth. According to this view, a society must make a host of inevitable and difficult choices – for example, between remaining internationally competitive and having a healthy natural ecology, between the productivity of labour and socially healthy workplaces and communities, between having enough energy to keep the economic engine running and the risks associated with energy production, between affordable municipal taxes and sustainable cities. In other words, our primary task is to promote technological and economic growth by maximizing desired outputs derived from requisite inputs in each and every endeavour and thereby to ensure that the national economy obtains the greatest possible 'output' of goods and services from the 'inputs' of communities and ecosystems. This strategy depends on maximizing gross wealth production, and using some of the generated wealth to pay for unavoidable costs. The latter are essentially imposed by regulations, which should, therefore, be kept to a minimum. The implication is that economic growth is essentially synonymous with human and social development. A world in which win–win situations exist for all parties, including future generations, appears entirely utopian.

From this perspective, policies that would put the quality of human life, sustainable communities, and a healthy biosphere first are unrealistic, given the many constraints in 'the system.' Such policies would push up the national deficit and thus negatively affect the very wealth from which they necessarily derive their funding. It seems, then, that the risk of increasing national deficits is supportable only to kickstart a sluggish and poorly performing economic engine, and not to create a more sustainable future.

This interpretation of our situation and the behaviour based on it is flawed for three reasons: technology is separate from society and the biosphere, its desired results are separate from the undesired ones, and technological values are separate from human values in assessing the results attained. Once this is recognized, alternative possibilities for the engineering, management, and regulation of technology, and thus for economic development, open up. These ways of separating what belongs together need a brief analysis. First, across the entire political spectrum, we are stuck with erroneous views of the role of the economy in a society and its dependence on the biosphere. Different sociopolitical groups behave as painters sitting side by side who decide to paint the same landscape but to put different things in the foreground and the background. Their paintings would provide us with no idea of what the landscape was really like. Similarly, in society there are some who put the economy centre stage; others, the community; and still others, the biosphere. Apparently, in our political, economic, and moral debates, we have a great deal of difficulty sorting out the reality of our individual and collective life. I am not talking about abstract issues, but of the understanding of the everyday reality that faces us all. Experienced reality does not come with labels indicating what is foreground and what is background: the economy, communities, and ecosystems are integral to it. There is no technology without a society, and no society without the biosphere. Since neither technology nor society can create or destroy matter and energy, both fundamentally enfold some part of the biosphere. Similarly, there is no technology without human beings, and no human beings without a society; therefore, technology is also enfolded into a society. This fundamental interdependence is distorted when the economy, society, or biosphere is placed in the foreground, as if it were separable from the others. Each contemporary sociopolitical interpretation reflects the fact that the symbolic and cultural processes by which we individually and collectively make sense of reality have increasingly come under the influence of technology. What is urgently required are interpretations of our situation that recognize that it is impossible to have a vital economy without a 'healthy' society, and that a healthy society requires a 'healthy' biosphere. Any other interpretation is as fruitless as attempting to demonstrate that a good family life depends, first and foremost, on the well-being of the father or the mother or one of the children.

There is a growing consensus across the political spectrum that, in a global economy, we have no choice but to do all that is required to ensure that the economic engine performs as well as possible. In this view, all contexts are stripped away and are represented only by the desired outputs they receive and the requisite inputs they must supply to the economic engine, in a manner analogous to the drawing of a free body diagram in mechanics. This abstraction has great utility in answering some questions, and none in answering others. Modern civilization has decided that technological and economic growth is the answer, but no longer seems to know what the question was.

Second, in our knowing and doing, in general and with respect to the engineering, management, and regulation of modern technology, and thus economic growth in particular, we have also separated the achievements of our civilization from the difficulties it has created. We no longer see the two as indissociable, stemming from the same way of life. The desired consequences of technological and economic growth are put into the foreground of our thinking and doing, while the unintended, and usually undesired, consequences are put into the background, as if they were incidental. Unhealthy communities and polluted ecosystems are as much the outcome of technical and economic development as are the latest consumer goods. The growing number of homeless people is as much an 'output' of society as is the rising GDP. Once again, what is required is a mode of thinking and acting based on the realization that modern technology is an integral part of human life, society, and the biosphere, with the result that none of the relationships between them can be marginalized without serious consequences. We are collectively behaving like an electrical engineer who refuses to accept that an amplifier cannot produce a signal without noise and fail to see that our accomplishments and failures are intrinsically related to the structure of our present world. This will become fully evident when I demonstrate that technological and economic growth are primarily guided by what I call 'performance values.' These output–input ratios, such as efficiency, productivity, profitability, cost-benefit parameters, and GDP, are used to measure success in maximizing the desired outputs obtained from requisite inputs. Undesired outputs are attended to only in so far as they violate applicable regulations and laws. Since performance values as output–input ratios are entirely mute on the question as to whether any gains are partially or wholly derived when the contexts of what is being improved (as a result of undesired outputs, for example) are degraded, what is really being maximized are gross performance ratios. The inefficiencies are not subtracted to obtain a clear picture of net performance. Furthermore, the meaning and value of desired and undesired outputs as well as requisite inputs for human life, society, and the biosphere are not being considered. Hence, the maximization of gross performance increases compares technical and economic alternatives only in terms of desired outputs and inputs, which amounts to measuring technology on its own terms rather than in relation to everything else. Because technological and economic growth are guided primarily by performance values rather than by human and social values, the structure of a modern society may, from time to time, produce genuine (i.e., net) wealth and human well-being, but this outcome is almost always accidental as opposed to intentional.

The plausibility of these claims may be reinforced by some preliminary considerations. The fact that ours is a world of scientific and technical specialization has profound implications for the engineering, management, and regulation of modern technology and economic growth. Specialists such as engineers may argue that they are competent in a particular area of technology and that any consequences of the decisions they make that fall outside of their domain of expertise are best left to other experts. Hence, problems tend to be created, and then fixed in what is usually referred to as an 'end of pipe' or 'after the fact' manner. Because of this division of labour, technology and economic growth are much costlier and much less effective than they might be. This is readily illustrated by our thought patterns. When I asked audiences made up of professionals to imagine themselves to be vice-presidents of a large corporation charged with the task of choosing among three alternative designs for a new production facility and to list in decreasing order of importance their decision criteria, most of them listed performance values first. Only marginal attention is given to the fact that the performance of a production facility fundamentally depends on the ability of its natural and societal contexts to provide matter, energy, capital, and personnel, and, in turn, to receive them in an altered state. This ability or lack of it determines whether inefficiencies exist that must be subtracted from measures of gross performance.

The inefficiencies of the 'best' design are determined by the answers to the following questions: Will the plant produce higher levels of nervous fatigue? If so, how will this 'byproduct' affect motivation, initiative, workmanship, and absenteeism? How could it affect the social relations the employees enter into at work, at home, and in their communities, and thus how might the entire social fabric of a society be affected? How will that society affect employees' daily ability to do their work? How will these and other sources of inefficiencies affect the ability of the economy to produce net wealth, in turn affecting the cost of capital, taxation levels, and those regulations that impose limits on the harm that can be tolerated? How will these and other inefficiencies affect the ability of local ecosystems and the biosphere to sustain the production facility by providing the necessary inputs of matter and energy and receiving unwanted outflows? Daily life may be thought of in terms of cycles of activities, some draining and others restoring people's energies and resources. The creativity, resources, and vitality of a workforce may be 'mined' if work is highly demanding in comparison with a society's ability to support regenerative activities (such as eating, relaxing, and sleeping) and growth (such as ongoing learning and active hobbies). If this goes on too long, the economy may no longer be able to adequately serve society. In the same vein, since the activities in the plant can neither create nor destroy matter and energy, they are necessarily part of long chains that must be supported by the biosphere. In the case of matter, this support involves cycles of transformations; in the case of energy, it involves linear chains powered by the sun. Once again, 'mining' occurs when the capacities of such cycles and chains are exceeded, thus threatening the capacity of the biosphere to sustain the human chain of activities in the long term. In sum, the human, social, and metabolic 'efficiencies' of a plant also depend on society and the biosphere, and 'inefficiencies' occur when these are unable to perform their supporting functions.

My point is that, when millions of engineering, business, and economic decisions are primarily guided by performance values, advances of one kind will increasingly be associated with problems of another kind. These problems arise from incompatibilities of one kind or another between technology, on the one hand. and human life, society, and the biosphere, on the other. This explains the current patterns associated with technological and economic growth. Our spectacular successes occur with respect to performance values, while our failures show how little weight is given to values related to context compatibility. Despite the fact that our homes, factories, and offices are filled with technologies that are supposed to save us time, many of us feel that we have less and less time for ourselves.[2] Although communications technologies facilitate human contacts, many people feel isolated and lonely, and experience difficulties in sustaining intimate relations.[3] The performance of our weapons systems has been improved to the point that their all-out use can no longer defend anyone, but only destroy everyone. The ongoing improvement in the performance of chemicals for certain tasks is undercut by their collective undermining of the health of the biosphere on which all life depends.[4] The benefits of computers may be offset by the negative impact they have on how we see ourselves as human beings, and how this, in turn, affects our relations with others and, through them, the entire fabric of society.[5] Ongoing agricultural advances permit an increased production of food, which is essential in a hungry world, but serious questions have been raised about the ability of the soils to sustain this kind of agriculture over long periods of time.[6] Our more efficient fishing methods are rapidly depleting fish stocks. Advances in automotive engineering have improved car performance, but the average speed at which cars are able to move people around many large cities continues to drop. The car is now one of the primary threats to the liveability and sustainability of our cities.[7] The advantages of automation are increasingly being offset by high levels of unemployment. In many cases, foreign aid introduces 'solutions' that are incompatible with the local way of life and the ecosystem, thus creating as many problems as it solves.[8] These examples, and many others, are instances in which increases in performance have been gained, in part, at the expense of context compatibility. The environmental crisis, the appearance of a mass society, and a variety of social and health problems are therefore directly rooted in the patterns of technological and economic growth in so far as these are primarily guided by performance values in the pursuit of something 'better.' Performance values separate desired from undesired outputs, the quantitative from the qualitative value of such outputs, gross wealth production from real wealth, growth from development, and living standards from quality of life. Yet all of these are intrinsically related to the structure of technological and economic growth.

You may say to me, 'This is an oversimplification! Engineers and business or government decision makers are swamped by regulations and constraints of all kinds, such as occupational health and safety regulations and environmental-protection legislation.' This is true, of course. Yet all specialists participating in the engineering, management, and regulation of modern technology work within job descriptions that relegate dealing with the unanticipated consequences of their actions to others who are more competent in those areas. Why risk one's professional neck when others can do a better job? From the individual's standpoint, this makes sense, but it leads to a 'system' that first creates a whole range of problems, and then sets out to fix them. This division of labour works because the tasks within the 'system' are guided primarily by performance values that assess them on their own terms, making it possible to reduce any human task to a sequence of separate and relatively independent steps. Thus, for example, engineers design a factory, a machine, or a product, or schedule deliveries, and leave any context issues that may arise to specialists (in areas such as human factors, occupational health and safety, environmental engineering, and industrial relations). This approach usually entails dealing with problems after the fact and only to the point required by law.

Of course, there are exceptions to this general pattern. Some context issues do fall within the domain of expertise of practitioners and are dealt with more preventively, but most context issues arising from a particular area of specialization tend to receive only peripheral attention or none at all.

From an individual and organizational perspective, the status quo is easily defended: 'Look here, I'm an engineer and I'll make sure the job will be done in a technically sound manner for the least cost, and the rest is up to others.' From an organizational perspective, the same argument is made: 'Sorry, this is not our business; we refer these matters to the environmental section, department, or ministry.' It is not difficult to understand how the conventional approach emerged soon after the Industrial Revolution. Rapid technological and economic development required a greater division of labour: no one could possibly stay abreast of all the new developments. People had to specialize and stick to what they were competent to do. Nevertheless, it is extremely expensive and ineffective to first create problems and then mitigate them. From the point of view of the individual specialist or organization, the conventional approach makes sense and can readily be justified. However, the overall effect of this approach to the engineering, management, and regulation of modern technology is to create a 'system' whose benefits are increasingly undercut by global problems. The quality of life of many people in the industrially advanced nations and the ecosystems they live in are in decline. Conditions in the rest of the world are even worse. What makes sense and can be justified on one level does not make any sense and cannot be justified on another. The 'system' is lost in the labyrinth of social and environmental implications of technology, which it constantly expands by rarely getting to the root of any problem. Each end-of-pipe 'solution' engenders other problems, and in our economic bookkeeping the resulting labyrinth is recorded as an increase in the output of goods and services. This makes it possible for gross wealth production to rise while net wealth production declines.

Here we encounter the fundamental contradiction in technological and economic growth. At the micro level, we find technical and economic rationality; at the macro level, technical and economic irrationality. The pattern is almost always the same: mesmerizing success in terms of performance values, and equally spectacular failure in terms of context compatibility. There is something fundamentally absurd about a 'system' that destroys the very foundations on which it depends – namely, society and the biosphere. Would we consider an electronics engineer competent if the circuit he or she had designed produced the desired signal along with so much noise as to make that signal almost useless? In fact, the 'noise' produced by technological development in the form of context incompatibilities (social and environmental pollution) is threatening many of its benefits. We should face the facts. We cannot separate the 'signal' from the 'noise' in technological and economic growth any more than we can do so for an electronics circuit. The only reason the 'system' worked for a century and a half was because the damage done to the human, societal, and biospherical commons was not economically priced or valued in any way that had an effect on the emerging 'system.' Gradually, these commons became so degraded as to require intervention through laws and regulations forbidding unacceptable harm. Unfortunately, the absurdity of the system is only gradually coming to light.

There is a third reason why the common view of technological and economic growth is inadequate. By emphasizing either performance values or human and social values, the interdependence among technology, society, and the biosphere is only partly recognized. The emphasis on performance values fails to recognize that these are not values in any traditional sense of that term. After all, what have we gained if, on an individual basis, many technologies are time-saving but we appear to have less and less time to ourselves? What does it profit us to endlessly improve the means of communication when many social relations appear to be undermined? None of the current sociopolitical views, and the policies based on them, are guided by a set of values that recombine performance and context values. The former are required to ensure the optimal use of scarce resources. The latter ensure that any technological and economic growth genuinely benefits human life and society. Together these two categories of values must ensure that growth does not occur at the expense of its contexts, which can easily happen because of the way technology, society, and the biosphere interrelate. If these three criticisms of the current views of technological and economic growth are correct, modern civilization 'steers' its most powerful creation – namely, technology – much like someone driving a car with the windshield covered over. In simple terms, it puts performance before context. The driver concentrates on the former, as indicated by the instruments on the dashboard: vehicle speed, engine r.p.m., oil pressure, engine temperature, and fuel economy. Once in a while, a scream is heard or a bump is felt. The driver responds by attempting to rip holes in the cover on the windshield. After a while, there are many small rips, but performance considerations continue to dominate. The situation is headed for a final crash that will bring the functioning of the vehicle to a halt. As a civilization, we now have many small rips in our windshield cover, allowing us the foresight to create a diversity of end-of-pipe measures, but these 'glimpses' have not resulted in our fundamentally altering the way we individually and collectively 'steer' technology into the future. Preventive approaches make a radical break with our past by attempting to remove the covers from the windshield and to steer technology according to both performance and context considerations.

I will conclude this brief introduction to preventive approaches by asking a question: if preventive approaches are superior to their conventional counterparts, why has it taken so long for them to make their piecemeal and embryonic appearance? The simple answer is that these approaches run counter to the fundamental beliefs and values of our age and of the cultures that generate and sustain them. These owe a great deal to the myth (in the cultural-anthropological sense) of progress that guided industrial societies in the nineteenth century. In essence, it made unthinkable the possibility that material advances would not automatically and inevitably produce social, and even spiritual, advances.[9] The events of the first half of the twentieth century have profoundly shaken our belief in progress, but no alternative way of making sense of our individual and collective existence has emerged. Progress is now associated with the development of high technology. Within this context, it makes sense to have technological values take priority over human and social values, since the former assure that progress. Preventive approaches challenge this prioritizing. By taking into account both the positive and the negative effects of technology, they imply a criticism of what modern societies, through their cultures, have assigned a very high value. An iconoclastic attitude to technology is essential for its regulation and control. Think of a thermostat regulating a furnace – an example to which we will return later. For the thermostat to function, it must 'criticize' the performance of the furnace in terms of the temperature to which it heats the room, and do so on an ongoing basis. This 'criticism' is converted into the useful actions of switching the furnace on and off in order to keep the temperature within a comfortable range as determined by the occupants. It is common to confuse this 'criticism' of technology with 'technology bashing' and a pessimistic outlook on life.

Modern cultures attribute one of the highest, if not the highest, of values to modern technology in recognition of the central role it plays in modern life. There is hardly a day-to-day activity that does not involve technology: waking up in time, fixing breakfast, getting to work or school, doing our jobs, amusing ourselves, staying in touch with friends and relatives, and keeping abreast of events in our communities and around the globe. So fundamental are the roles of many individual technologies that their failure or withdrawal from life poses serious threats. Our most serious problems are seen as crying out for more technology: global warming, ozone depletion, unemployment, stressful lifestyles, skyrocketing health and social costs, a growing gap between technology-rich and technology-poor regions and nations, and the breakdown of large technological systems that provide us with the necessities of life. Nevertheless, the World Commission on Environment and Development has declared that our current way of life is unsustainable because of its effects on human life, society, and the biosphere.[10] It is not surprising, therefore, that many people experience some level of ambivalence with respect to our most fundamental and important creation. Clearly, it is not omnipotent – capable of doing all things for all people in all circumstances. Technology is like any other human creation – good for certain things, useless for others, and irrelevant to still others, and we must develop the most realistic attitude towards and expectations of it. However, doing so is very difficult in a culture that has declared technology to be possibly the greatest good we know today. Hence, our deepest hopes, fears, expectations, and dreams rise up to thwart our discussions of preventive approaches. When controversies arise, we must learn to hear one another rather than to write others off as pessimists, tree-huggers, or technology bashers, and thereby legitimate the avoidance of our problems.

'Technology,' as it is used here, requires a brief explanation. During the twentieth century, technologies were developed for almost every sphere of human life. The early phases in this development of a much larger phenomenon were described by Max Weber, who called it 'rationality.' Its later phases were described by Jacques Ellul, who called it 'technique.'[11] How we engineer, manage, and regulate particular technologies within this larger phenomenon forms the paradigm for this larger constellation of technologies, much like physics constitutes the model discipline for the sciences. In other words, although our focus here is on the engineering, management, and regulation of technology in the more narrow sense of the term, our findings are directly applicable to all technologies, regardless of their current stage of evolution. The principal thesis in this book is that science and technique represent an approach to knowing and dealing with the world that makes minimal use of context, in contrast with their traditional counterparts, which, being integral to a particular culture of a particular society, make maximum use of context. It is in terms of this schema, simple as it may appear, that many aspects of modern civilization can be explained.

## 1.2 Preventive and Conventional Approaches

The conventional approach to the engineering, management, and regulation of modern technology is essentially non-preventive. This is exemplified by an ongoing study of engineering education[12] that asked two questions: (1) How much do we teach our students about the way technology affects human life, society, and the natural ecology?; (2) To what extent are they expected to use this knowledge in a negative-feedback mode to adjust engineering methods and approaches to achieve a greater compatibility with these contexts? These questions were applied to the formal curriculum through an examination of course outlines, textbooks, class hand-outs, supplementary readings, student lecture notes, project exercises, examinations, tutorials, laboratory manuals, field trips, and audiovisual materials. During the first phase of the research, only the quantitative, and not the qualitative, aspects of these questions were studied. The second phase is currently under way; in it, we are examining the quality of that information in terms of what is known at present, and the extent to which the information could have been used to modify engineering practice.

The formal undergraduate engineering curriculum comprises two components – namely, the technical 'core' and complementary studies. Examining a wide range of the most commonly used textbooks in these two components reveals that technical ones rarely refer to human life, society, and the biosphere, while the texts used in complementary studies make minimal reference to science and technology. The latter works appear to be able to describe most aspects of modern societies, such as their economic, social, political, or religious institutions and frameworks, with little or no reference to modern science and technology, even though such aspects of society are unimaginable without them. The disciplines in both components of the engineering curriculum form two groups: one in which technology almost entirely fills the intellectual field of vision; another in which technology has almost disappeared from it. Attempts to restructure the curriculum around a core of design courses that involve the synthesis of everything students have learned have had significant limitations, because the 'world' of the technical courses is incommensurate with the 'world' of the courses in complementary-studies subjects. Here we encounter in detail the fundamental problem of modern civilization described above with the analogy of the painters.

The study of engineering education (discussed in detail in chapter 3) found that almost no context information is used in the technical component of a typical North American undergraduate engineering curriculum, and that little or no reference is made to science and technology in the complementary-studies component. The former is equally true for the journal articles and books published by the faculty members teaching the curriculum. A longitudinal study of undergraduate and masters theses from 1981–92 and 1950–90, respectively, demonstrated no significant trends. The detailed quantitative study shows that engineers acquire non-prevention-oriented mindsets and professional approaches that will, in turn, influence the ethos of the profession and the 'organizational cultures' of the institutions in which they exercise leadership roles. The research instruments used in the study were also related to the goal of sustainable development. Similar results would emerge for studies of other professions, including management science, business administration, accounting, medicine, and law.

The study also shows that there is a qualitative difference between conventional engineering theory and practice, with a low context score, and current frontier methods for dealing with the social and environmental implications of modern technology that receive high context scores. In other words, the range of values measured by the research instruments does not correspond to a continuum of theory and practice. At the high end of the spectrum, we encounter what I call 'preventive' approaches for the engineering, management, and regulation of modern technology, and, at the low end, their conventional counterparts. This finding was further confirmed by an extensive analysis of the methods used for dealing with the social and environmental implications of technology in four areas of application: materials and production, energy, work, and a built habitat. The literature associated with each of these areas of application was searched, and the approaches found were scored by means of the same research instruments and were classified accordingly. The quality of the context information used was also assessed in terms of what is currently known about the subject in other disciplines. The results have been compiled in four annotated bibliographies.[13] It became apparent that conventional approaches tend to separate what I call the 'economy of technology' from the 'ecology of technology' because they generally take the form of a two-stage approach. The economy of technology strips away the contexts of human life, society, and the biosphere, leaving only the inputs and outputs that connect these contexts to the technology. The ecology of technology includes the consideration of undesired outputs, and the meaning and value of all inputs and outputs by means of which technology is embedded in, depends on, and interacts with its contexts, in so far as this is relevant to the particular issue at hand. The first stage of conventional approaches concentrates on the economy of technology on the assumption that what is perceived as the technological and economic core of any task must be dealt with first. Success is measured in terms of obtaining the maximum desired outputs from inputs (such as materials, energy, capital, labour, and expert knowledge) as measured by performance values. This accounting of desired outputs and inputs (from the perspective of the economy of technology) is necessary for the effective use of scarce resources, but it is not sufficient to ensure genuine technological and economic development. Performance values are entirely mute on whether any gain in output has been partially or wholly achieved at the expense of the compatibility between whatever is improved and human life, society, and the biosphere. This first stage does not deal with undesired outputs, nor with the consequences of inputs and outputs for human life, society, and the biosphere. These are put into the background during the first stage of conventional approaches, to be dealt with as secondary matters during the second stage.

The second stage preoccupies itself with the ecology of technology only to the point of ensuring that the undesired outputs and their implications are within the acceptable limits set out by the community or nation where implementation is to take place. Any residual implications are considered 'soft' and out of reach in a globally competitive situation. Success is measured in terms of compliance with applicable regulations. From an economic perspective, the second stage internalizes the prohibited effects on human life, society, and the biosphere, but does not question the technological and economic optimum arrived at in the first stage. Hence the two-stage process optimizes gross rather than net (i.e., real) efficiency, productivity, and profitability.

The two-stage approach is founded on a number of assumptions. The optimum achieved in the first stage is seen as unaffected by whatever regulatory end-of-pipe additions to a 'system' may be required during the second stage. These second-stage additions should be 'economic'; that is, mitigation should be required until benefits no longer exceed its costs. However, adding mitigation devices or services frequently undermines the optimum achieved in the first stage. Preventively redesigning the overall 'system' so that it no longer first creates problems and then mitigates them transforms the cost and benefit curves, now pointing to a real optimum. The first stage is assumed to focus on the creation of wealth while the second minimizes 'unavoidable' costs in the production of that wealth, which is the case only if end-of-pipe approaches are used. The production of gross wealth is thereby substituted for the generation of real or net wealth. The first stage is largely of an applied scientific nature and is assumed to be essentially objective and value-neutral. Performance values are treated as technological and economic, representing an objective system of accounting of desired outputs and inputs. The second stage minimally applies human values and political judgments mostly to the undesired inputs and outputs, thus far kept entirely in the background. These take the form of context values, such as freedom, quality of life, health, sustainability, and integrality. These should be interpreted in terms of a reciprocal interaction between any whole and its contexts without any significant damage being done to jeopardize the long-term viability of this interdependence. It is assumed, therefore, that the best way of dealing with the social and environmental implications of technological and economic development is an end-of-pipe approach, which amounts to an end-of-pipe professional ethics.[14] Implicit in this approach is a hierarchy whereby performance values are more important than context values.

I suggest that, to the contrary, the state-of-the-art methods for dealing with the social and environmental implications of modern technology try to reintegrate the economy of technology with the ecology of technology, if only with respect to one or more issues. This reintegration becomes possible, however, only when the ecology of technology is fully examined. It is this examination that informs us of where technology is within society and within the biosphere, and where it is headed, thereby enabling us to compare its future path with where it ought to be according to our goals, values, and aspirations. These frontier methods use the insights obtained from the ecology of technology in what I have elsewhere designated as 'negative-feedback mode' to adjust design and decision-making processes in order to prevent or greatly reduce its harmful effects to human life, society, and the biosphere.[15] Examples include pollution prevention, industrial ecopark design, industrial ecology, design for disassembly, design for the entire life cycle, design for environment, green product design, total environmental management, total quality management, energy efficiency, energy-demand management, integrated resource planning, healthy workplace design (based on the findings of social epidemiology), sociotechnical approaches, and healthy/sustainable city concepts. In essence, these methods 'optimize' technological design and decision making in terms of *net* measures of performance values.

Preventive approaches resemble the way we perform day-to-day activities, such as driving a car. The concept of negative feedback can be used to explain how we safely reach our destination. The skill of driving combines our knowledge of where we should be on the road with our observations of where we actually are and where we are heading, enabling us to close the gap between the two by making steering and speed corrections. In the same vein, individual and collective life in a modern society may be regarded as 'driving' our way of life into the future – an activity that frequently involves technology either directly or indirectly. Preventive approaches use the ecology of technology to 'steer' technology in a way that avoids 'bumping' into human life, society, or the biosphere. No comprehensive approach to dealing simultaneously with all these implications has yet been developed. Nevertheless, my comparative study clearly points to an emerging alternative paradigm in engineering theory and practice to be based on a comprehensive preventive approach that implies an alternative to the economic world-view described above.

## 1.3 Comparative Advantages

Three growing bodies of evidence are converging to suggest that taking a proactive approach to dealing with the human, social, and environmental implications of modern technology is more cost-effective and frequently avoids a collision with human life, society, and the biosphere. These bodies of evidence challenge the conventional engineering and economic wisdom that improving the health and viability of human life, society, and ecosystems necessarily reduces the competitiveness of corporations, communities, and nations and increases national deficits. This outcome is inevitable for the two-stage approach, but not so for preventive approaches. While this fact is acknowledged for pollution prevention, its recognition has not spilled over to other domains. Nor have the implications been recognized for the general patterns of development of the industrially advanced nations. As noted previously, the successes and failures of these patterns can readily be interpreted in terms of the conceptual framework outlined above. Most of the successes of the industrially advanced nations have been in the domain of performance values, and most of the problems fall in the domain of context values as technology 'bumps' into human life, society, and the biosphere. The argument that economic constraints necessitate the priority of performance values over context values must now be questioned.

The first body of evidence is constituted by the comparison of preventive approaches and conventional ones, and shows that the former are generally more cost-effective. For example, many pollution-prevention projects have payback periods of less than two years, and electric power can be saved (i.e., negatively generated), often for half the cost of producing it.[16] The evidence in the area of healthy workplace design and healthy cities is still scarce, largely because few attempts have been made at preventive approaches, but it would appear that similar results can be anticipated. This possibility is explored later in this work.

The second body of evidence comes from new thinking in modern economics. The patterns of development of the industrially advanced nations are justified by modern economics, which is dominated by input–output methods and performance values. This justification is increasingly coming under attack because it is recognized that its primary indicators, such as the GDP, are flawed. These indicators ignore undesired outputs, such as the costs associated with 'bumping' into human life, society, and the biosphere; as well, because they fail to subtract such costs from the wealth produced, they do not reflect true wealth. The result has been a sharp decline in human, social, and natural 'capital' because we have not been living off the sustainable 'interest.' Ecological economists and others have developed better indicators, which are only slowly gaining acceptance by international agencies, governments, corporations, and, least of all, conventional economists. Ecological economics is the essential intellectual complement to preventive approaches for the engineering, management, and regulation of modern technology. Without it, the real benefits of preventive approaches cannot be assessed. These new kinds of indicators also partly reintegrate the economy with the ecology of technology. For instance, one study shows that, since 1970, although the GDP has consistently risen, other indicators show a marked decline, thus confirming the experiences of many people, communities, and nations. For example, the 'genuine progress indicator' (GPI), developed as an alternative measure in this study, shows a gradual decline of roughly 45 per cent since about 1970.[17] We may well have reached a situation where the rate of increasing wealth production, resulting from technological and economic development, is now outstripped by the rate of increasing costs incurred in the production of that wealth. This would go a long way towards explaining our present structural economic difficulties.

The third body of evidence derives from studies of professional education such as the study of engineering discussed above. The patterns of development of the industrially advanced nations and those of the professions involved in the evolution of technology reflect and support one another. This has obvious positive consequences, but also some negative ones. Professional education is less interesting and relevant than it might be; women are underrepresented; and, in the eyes of the public, prestige and status are slipping. The previously discussed examination of undergraduate engineering education shows that it continues to be largely based on the two-stage approach, thus leaving the vast potential for preventive approaches largely untapped. It is not a question of conventional approaches having an as yet undeveloped area that can now be filled with pollution prevention or other prevention-oriented courses. Conventional approaches deliberately exclude social and environmental considerations and delegate them to other specialists, who then have no choice but to treat them in an end-of-pipe fashion.

## 1.4 The Economy and Ecology of Technology

Our findings thus far may be summed up in terms of the urgent need to make our present way of life more sustainable. Despite the controversy over the concept of sustainable development, there exists a wide consensus that modern technology imposes too great a burden on its contexts. The process of sustainable development may thus be operationalized as one that continuously reduces the burden individual technologies impose on human life, society, and the biosphere. Preventive approaches appear as the common-sense answer to our current predicament: they are able to make corporations more profitable while reducing the social and environmental problems faced by the societies in which they operate. How was it possible not to see this for nearly two centuries?

To answer this question and institute preventive approaches, we have to examine the split that occurred some time ago between the *economy of technology* and the *ecology of technology.* The concept of an economy has long historical roots that go all the way back to the beginning of Western civilization in Greece. It focused on those aspects of individual households, or even the collective human household on earth, that could be counted. The concept takes the perspective of an accountant quantifying, for example, the basic needs of a household in terms of inputs (e.g., the number of bags of wheat or corn, items of clothing, linens, pots, or furniture) or the outputs of productive work (e.g., the number of metres of woven cloth or litres of olive oil). Managing these aspects of a household was not regarded as very honourable work by the Greeks, and was thus frequently left to slaves. What was more important was the role these inputs and outputs played and should play in a household, and in the lives of its members. After all, the quantities of different foods consumed depended on more fundamental questions, such as the kinds of foods that should be eaten to remain healthy or to mark certain occasions. In the same vein, the quantities of clothing required reflected what constituted appropriate dress for the different events in human life. These and other such matters helped constitute individual and collective lifestyles, in turn embedded in customs, traditions, and the way of life of Greek culture. All of this refers to the *ecology* of the material necessities and the role they played in enabling people to live good lives. The same is true for contemporary society. It is important to know how many technical products such as VCRs are being produced, but it is equally important to know how such products influence our habits and, by extrapolation, our lives.

The economy of technology focuses on inputs and outputs, and thus on performance, while the ecology of technology focuses on connections and on relationships, and thus on context compatibility. Throughout prehistory and history, most cultures believed that living good lives had more to do with relationships, leisure, appreciation of habitat, involvement in one's community, and the like, rather than with performance.[18] As a result, throughout human history the economy of technology was always subject to the ecology of technology. For the last three centuries, however, a gradual reversal of this situation has taken place. Priority has shifted to the economy of technology as Western cultures became convinced that this was the best path to the improvement of human life and society. The twentieth century, in particular, saw technological and economic development being increasingly guided by performance values, thus emphasizing the accounting of inputs and outputs, and the economy of technology over the ecology of technology. During this era, the trends and what was expected to be their eventual results seemed evident to everyone. Thanks to industrialization, the quantity of goods and services produced was increasing every year. Soon, it was thought, everyone's needs would be met and the age-old struggle for survival would come to an end. With scarcity no longer a threat, humanity could apply its energies to social and moral development. Hence, focusing on performance values was also supposed to lead eventually to improved social relationships. This view generally took for granted that human needs would not change under the influence of the new conditions created through industrialization. No one anticipated the emergence of a consumer society. It was also assumed that the negative social and environmental consequences were nothing more than the childhood diseases that maturing industrial societies would soon outgrow. With a very few exceptions, no one appears to have thought of the possibility of an environmental crisis. From a common-sense point of view, the reversal of the economy of technology and the ecology of technology amounts to covering over the windshield of a car and concentrating exclusively on the performance as indicated by the instruments on the dashboard. The consequences are inevitable but appear obvious only in highsight. The economy is in an 'upward trend,' but humanity and the biosphere are spiralling 'downward.' To live good lives, we need to know the significance of technology for human life as lived in a society located in the biosphere. This is what traditional human values expressed.

Preventive approaches thus introduce (or perhaps I should say restore) a level of common sense that avoids a number of frequently made mistakes. At least two interdependent but distinct levels of preventive approaches must be recognized – namely, the macro level, which deals with the technology of a society as a whole, and the micro level, which deals with specific individual technologies. The whole is more than the sum of the parts, and failing to recognize this is equivalent to attempting to understand the effect water has on paper in terms of the influences of hydrogen and oxygen. Unfortunately, this is rarely recognized in the literature, where works on the subject of technology and society immediately divide that subject into categories, such as production technology, information technology, mass media, and biotechnology, among others.[19] The present work develops a conceptual framework for the engineering, management, and regulation of specific technologies through preventive approaches. The discussion of a conceptual framework for examining the relationships between technology as a whole and its contexts of human life, society, and the biosphere is undertaken elsewhere.[20]

A second common mistake may be recalled from a previous discussion. Electronic components can be wired together to create a variety of devices, such as amplifiers and oscillators. The positive and negative characteristics of these devices are inextricably linked because they are rooted in the 'structure' of the circuit. Similarly, the Industrial Revolution began an ongoing process in which people change technology, and technology changes people and society. The result was dynamic 'structures' that linked its desired and undesired effects. Preventive approaches seek to understand these 'structures' and how, through preventive approaches, a restructuring can take place to make our modern way of life more sustainable with respect to human life, society, and the biosphere. I suggest, therefore, that the negative implications of technology can frequently be eliminated only through structural changes. Another common mistake is to believe that, when the ownership of such structures changes from the public to the private sector or vice versa, their characteristics are somehow transformed. However, life is not so simple. There is no question that the effect any technology has is a function of its own characteristics and those of the environment in which it operates. For example, two identical bombs, one exploded in the downtown core of a city and the other in the middle of a desert, clearly have very different effects. Nevertheless, the psychosocial effect of the assembly line on workers appears to be essentially the same whether the plant is publicly or privately owned. The soils and local ecosystems are similarly affected by industrial-agricultural practices whether the agro-business is privately or publicly owned. Examples can be multiplied, but the ideology of private and public ownership comes from an age in which modern cultures declared technology as good in itself, and thus the 'cause' of any problems must be found elsewhere since they have nothing to do with technology itself.

High technology continues to be based on a separation of the economy of technology from the ecology of technology. Consequently, any hope that expects it to deliver us from our present woes is probably misplaced. Can it rescue us from the problems created by smokestack industrial technologies? Can it deliver us from unemployment by improving productivity and competitiveness, and thus creating jobs? Will it be able to reduce the pressures generated by the spiralling costs of health care and the social safety net by making them more efficient? Will it be able to resolve the environmental crisis through dematerialization and generally doing more with less (performance-value-related achievements)? High technology can undoubtedly do some of these things, but will the gains not be undermined by the losses incurred from an ongoing separation of the economy of technology from the ecology of technology? Will it be able to help us meet our human rather than our technological (performance-value-related) hopes and aspirations, or will it force us to look elsewhere for deliverance? I will argue that preventive approaches have a better chance of success, and that high technology should be brought under this umbrella.

## 1.5 Prevention and the Ecologies of Individual Technologies

The design and characteristics of individual technologies can vary considerably in terms of the extent to which their contexts are factored into the assessment. This can be illustrated by several examples. First, consider the prototypical element of a technology – namely, a machine used for production. Its ecology includes the relationship between the machine and its immediate context, between that context and a still larger context, and so on, until the largest system or whole is reached. In terms of human life and society, the immediate context is the operator of the machine. The next context is the shop-floor organization, with its supervisors, who are, in turn, embedded in the organization of the plant, which is embedded in the organization of the corporation, the local community, and society. The design and utilization of the machine are directly affected by the extent to which these contexts are taken into account. It is possible to design the machine to produce the maximum output from the available inputs. It is possible to do so and also take into account the fact that human beings have to operate it. Obvious as this may appear, human-factors specialists have recorded numerous incidents where this had to be done in an after-the-fact manner. It is possible to design a machine for optimal performance, to be operated by human beings, and to function in a certain type of industrial organization. For example, in a Taylorist organization, a computerized, numerically controlled machine will be designed to be programmed by people in the office, and the role of the operators will be reduced to loading, unloading, and minimal supervision. In sharp contrast is that same machine designed for a work organization based on sociotechnical principles, whereby skilled machinists on the shop floor program the machine by manually operating it while each step is recorded, monitor the machine as it executes the recorded steps, and intervene to make any adjustments necessary before a more automated mode is utilized.[21] It is also possible to design a machine to give optimal performance, to provide a well-designed interface with the operators, to function in either a Taylorist or a sociotechnical organization, and to meet the demands of very stable markets permitting long production runs or respond to highly turbulent markets with considerably shorter runs. In the former case, highly specialized machines are designed for particular operations; in the latter case, 'flexible' machines are the answer. It is possible to add yet another context by designing machines to work with human beings in either kind of organization in either kind of market and anticipate environmental problems (including present and potential regulations). In other words, the machine may also be contextualized with respect to the biosphere.

The immediate context of the machine with respect to the biosphere is the flows of materials and energy into and out of the machine. These flows are a consequence of other activities, similarly depending on inputs and outputs of materials and energy, and constitute the next immediate context. The interconnected contexts form chains of human activities that bind the machine to the biosphere as the source and sink of all flows of matter and energy directly and indirectly involved in the machine. This approach is used in industrial ecology to improve the compatibility between technology and the biosphere. To sum up: the ecology of the machine consists of how it is embedded into its immediate contexts, how these in turn are embedded into larger contexts, and so on, until its ultimate dependence on society and the biosphere is clarified. This approach informs design and decision making and can result in substantial differences between machines that perform identical production steps.

The next example is drawn from the materials involved in the technology of a society. Until recently, the design and production of materials concentrated on the most efficient extraction of resources, refining, processing, and manufacture. The emphasis was on engineering performance and cost, which are entirely determined by the characteristics of the material and its production. Such 'intrinsic' properties were joined by an 'extrinsic' one – namely, its market-determined price. Gradually other 'extrinsic' properties, such as toxicity with respect to human and ecosystem health, were added, necessitating the creation of regulations for the manufacture, use, and disposal of those materials that proved particularly toxic. These developments took place in the context of a linear economy, which depended on unsustainably high rates of resource extraction and waste disposal. It is increasingly recognized that modern economies must be made more circular, in the sense that materials will have many 'lives,' beginning with a first application demanding high performance characteristics and gradually being degraded through successive uses until they eventually leave the economy. In other words, the revolution in materials that will occur over the next few decades will be one in which their intrinsic properties are complemented by extrinsic ones that manifest their suitability for a more sustainable society in terms of principles of industrial ecology. The new extrinsic properties will show how suitable a material is for open- or closed-loop recycling, remanufacturing, and eventual safe disposal. In other words, the ultimate context of sustainable societies and a healthy biosphere capable of supporting them will, over the next few decades, be operationalized in part as extrinsic properties of materials, thus constituting a significant revolution in materials' design, manufacture, use, reuse, and eventual disposal.

Germany's automobile industry is leading the way in attempting to close the materials loop by disassembling worn-out vehicles to retrieve useful components and materials from which new ones will be made. Manufacturers have become responsible for disposing of their products after they have served their useful life. The result has been a revolution in the way the cars are designed and manufactured. As disassembly has to be cost-effective, the way components of vehicles are attached and installed is changing, which is also making them easier to repair. The diversity of materials used in a vehicle is constrained as much as possible, and the kinds of materials used are identified with bar codes to facilitate closing the loop. Other products, such as computers, are expected to follow this pattern, and it will undoubtedly have equally revolutionary effects in the electronics industry.

A third example of the growing importance of context issues is drawn from the engineering and management of electricity grids. Until recently, the engineering, management, and regulation of energy systems focused almost entirely on energy supply. Somewhere along the way we forgot that energy is not an end in itself; that is, energy itself is not what people want, but, instead, the services that that energy makes possible in different contexts. The distinction may appear trivial, but putting energy use in context makes a fundamental difference to the design of these energy systems. People simply want to be able to do things such as cook and refrigerate their food, heat and cool their homes and work-spaces, shower and bathe with warm water, and listen to music and watch television. It is not difficult to understand how we became 'hooked' on a supply-side focus. The only cost-effective way of supplying electricity and natural gas is through regional or national monopolies. In a free society such monopolies can be justified only if consumers are assured of a reliable supply at a reasonable price, since they have no competitor to turn to if they are dissatisfied. Hence, electrical utilities and gas companies that were granted such monopolies became preoccupied with meeting these conditions and planning for ongoing growth in demand.[22] Several decades of rapid economic growth after the Second World War reinforced the cultural myth that growth was the answer to all our problems. For energy, this meant focusing all technological and economic efforts on ensuring adequate supplies because, without them, progress and human well-being would be threatened. No problem could possibly be significant enough to justify halting growth. The report of the Club of Rome that there were limits to growth on a finite planet, and the growing awareness of the environmental crisis, initially had little impact.[23] The consequence was the building of energy systems that were technologically sophisticated in terms of energy 'production' and distribution and much less so on the energy end-use side, making them entirely unsustainable in the long run.

The evolution of these energy systems stands in sharp contrast with the original idea that Edison had for the supply of services requiring electricity, rather than the supply of electricity itself. For example, the provision of an illumination service would ensure that the costs of providing that service would be kept to a minimum.[24] This shift in focus would not have affected the supply and distribution side, but it would have significantly altered the end-use side by making necessary the provision of efficient light fixtures, long-lasting light bulbs or fluorescent tubes, and durable and repairable components in order to remain competitive. In the present system, suppliers of electricity, light bulbs, or light fixtures seek to sell as much as possible, with the result that there is no pressure to maximize end-use efficiency. In sum, current energy systems are imbalanced in terms of their levels of technological and economic sophistication. This discrepancy creates a choice between two possible strategies for meeting increases in energy demand over the next few decades. The first is to continue the focus on supply, symbolized by extrapolating the exponential-growth curve to project the number of power stations that must be built. The second is to negatively generate increased capacity by investing in energy efficiency, which will free up time to develop alternative sources of energy that are more sustainable. Although both strategies are able to match supply with demand, their economic, societal, and environmental implications are vastly different. This is particularly evident when the social costs are added to the private costs, because the externalities in the supply-oriented strategy are much greater than in the energy end-use efficiency strategy. Consequently, the market cannot be relied upon to steer the development of energy systems along a path to increased sustainability. Nevertheless, utilities are now beginning to recognize that energy can be negatively generated by investing in energy end-use efficiency, often for half the cost of producing it. It is estimated that the design of energy-using equipment that takes full advantage of its context could permit modern societies to do everything they do now with half, or even one-quarter, of current energy consumption. This would greatly increase competitiveness, as well as vastly reduce the burden imposed on the biosphere. An energy-efficiency strategy could accomplish all this and create a great many new jobs in the process.

From the above examples, it is clear that to primarily guide the design and decision making related to a particular technology by means of performance values is to selectively focus on a very small portion of its ecology. Increasing the context under consideration results in a different technology that may be identical in terms of performance values but vastly different in terms of undesired outputs, such as waste streams, waste heat, products less well adapted to their intended use, and less healthy human beings. These lead to less viable ecosystems and communities, which, in turn, affect their contexts, and so on. By primarily relying on performance values, we are collectively engaged in a kind of magic. But having the undesired outputs of a technology disappear from our minds does not mean they have gone away. To be realistic is to recognize that every technology has both intrinsic and extrinsic properties, measured by performance values and context values, respectively.

## 1.6 The Preventive Management and Regulation of Technology

How is and how should technology be managed and regulated? A modern industrially advanced nation utilizes its technology to achieve certain goals. To make sure that things develop in the desired direction, the consequences of the utilization of that technology must be determined and compared with these goals so that corrective action can be taken as required. On a smaller scale, this situation is not unlike the one in which the occupants of a home keep themselves comfortable in winter by means of a hot-air furnace. It must be regulated to ensure the delivery of the right amount of heat to the house. This is usually accomplished by means of a thermostat, which monitors the temperature in the home and compares it with a reference temperature set by the occupants. The thermostat switches the furnace on when the temperature in the home falls below the set point, and switches it off when the reverse is the case. Such a control is called a 'negative-feedback system' because the action of the control is counter to the direction of the influence to be controlled. This mode of control is distinguished from 'positive feedback,'where the control does the exact opposite. These two modes of control are in turn distinguished from 'end of pipe' or 'after the fact' approaches, which do not regulate a system directly, but create one or more additional systems to control one or more of the outputs of the original system.

Consider the following three scenarios in which the furnace is regulated by means of positive feedback, faulty negative feedback, and an end-of-pipe solution. First, suppose that the thermostat had been wired incorrectly so that the furnace turned on when the desired temperature was exceeded and turned off when the temperature fell below the set point. The system is now unstable, and is an example of positive feedback. It should be pointed out, however, that positive feedback is not simply the result of a faulty negative-feedback system. There are situations when positive feedback is desirable, for example when babies and children are maturing physically. When they have grown up, negative feedback seeks to maintain their ideal body weights. Positive feedback is therefore required in some systems to achieve growth. Such phenomena may also be interpreted as being under the control of negative feedback, as when an individual's body seeks to attain normal height and weight.

Next suppose that the thermostat had been mounted in the hot-air plenum, which distributes the air heated by the furnace to the air ducts, which in turn transport it to the different rooms in the house. In the original system, the temperature in one of the rooms is selected as an indicator of the temperatures in the other parts of the home. As the temperature in the hot-air plenum is an inaccurate indicator of the temperatures in the home, effective regulation now becomes impossible. The moment the furnace comes on, the air temperature in the plenum quickly exceeds the set point, leading to a shutdown. The temperature in the hot-air plenum then begins to gradually decrease until the set point is reached, causing the furnace to be momentarily turned on again. Thus, the system switches the furnace on and off without appreciably affecting the temperature in the home. The regulation has become ineffective, because the indicator of the temperature in the home no longer reflects the desired comfort of the occupants. A similar malfunctioning of the negative-feedback control system would occur if the furnace could affect the set point on the thermostat. This is not as silly as it may appear at first glance. Suppose, for example, the occupants were anxious about their old furnace running constantly during a cold spell and figured (rightly or wrongly) that, by turning the thermostat way down, they could prolong the life of the furnace. The latter has now in effect 'influenced' the set point as opposed to its being related to the comfort level of the occupants. Regulation now occurs by means of a technological value rather than a human one. Effective regulation can occur only if the occupants alone can alter this set point. It follows that a negative-feedback control system, to function properly, requires adequate indicators of the output to be controlled and of the desired goals the system must reach and maintain.

Finally, consider an end-of-pipe or after-the-fact approach to keeping the occupants in the home comfortable. The furnace now runs continuously and is no longer directly regulated. All windows have been equipped with electric motors controlled by a thermostat, which opens the windows when the set point is exceeded and closes them again when it gets too cold. The problems created by a first system are now compensated for by a second system, rather than being regulated at source. Such an after-the-fact method of regulating the temperature in the house is more expensive to build, incredibly wasteful of energy, and much less satisfactory in terms of comfort because whenever the windows are opened additional drafts are created.

It is clear that a properly designed and installed negative-feedback system is the most effective among the alternatives described above, because it controls any problem of overheating or insufficient heating right at the source, by regulating the very system that produces it in the first place. The other three approaches do not work at all or perform unsatisfactorily or get the job done with much greater expense and lower satisfaction. Surprising as it may appear, all four approaches reflect ways in which modern industrially advanced societies control and regulate their technologies, and the negative-feedback approach is the least common. A few examples will illustrate the point.

Consider what happens when a modern city continues to expand, causing the number of motor vehicles to become too large for its network of roads, particularly in the downtown area. As the average speed for getting around the city declines, public pressure escalates to build expressways, widen streets, and in other ways improve the flow of traffic. To the extent that this is successful, it becomes easier to get around the city, and more people are encouraged to use their cars to go downtown. Soon, traffic becomes snarled again. The outcome of several such cycles is all too evident in many large North American cities. The average rate of speed for getting around these cities is often less than that of a bicycle, noise levels are high, pollution levels have become a threat to human health, a great deal of valuable space is now taken up by automobiles – all these factors contributing to making the cities less habitable for people. Little has been accomplished in terms of better transportation, and much has been lost in terms of the livability of cities. But it appears that we may be on the verge of yet another cycle in this process, based on the building of a smart road system designed for smart cars. Under these circumstances, it is essential to ask whether the development of transportation systems based on the automobile should continue to be guided by positive feedback. While this may be useful during a certain growth period, there clearly comes a point beyond which the scale of such a system is incompatible with that of the city. At this point, the system must be guided by negative feedback.

In this example, we encounter the limitations of applying the concept of negative feedback to more complex situations. A city does not have a single primary goal such as the accommodation of a certain number of cars. It is, first and foremost, a life milieu for its inhabitants that promotes, facilitates, inhibits, renders impossible, or is neutral in regard to a host of activities these people engage in every day. Just as their language names everything, their culture values everything; and, although individual diversity exists within a cultural unity, any activity is similarly valued by each person. The regulation of cars, therefore, has to reflect an accommodation of all the activities affected, keeping in mind that these are not equally valued in a society. Such regulation would require a multidimensional negative-feedback control based on a multidimensional set point determined in relation to a diversity of hierarchies of values. Clearly, no technological approach based on the principle of negative feedback is applicable in these more complex situations, yet the fundamental principle can help us better understand their control aspects.

Consider a second example related to the regulation of a modern economy. Typically, indicators such as the Gross Domestic Product (GDP), growth rates, and other indicators of specific economic outputs (such as tonnage of steel, coal, or goods transported) guide economic decision making. It is assumed that the more goods and services an economy produces, the better off we will all be. These indicators are assumed to represent what free and democratic societies aspire to. Yet some economists have pointed to the inadequacy of such indicators. Consider a typical suburban family with two children. The family car is destroyed in an accident, fortunately with only minor personal injuries. When a new car is purchased and the family receives some medical services, the GDP increases. Apparently, accidents boost economic output but negatively affect the quality of people's lives. Have we done the equivalent of putting the thermostat in the hot-air plenum?

The same family finds life in the suburbs very stressful. With both parents working to make ends meet and the teenagers dealing with a complex time in their lives, family life suffers. The parents decide to seek the services of a marriage counsellor to help them work out some problems, and one of the kids is seeing a psychologist. Finally, they decide to do something dramatic, and move to a smaller community in search of a simpler and less stressful lifestyle. Once they move to the country, they begin to grow vegetables. They no longer need special services or sleeping pills, and they give up smoking. They also drive their automobiles less, and as a result of all these lifestyle changes the GDP decreases. After a year, they decide that this lifestyle is not for them after all, and they return to the city. Some of their earlier problems reoccur, and as a result the GDP increases again. These examples bring home the fallacy of making economic policy decisions to speed up or slow down the economic engine, or to restructure it primarily in terms of performance indicators. All this is well known, yet even under the present circumstances little change appears to be forthcoming. From a control system's point of view, however, it is easy to understand why we don't seem to be able to make much headway with problems such as unemployment, spiralling health costs, and controlling large deficits. Many observers now recognize that more is not necessarily synonymous with better, and that a distinction must be made between growth (more) and development (better). It has long been recognized that unlimited growth sustained through a kind of positive-feedback control is incompatible with a finite planet. A negative-feedback type of control will have to be created to ensure that the scale of our economic activities is compatible with the scale of the biosphere.

The multidimensional 'set point' based on the hierarchy of values of the cultures of modern societies has been profoundly influenced by technology. Technological or performance values have risen to the top of the hierarchy, taking priority over traditional human and social values. The reasons for this are complex but can be simply explained: As people change their technology, technology simultaneously changes people. The influence of technology on people becomes significant when their surroundings primarily consist of technological objects and systems. This is the case because the experiences of those surroundings modify the neural connections in the brain to symbolize the experiences of a person's life in the mind. The influence of these surroundings on people's minds,and thus on their culture, can be expected to be as great as the well-understood influence nature had on prehistoric cultures. If the influence technology has on people is more decisive in shaping the evolution of a society than the influence people have on their technology, a situation arises that is analogous to the furnace having affected the set point on the thermostat. Here we encounter the limitations of preventive approaches. They are a necessary but not sufficient condition for a society to effectively manage and regulate its technology. These three examples show how the management and regulation of modern technology have broken down in ways that are analogous to those discussed for the thermostat-controlled furnace.

## 1.7 Preventive Approaches Oriented to the Biosphere

Preventive approaches attempt to overcome the pitfalls of the conventional two-stage approach to the engineering, management, and regulation of modern technology. They use information about the human, social, and natural implications of technology, along with technical factors in the engineering process, in a negative-feedback mode so as to reduce or entirely avoid negative implications. Mitigation technologies and services will be added only if unacceptable negative implications remain. The concept may be illustrated by rethinking the environmental strategy of a corporation. With the end-of-pipe approach, each time a government passes stricter environmental regulations, additional mitigation technologies must be installed, operated, and maintained. This situation is analogous to the one in which motors were installed on windows to control room temperature. There is no reason to believe that this pattern of events will ever change. It may be accelerated under some governments and decelerated under others, but the company expects no relief in sight, and hence lobbying and uncooperative behaviour have been the rule. The situation appears hopeless, particularly when companies in other parts of the world can operate with much lower environmental standards.

The situation appears quite different, however, if we go back to the root of the problem. Why are these pollutants produced in the first place? How can products, processes, and operations be redesigned and adjusted so as to eliminate or significantly reduce them? Conventional wisdom would have us believe that trying to answer these questions is a waste of time because it would make the company less efficient, less productive, and less profitable. However, even if a reduction in performance measured in terms of output/input values occurs (which often is not the case), might this not be offset by other factors? These may include a reduction in maintenance and energy costs, as the pollutants would not have to move through the production equipment; savings on the costs of installing, operating, and maintaining mitigation technologies; a reduced need for occupational health and safety measures, since the plant would no longer present workers with the same risk to their health; a reduction in the risks flowing from liabilities associated with an accidental spill or discharge; the ability of management to engage in more effective long-term planning, since the plant is now operating well within local environmental standards; improved worker morale as workers take pride in their 'green' company; and the competitive advantage derived from producing a greener product by means of cleaner processes, favourably affecting consumer choices. Such a company is creating a new corporate culture able to deal with challenges more effectively by comprehensively rethinking strategy based on negative feedback – a feature that may be critically important to operating successfully in an increasingly turbulent global competitive climate. These efforts can also help reduce a nation's 'expenses,' such as health care and environmental costs, thus reducing deficits and tax loads.

It is clear that innovative approaches in industry are beginning to contradict the end-of-pipe or 'do it only if forced to' approach. Pollution prevention is, of course, the best-known one. A considerable number of case studies demonstrate that, when environmental factors are incorporated into the design process right from the beginning (what I call 'internal environmental engineering' as opposed to end-of-pipe or external mitigation), the typical outcome is products and processes that are significantly less costly, healthier, and more sustainable. Some widely publicized examples are the results achieved by the 3M Corporation and Northern Telecom.[25] The chemical industry has adopted a somewhat more prevention-oriented Responsible Care Program across North America.[26] The German automobile industry is beginning to design cars for the entire life cycle, including the recycling or reusing of as many parts and materials as possible at the end of a car's useful life. The Germans are also enacting some of the most stringent legislation with regard to packaging.[27] The white goods industry is beginning to design for disassembly and reabsorption of materials into the production and consumption cycle.[28] Economists have begun to realize the potential of these strategies.[29] From these examples, it is clear that pollution-prevention efforts must not be limited to the production process but must deal with a product's entire life cycle, including its disposal.

The realization of this potential, however, is in part limited by the considerable gap that exists between innovative industrial practices and supportive government policies, on the one hand, and the current view in engineering and business culture, on the other. For example, what I call 'internal environmental engineering' would require the incorporation of environmental considerations along with technical ones from the beginning in the engineering process, so as to achieve the most compatible technology. It is only when this potential has been exhausted that 'external' environmental engineering should come into play and add on the mitigation technologies as required. 'Internal environmental engineering' should be an integral part of any engineering department and specialty, while mitigation approaches can continue to flourish, typically within civil or chemical engineering. Both internal and external engineering are required, although the potential of the former must be exhausted before external solutions are implemented.

We are really dealing with what is called a 'paradigm shift.' During the transition period, factual evidence is regarded differently by competing perspectives. In the engineering and business community, the established view for dealing with context issues is very strong, and probably constitutes the primary stumbling block in the development of a technological strategy able to create win–win situations for all parties. I have difficulty imagining a greater potential resource for helping to deal with our present economic difficulties, which are now evident as being structural in nature, than the preventive paradigm I am proposing. By reducing the costs associated with the creation of wealth through the application of preventive approaches to economic development, much can be done.

Despite considerable activity in the domain of pollution prevention, modern societies have barely begun to exploit the potential of preventive approaches. Most engineering, management, and business schools teach no courses in pollution prevention; in those schools that do, the rest of the curriculum remains entirely non-prevention-oriented. Hence, the curricula of these schools continue to be an unfriendly environment for such activities. Because no school as yet has begun to recognize pollution prevention as the tip of the iceberg of a new paradigm for the engineering, management, and regulation of modern technology, it is not surprising that many graduates acknowledge the successes of pollution prevention but are sceptical about its long-term potential. When the potential of pollution prevention approaches comes up against economic limits, the end of the road is not even in sight. Industrial ecopark design, closed-loop and open-loop recycling, product stewardship, design for environment, resource recovery from waste streams, and lifestyle changes through the redesign of our technologies and urban habitat are an integral part of the new preventive paradigm. The universities, corporations, and government departments most advanced in the domain of pollution prevention still largely treat it as an addition to the conventional paradigm for the engineering, management, and regulation of technology, as opposed to the beginning of a paradigm shift. As a society, we fail to recognize that technological development, mostly guided by performance values, is itself as significant a factor in the environmental crisis as are population growth and increased consumption. Technological development, guided by a preventive paradigm and the corresponding creation of a more circular economy powered by sustainable-energy practices, can vastly reduce the throughput of matter and energy in modern societies. Thus, restricting the role of the biosphere as their ultimate source and sink will enormously reduce the pressure on all living systems.

The preventive approach is the opposite of the one advocated by modern economics, which suggests that end-of-pipe abatement equipment be added until the cost of installing and operating it equals the benefits derived from it. The assumption is that the industrial process under consideration is optimal, and therfore that mitigation is the only option. Preventive approaches seek to modify the process to make it optimal in terms of both performance and context values, thus shifting the cost and benefit curves.

## 1.8 Preventive Approaches Oriented to Society

Preventive approaches must also be developed for the human and societal contexts of technology. Consider how we organize work in industry as an example. The development of the production system based on Fordism and Taylorism helped make the United States the pre-eminent industrial nation in the middle of the twentieth century. These systems were in part a response to the unique American situation characterized by an abundance of resources but a scarcity of the highly skilled labour required to manufacture complex technological products. This problem was overcome by breaking down the production process into a sequence of steps performed by machines or workers connected through an assembly line and by providing management with an unprecedented control over that work because of its ability to determine precisely the 'one best way' to carry out and organize it. Even more than before, production systems began to resemble gigantic machines into which human beings were fitted as parts. These systems utilized the hands of the workers, but not their brains – workers were meant to leave them at the plant gate.

These new production systems excelled in the domain of performance values by improving productivity, profitability, and, in many cases, wages. Improved performance was, to a significant extent, achieved at the expense of society. Socio-epidemiology has produced considerable evidence that this form of work is incompatible with healthy workers. It leads to a less creative and productive workforce, a serious deterioration in the relationship between management and labour, a diminished capacity of the workers to participate in family and community life, all of which adds up to a troubling degradation of the social fabric of a society. It has also placed additional restrictions on the development of democracy. The design of these production systems overlooked several important context issues. The abilities of human beings and those of machines are generally opposites.

Machines thrive on repetition, while research shows that human beings are destroyed by it. The suppression of the self by work that necessitates behaviour patterns that more and more resemble those of robots requires mental energy; thus, physical fatigue is replaced by nervous fatigue, which negatively affects physical and mental health. Socio-epidemiological studies are finding that, at any given age, people are more likely to die prematurely when they have low-paying, monotonous jobs with low decision latitude as to how their work is to be performed. It is increasingly evident that psychosocial conditions at work produce a variety of dysfunctions and disabilities. The consequences for human lives, families, communities, and corporations, and, by extention, for the social ecology (i.e., society) are simply not measurable.[30]

The emergence of the lean production system is once again a story of necessity being the mother of invention, making it possible for a small Japanese automobile manufacturer to compete with much larger firms. An important ingredient in its success is the fact that improvements in performance over the Fordist/Taylorist system derive in part from a much more tightly coupled system and the shortening of negative-feedback loops.

Human capabilities are more fully integrated into lean production systems, creating in a variety of benefits for all stakeholders. A definitive verdict on the impact on workers will have to come from socio-epidemiological research, but the advantages for management are widely recognized. Commenting on American business, a Japanese industrialist claimed that

> we will win and you will lose. You cannot do anything about it because your failure is an internal disease. Your companies are based on Taylor's principles. Worse, your heads are Taylorized too. You firmly believe that sound management means executives on the one side and workers on the other. On the one side men who think, on the other side men who can only work. For you management is the art of smoothly transferring the executive's idea to the worker's hands.[31]

Recent developments in Japan show that the Japanese, too, face growing problems.[32] This is not surprising, since their approach is also ultimately based on performance values and measures. Nevertheless, Japan and Scandinavia, in particular, show us what we ought to have realized a long time ago – namely, that the effectiveness of a particular production system that involves people and machines depends greatly on how people are treated. Employers depend on a healthy and creative workforce, while the community requires healthy work. If employers are to reduce costs associated with high turnover rates, absenteeism, poor-quality work, and accidents, and society is to get its health (including Workers' Compensation) and social costs under control, the design of healthy workplaces must become a high priority. We may have to rethink much of the conventional wisdom, adjusting what we do on the basis of what we know about modern work and workplaces. Corporations that have scratched the surface in this area appear to be finding that healthy work can create win–win situations for all parties. However, based on what we are learning about healthy work and healthy workplaces from socio-epidemiology, one thing is certain: lean production is not an answer to the problems created by the Fordist/Taylorist precursor.

A comparison of the performance of Silicon Valley with its equivalent in the Boston area during the economic downturn of the last few decades shows that the former did much better than the latter. The two industrial areas are comparable in almost all respects except in the way 'human and social capital' is organized. In Silicon Valley, there is a distinct absence of rigid organizational hierarchies and well-defined reporting lines. People form teams that emerge, evolve, and reorganize according to the tasks that need to be carried out. There is also more cooperation between firms. In contrast, the Boston industrial area characteristically exhibits a more conventional approach.[33]

Preventive approaches for the design and organization of human work may well unlock the same kind of success as approaches oriented to the biosphere. Much less effort and research have gone into approaches for the engineering, management, and regulation of modern technology that are preventively oriented to human life and society than into those oriented to preventing harm to the biosphere. Since lean production systems also involve significant implications for human life, much work needs to be done, but too many corporations still regard any explorations in this area as luxuries they cannot afford in times of intense competition. There is nothing new here. Preventive approaches oriented to the biosphere continue to be received with the same levels of scepticism by many engineers, managers, and regulators of technology steeped in the conventional approaches. This book expresses a cautious optimism that preventive approaches oriented to human life and society can unlock benefits for all stakeholders and make a substantial contribution to human sustainability and our common future.

The viability of preventive approaches oriented to the biosphere, and those oriented to human life and society, may be further enhanced if their development is synergistic, that is, if it is ensured that the implications of each one enhance those of the other. Preventive strategies ought to simultaneously consider the human-social as well as the natural contexts of technology. Energy production and distribution is a good example. Traditional approaches have created a societal addiction to ever more energy, making us highly energy-inefficient. Less convivial communities, a high debt load, and considerable environmental damage have been the result. Preventive approaches not only make us much more energy-efficient and environmentally responsible, but also more competitive, since the cost of saving power is usually less than that of producing it. In addition, a less energy-intensive lifestyle will require a more convivial urban habitat. Here again we have only begun to scratch the surface. Co-generation is the best-known and most obvious example of the gains that can be realized by creating a greater synergy between an energy system and its contexts. These synergies must be created on both the micro and the macro level.

The increasingly recognized need to make our cities more liveable, healthier, and sustainable also has significant implications for the practice of engineering. Any engineering undertaking contributes to the shaping of the urban habitat and the kinds of stresses created between it and the people living within it. Many of these, such as sensory overload, social overload, crowding, noise, pollution, violence, and crime, have been extensively examined and correlated with various characteristics of that habitat. This kind of information can be analysed from an engineering perspective for the purpose of adjusting engineering theory and practice, so as to help create healthier and more sustainable urban habitats. Adherence to the 'healthy city' concept is gaining momentum, and it is evident that engineering practice can make a significant contribution in this area through preventive approaches. This is spilling over into the management and regulation of technologies to help make our cities more sustainable. It is difficult to exaggerate the enormous benefits that will accrue from preventive approaches oriented to society.

Finally, all of the ideas discussed above have implications for technology transfer to remote regions and the so-called third world, as well as for foreign aid. The design of contextually appropriate technologies, once established in the industrially advanced nations, will create a welcome spin-off for the poorer nations, provided we do not export to them the previous generation of non-prevention-oriented technologies.

### 1.9 A Political Framework for a Preventive Strategy

The former premier's Council of Ontario formulated an integrated approach to ensure the long-term viability of the province and had just begun to recognize the role of preventive approaches at the time of its disbanding.[34] Its approach was based on the recognition that everyone has a perspective on what is happening, but that perspective may be different from those of others. Setting up a round-table discussion of how we all live in one world and will share one future will create a more comprehensive understanding of the situation, thus opening the way to more integrated solutions. The approach may be summarized as follows.

Consider the debate over how best to allocate public resources. You would almost think there were three separate and distinct worlds with their own life-support systems. The first and most influential is the economic world, which includes manufacturing and the distribution of scarce resources, products, and services. It encompasses much of science, technology, industry, and business. Its proponents regard the creation of wealth as the primary activity on which all others depend. How else would society create jobs, and how would it pay for necessary activities such as education, health care, social services, leisure-time pursuits, and environmental rehabilitation? From this perspective, the most important things are those that either enhance or interfere with the ability of society to create wealth. Large deficits, high taxes, and excessive public spending are seen as the primary threats to wealth creation, and are therefore placed at the top of the political agenda.

A second and much less influential world is that of society itself. Its proponents include community groups, social agencies, religious groups, and labour movements. The primary activity in this world is ultimately the raising of the next generation, without which there would be no society, no economy, no deficits – nothing. What is most important from this perspective is whatever helps create well-educated, physically and mentally healthy, socially responsible, and morally caring human beings – that is, a healthy community. Anything that threatens this undermines the viability of society itself, and thus of everything else. As a result, at the top of the political agenda are issues such as social justice, equality, full employment, accessible health care, adequate social services, and a social safety net for those who are disadvantaged for reasons beyond their control.

A third world, even less influential than the two previous ones, is the community of all life constituting the biosphere. It cannot speak for itself except through environmental groups, which are its only proponent. From the perspective of this world, the life-support provided by the biosphere is more fundamental than the creation of wealth and, for some, more fundamental than human life and society as well. Everything that threatens life and the life-support systems moves to the top of the political agenda. These include global warming, ozone depletion, the extinction of species, and all forms of pollution.[35]

The proponents of each of these three worlds tend to behave as if each world had its own separate and distinct future. Debates on apparently insurmountable differences are rarely mediated by the recognition that in fact we all share one common future, particularly as a result of growing global integration. Instead, proponenets of the first world regard the priorities of those of the second world as well and good in so far as we are able to pay for them through wealth creation. It regards the demands of the proponents of the third world as a direct threat to wealth creation, and thus to environmental rehabilitation. Second-worlders regard the demands of first-worlders as anti-human and insensitive, and those of third-worlders as important but secondary constituents of a healthy biosphere, seen as a prerequisite to a healthy community. Third-worlders regard the demands of first-worlders as a threat to present and future generations, as well as to all life, and those of second-worlders acceptable only in so far as population growth and resource consumption can be rolled back to levels the planet can support. Knowing, doing, and controlling in each of the three worlds is severely out of context with respect to the other two, a situation that spells disaster unless it is fundamentally changed. It is urgent that the three worlds come together in the recognition that, in emphasizing their different value priorities, they have put different aspects of reality in the foreground and background, but, nonetheless, we continue to share one common future. The premier's council sought to reintegrate these perspectives through constructive dialogue. Avoiding 'us and them' distinctions recognizes that, over generations, a system is built that rewards certain kinds of behaviour and penalizes other kinds. To varying degrees we all respond to these signals, and in any case we are all contributing to the evolution of the system. As we are people of our time, place, and culture, even our moral judgements are affected, and few of us have the moral authority to cast the first stone.

I acknowledge the changes that have taken place in the right direction. Some decision making in each of the three worlds is now being contextualized in terms of the others. For example, some public policies now attempt to balance economic and environmental concerns. Others attempt to balance economic development with health, productivity with occupational health and safety, community well-being with controlling deficits, delimited social inequality with poor health and crime, and so on. Yet the contextuality of worlds in relation to one another still occurs primarily only in pairs. A more integrated approach is required. This brings us to the subject of barriers against preventive approaches in professional education and in the institutions playing a key role in the engineering, management, and regulation of modern technology.

## 1.10 Barriers to Preventive Approaches

To explore how the after-the-fact approach is both cause and effect of the contexts in which it operates, I offer three examples: (1) the intellectual division of labour within an engineering faculty, a corporation, a government, and a university; (2) the broader economic context within which these operate; and (3) the cultural context created by a society. First, consider the intellectual division of labour in an engineering faculty. Since environmental engineering is typically the purview of civil and chemical departments, it can deal only externally and after the fact with environmental issues occurring in other engineering specialties. No equivalents to preventive medicine and public health exist in engineering. Preventive approaches are virtually absent in engineering textbooks, including those on environmental engineering.

Next, think of an environmental department in a corporation. It is usually restricted to dealing with end-of-pipe solutions because these issues are not dealt with 'internally,' by other departments, and this division of responsibilities is sanctioned by the organization chart and the corporate culture. Because after-the-fact approaches to environmental problems cannot provide long-term solutions, corporations faced with growing competition have little choice but to lobby for free-trade agreements that permit technologies to flow to areas where greater negative social and environmental impacts are tolerated. Many more companies would have an alternative option – namely, to develop preventive approaches – if their professional staff had received a different kind of education.

A similar situation commonly occurs in government. The division of labour between the different ministries has, until very recently, favoured end-of-pipe approaches by whatever ministry is responsible for environmental issues. The situation is beginning to turn around, but the resistance to more preventive approaches as a result of organizational design is still considerable. Experience with more preventive Quality Management Systems and Environmental Management Systems has shown that these concerns cannot be an afterthought in organizational terms, but that they require that quality and environmental issues be internalized to apply to all aspects of the corporation or government, thus necessitating a change in organizational 'culture.' Hence, there continue to exist considerable institutional barriers to more preventive approaches, which must be reckoned with when initiating fundamental changes. Although the International Standards Organization (ISO) has developed a number of new environmental provisions, they are not deliberately prevention-oriented. If these new environmental standards lead to more end-of-pipe solutions, the long-term benefits to corporations and society may be outweighed by the additional business costs incurred. It is not a question of better managing the quality and environmental aspects of the operations of a firm, but of transforming them through a prevention orientation.

These patterns are both cause and effect of the kind of knowledge base that supports engineering, business, and economic decision making. The university educates the decision makers, which brings us to the issue of scientific specialization and the scientific approach to the study of reality. Unlike culture, which deals with the complexity of the world on the level of daily experience, science behaves as if complexity is unmanageable. Whatever is scientifically examined is intellectually, if not physically, abstracted from its context. For example, when something is studied in a laboratory, the context is not complex, as it is in nature, but highly simplified so that the few variables can be controlled individually. In this sense, science knows things out of their usual context and in a much simpler one. As a result, science produces knowledge of one kind and ignorance of another kind. This ignorance cannot be eliminated by piecing together the findings of the many scientific disciplines into one coherent scientific image of the world. There is no science of the sciences capable of providing this synthesis. Thus, science has greatly simplified the task of knowing our world, making it extremely efficient, as the exponential growth of scientific knowledge demonstrates. This much-admired human accomplishment is offset, however, by the simultaneous exponential growth of ignorance of how everything in our world relates to everything else. This is increasingly a serious weakness in our modern knowledge base, as modern societies seek to make their ways of life more sustainable, which requires the understanding and improvement of the compatibility between technology and its contexts. Technology must be understood in its real-world context. This is not a call to abandon science; on the contrary, what is required is a recognition of the limitations of modern science that, like any other human creation, is good for certain tasks, useless for others, and irrelevant for still others. We have brushed aside all other forms of knowing in the belief that science, and science alone, can provide reliable and objective knowledge, valid for all times, places, and cultures. However, science, like any other human creation, is not omnipotent.

The limitations of science become particularly evident when we seek to understand a phenomenon such as modern technology, which affects many aspects of our world, and in the application of scientific knowledge to the solution of real-world issues. Textbooks show that the social sciences and humanities appear to be able to describe many aspects of our world with minimal reference to the roles science and technology play within it, while the engineering and management sciences proceed as if the context implications of technology can be dealt with largely outside of the core of the undergraduate and graduate curricula. Is it possible to describe a modern economy with a brief reference to science and technology on one page and in one chapter near the end of the book? Is it possible to describe the social organization of an industrial society without ongoing discussion of how technological and social change are linked? Is it possible to understand modern democracy without considering how the computer and associated intellectual techniques have changed the workings of governments and political parties, the relationships between government ministers and the experts in their departments, the role of the mass media, or the condition of the citizen in a mass society? Is it possible to understand the health of the members of a modern society without taking into account the influence technology has on their lifestyles in general, and their work in particular, not to mention the effects on food and drink? Is it possible to be an effective engineer and not know how particular technologies within one's domain of expertise affect human life, society, or the biosphere? Surely these textbooks distort the world: in some, technology entirely fills the intellectual field of vision; in others, it is almost entirely absent. In neither case is technology intellectually proportional with respect to everything else. We have come full circle, and encounter once again the situation illustrated by the analogy of the painters depicting a landscape.

The application of specialized scientific and technical knowledge to real-world issues manifests the same difficulties encountered in the absence of a science of the sciences. Consider the following example, which is an extrapolation of a real situation in which international attention was focused (through a United Nations study) on the problem of malnutrition in a Colombian valley.[36] Many experts participated, but each looked at the situation through the lenses of his or her own area of specialization. A nutritionist argued that the problem was a question of diet. An adequate diet should be designed from the foodstuffs available in the valley, supplemented if necessary by others that could readily be imported. A health expert suggested that an adequate diet would not be sufficient to remedy the problem because the inhabitants suffered from intestinal parasites that caused diarrhoea, and therefore would remain malnourished. The water supply had to be improved by digging new wells, open sewers should be eliminated, and basic health care provided. An economist argued that all this was well and good but ultimately the problem was that these people were poor. If they were not, they could afford an adequate diet, and better wells, sewers, and health care. What was required, therefore, was some kind of development, possibly in the form of cottage industries or a stop-gap measure such as persuading one or more members of each family to go and work in the city and send money home. An agronomist did not agree, stating that the fundamental need was to teach the inhabitants better agricultural techniques, including the use of high-yield crops, fertilizers, pesticides, and modern machinery. They would then be able to grow not only enough food for themselves, but a surplus to produce income. This would solve the problem of poverty and everything else that comes with it. A political scientist had a different view: It was well and good to talk about an adequate diet, better hygiene and health care, economic development, and agricultural development, but all this overlooked the fundamental point – the reason why the inhabitants of the valley did not have any of these things was because they were not empowered. A few wealthy families controlled all the good land on which they grew food, not to meet local needs, but for export through agribusiness-type operations. They belonged to an elite that included the military leadership and many of the politicians. Hence, to resolve the fundamental problem, the poor inhabitants of the valley should mobilize and form a political party to empower them, gaining political influence so that they could press for land reform. For a sociologist, all of this overlooked yet another issue rooted in the custom of families to divide their land when the oldest son married. Since this had gone on for generations, many families now owned plots of land too small to meet their own needs, let alone produce a surplus. According to this specialist, what was needed was the creation of collective farms, cooperatives, or larger farms jointly operated by several families. The demographer challenged all these viewpoints by noting that thirty or forty years ago the problem did not exist. After the eradication of malaria and better control of several other diseases, the population numbers rose, eventually reaching a level that the valley could no longer support, at least not sustaining the current way of life. It was necessary to stabilize the population or to encourage people to move out of the valley. A wise person pointed to an additional issue: The valley did not have a 'problem' in the usual sense of that word. Genuine change in people's lives has to come from within. It may be guided, inspired, or challenged by wise counsel, but that is all. Implementing expert advice always raises a moral dilemma. Somebody decides how other people must live. Furthermore, because there is no science of the sciences, no comprehensive diagnosis of any situation can be made, with the result that the chances of good counsel coming from experts who are divided among themselves is not very great. Each expert approaches the situation with the prejudgment of the world that comes from having grown up in a particular culture, professional training, and experience. The resulting dilemma may be compared to their observing the situation from very different intellectual vantage points or through different 'mental lenses.' Gaining a discipline-based knowledge of the valley is clearly an irreversible process in the sense that it does not permit a return to the daily-life experience of life in the valley, which is integral and not segmented. Scientific and technical knowing are separated from daily-life knowing, without a common denominator.

The predominant mindset, accompanying behaviour patterns, and institutional forms referred to above are further rooted in the common view of the relationship between society and its economy. This creates a second barrier to preventive approaches. The economic view of human life and society with which we began this chapter implies that everything has an 'end of pipe' or 'after the fact' relationship with the economy, thus blocking the emergence of a preventive paradigm. This view of human life and society is supported by a range of other beliefs about how our human world functions and evolves.

Much of what constitutes the practices of modern engineering and business is explicitly articulated in the classroom, on the job, and in the literature. However, a great deal of it is only implicit in our behaviour. This brings us to the third barrier. For a long time, cultural anthropology has understood that much of a society's way of life and culture is transmitted implicitly, and what is not said at a particular instance is as important as what is specifically communicated. Any way of life is based on deeply held beliefs about the nature of reality and human life within it, as well as on values to orient human behaviour in that world. These can be taken for granted by the members of a society because they appear self-evident and so obvious that any radical alternative is viewed, at best, as an interesting curiosity, but not something one could actually live. In expressions such as 'Time will tell' or 'History will judge,' we are hinting at profound conceptions about how as a society we arrange our lives within time. Other cultures have done this in very different ways.

Of course, engineering is not a culture, but it constitutes a set of practices that cannot be divorced from profound conceptions about ways of dealing with and existing in the world. Some historians and sociologists of science have persuasively argued that a scientific community shares much more than what is explicitly transmitted in the classroom, the laboratory, and the literature. This is also true for the business community. In other words, I am assuming that my argument thus far has identified only the tip of the iceberg but that much of it is submerged in the deeper institutional and cultural waters. I return to these matters in chapter 2.

### 1.11 Conclusion

In this chapter, we have argued that the engineering, management, and regulation of any technology or undertaking can be done in either a conventional or a preventive fashion. The conventional approach gives priority to ensuring the efficient use of scarce resources by constant reference to values and measures of efficiency, productivity, cost-effectiveness, and profitability. Issues of compatibility between the technology or undertaking and its context are dealt with as constraints set out in regulations or laws and as issues to be managed after the fact. The preventive approach is guided both by performance and context values and by measures aiming to ensure that no gains in performance values are undermined by negative effects on human life, society, and the biosphere. Such approaches make use of the concept of negative feedback by constantly monitoring potential context implications and adjusting designs and decisions accordingly. They recognize that the optimal allocation of scarce resources is a necessary but not sufficient condition for successful engineering, management, and government efforts.

This becomes evident from a variety of new approaches developed primarily in industry and government, which are preventive with respect to some particular issue, the human-societal context, or the biospherical context. In other words, none of them is comprehensively preventive but each in some way does point in the right direction. These include pollution prevention, industrial ecopark design, industrial ecology, green product design, design for disassembly, design for the entire life cycle, design for environment, total quality management, environmental management systems, clean production, energy efficiency, energy-demand management, integrated resource planning, healthy-workplace design, and healthy-city concepts. They challenge the conventional idea that preventive approaches cannot be cost-effective and profitable. Their diffusion appears to be inhibited not so much by technical or economic constraints as by the fact that the intellectual division of labour in universities, corporations, and government bureaucracies has, over the past 100 years, had a cause-and-effect relationship with traditional approaches. This is further reinforced by a constellation of beliefs and values integral to the ethos of the professions involved and to the culture of society.

The purpose of our inquiry is to examine how engineering, management, and government regulation can contribute to making our way of life more sustainable by introducing preventive approaches. Much of the debate related to the meaning of sustainable development misses the fundamental point. Almost all living social and natural systems are in decline. What is required is that we make our way of life more sustainable by reducing the pressures we place on these systems. Operationalizing this is not difficult: each year we should reduce these pressures, compared with what they were in the previous year. Gradually, in the course of decades, we will begin to discover what a more sustainable way of life looks like. The present inquiry is concerned with the art of the possible; that is, it begins with current practices and the constraints on those practices, and, through the introduction of preventive approaches, seeks to reduce the harm modern technology does to human life, society, and the biosphere. The application of preventive approaches to individual technologies or undertakings may not be sufficient, but it is certainly a necessary condition for making our way of life more sustainable. Whether these undertakings will add up to a sustainable technology and economic framework is a subject discussed elsewhere.[37] It involves extending the current thesis beyond the engineering, management, and regulation of modern technology.

We found that the prerequisites for developing preventive approaches include the following:

- a comprehensive understanding of how the technology of a society is embedded into, interacts with, and depends on human life, society, and the biosphere
- an understanding of how a particular technology or undertaking is part of and contributes to this larger picture
- an understanding of past, present, and potential future issues related to context compatibility
- an ability to use the understandings identified above to adjust engineering, management, and regulatory processes in order to improve the context compatibility of technology by preventing or reducing harmful impacts
- the ability to use a set of values for guiding and assessing such adjustments from a context-compatibility perspective. Such values and measures cannot be technological in nature.

The first three items constitute what I call the 'ecology of technology.' This body of knowledge uses the methods and approaches of the social and environmental sciences as well as the humanities, but applies them to issues relevant to engineering, management, and government activities related to modern technology. These prerequisites for the development of preventive approaches are examined in some further detail in the next chapter.

2

# Individual Prerequisites for Preventive Approaches

## 2.1 Four Prerequisites

Society guides technology down the road of human goals and aspirations by means of countless decisions involved in the engineering, management, and regulation of modern technology – decisions made by many people working in a variety of organizations and institutions. These decision makers collectively 'steer' technology into the future, much like driving a car. Since technology does not exist in a vacuum, it shares the 'road' with people and societies who, in turn, fundamentally depend on the biosphere. To avoid 'bumping into' those who 'travel' in the company of technology, the decison makers must employ negative feedback; they need to look ahead to see where they are going, compare this to where technology should be on the 'road,' and correct their designs and decision making accordingly.

In the previous chapter, I suggested that the two-stage approach to steering technology (in which the economy of technology takes priority over the ecology of technology, and performance values are more decisive than context values) is highly unusual from a historical perspective. I compared it to driving a car with the windshield covered over, thus having to rely on the dashboard indicators to measure performance and being aware of ecology only when it intrudes from outside. The plea for preventive approaches amounts to an invitation to uncover the windshield in order to establish effective negative feedback. To properly steer its technology, society must educate professionals, equipping them with the following driving skills:

1 The ability to intellectually look beyond our specialty to learn what

others (specialists and non-specialists) know about technology that is essential for preventive design and decision making. In this way, a specialist's knowledge of the economy of technology is complemented by a knowledge of the ecology of technology.
2 An awareness of the limitations of our specialty to prevent intellectual blind spots. As is the case when we are driving a vehicle, an awareness of such blind spots is essential if we are safely to reach our goals.
3 A critical awareness of the prejudgments we bring to any situation, resulting from a previous experience. Growing up in a society that makes extensive use of technology in all aspects of its way of life involves the acquisition of an informal ecology of technology that can impair our ability to steer technology preventively. The situation is analogous to a self-taught driver who has to unlearn some bad habits when later taking training in defensive driving.
4 The ability to apply context values to complement performance values, so that design and decision making may be guided and assessed and, if necessary, redirected towards a future that is human and sustainable.

These four skills are the subject of this chapter.

## 2.2 A Map of the Ecology of Technology

A first 'driving skill' that is necessary for the development of preventive approaches in the engineering, management, and regulation of modern technology is the ability to think about technology in terms of patterns of relationships. It involves breaking with the intellectual habit of thinking about technology as a means to accomplishing certain ends. This is true, of course; however, technology simultaneously structures and restructures relations in the fabric of individual human lives, society, and the biosphere. To continue our previous metaphor, thinking about technology in terms of relations tells us where it is on the 'road' and where it should be, according to a society's values and aspirations.

Imagining the plausible situations outlined here may help us recognize the extent to which technology establishes ways of connecting with fellow human beings, society, and the biosphere. What would our world be like if we woke up one morning to find that a particular technology had vanished? If there were no cars, would we live as far away from work or school, and how would our social relations be affected? Would it be possible to sustain friendships and close ties with relatives who live far away? Is it not likely that we would live much closer to school and work, and that most of our friends would be from our immediate neighbourhood? Would there be any suburbs, and how would that affect the shape of cities? Suppose that tomorrow morning there were no elevators: How would that affect the people who live and work in downtown high-rises, and what would have to be done to make the core of cities habitable again? What would happen if we woke up tomorrow and there were no telephones? These kinds of hypothetical explorations help us understand how the technology of a modern society assists us to establish patterns of relationships that connect the members of a society to one another, and to local ecosystems, and, via those ecosystems, to the biosphere.

Such thought experiments reveal the difficulties many of us have experiencing technology in terms of patterns of relationships. We lack reference points, not having lived in a world without cars, elevators, or telephones. Even when we do have a reference point, rarely do we experience directly the influence of particular technologies on the patterns of our lives. A number of technologies are directly and indirectly involved in each moment of our lives; however, they cannot be experienced one at a time, and their influences often extend beyond these moments. Many of their effects are so small and subtle that they would be negligible were it not for the fact that other technologies have similar effects. This synergy of technologies may have a significant influence on people's lives, their communities, and their relations with local ecosystems, but is almost impossible to experience directly. The obvious effects of technology as means are therefore complemented by effects best perceived in terms of technologies that mediate relationships in a non-neutral manner. When the density of technology-mediated relations increases in individual and collective human life, fundamental transformations occur that are almost impossible to experience directly.

This shift in thinking represents another way of getting to know technology, which is somewhat analogous to getting to know people at a party. As we are being introduced, we first relate to others in terms of their physical appearance, which, in our culture, quickly translates into assumptions about who they really are. To get to know them is to go beyond these initial impressions and to become involved in their lives, that is, in the fabric of relations constituting their being in the world.

We may enquire what they do for a living or whether they enjoy the music being played, or we may endeavour to connect to some other aspect of their lives. In a society, the fabric of an individual's life relations is seamlessly interwoven with those of others to make the whole.

It is helpful to think of technology as having a 'body' and a 'life.' The former is evident in the physical presence of the technologies all around us. The latter is generally given much less attention, yet, for our purpose, it is much more fundamental. The 'life' of a technology reveals the way it depends on, interacts with, and is embedded in its contexts. The study of this 'life' leads to an understanding that we have called the 'ecology' of a technology.

The study of the ecology of an individual technology has a static and a dynamic component. The former involves an examination of the hierarchy of relations in which it is currently involved. These range from those within the technological 'black box' to those connecting it to human activities, and, via them, to the fabric and way of life of a society, in turn existing within and depending upon the biosphere. For example, the internal structure of an engineering material may yield certain properties that are essential for a part made from it. A subassembly made up from many different parts can be designed to carry out a range of functions that no single part is capable of performing. Several subassemblies, together constituting a device or process, can be designed to perform various tasks associated with human activities. These may, in turn, contribute to still larger networks of activities, such as those related to running a household, a plant, or an office.

The second component in the study of the ecology of an individual technology involves a tracing over time of the network of relations within which a technology is embedded. Any technology has a technological cycle, which in its simplest form comprises five phases: invention, innovation and development, application, diffusion, and displacement. This dimension traces the 'birth,' 'life,' and 'death' of a technology in a society. The ecology of an individual technology reveals how that technology simultaneously participates in the order of society and in the natural order. The former depends on the latter for all matter and energy, which it can neither create nor destroy. Hence, the ecology of a technology maps the interpenetration of the two orders.

Studying the ecology of a technology brings to light the fabric of relations in which it is involved, and reveals how its positive and negative effects are interrelated. Preventive approaches can then be used to explore the possibilities of redesigning the technology to obtain a more favourable balance between its positive and its negative effects. Once again, this involves negative feedback to guide us from where we are to where we wish to go by means of a map of all the relevant relationships involved in the technology under consideration.

It is by examining the ecology of an individual technology that we will seek to introduce negative feedback about context compatibility into the engineering, management, and regulation of modern technologies. Such an ecology tells us where we are and what will happen if certain 'steering corrections' are made over time. Some aspects of the ecology of an individual technology are currently being examined through procedures including life-cycle analysis, design for environment, input/output analysis, environmental assessment, and technology assessment. The first two methods are limited to considering how a technology is connected to everything else via flows of matter and energy. Input/output methods have been adapted to modelling the flows of matter and energy in an economy. Environmental assessment and technology assessment also include the consideration of human and social relations in their examination of the ecology of a technology in addition to the environmental implications. Such tools are changing the way decisions are being made in some corporations.

In order to ensure that a map of the ecology of a particular technology is an adequate model for supporting a particular task, three things must be considered – namely, the choice of the level of analysis (macro or micro), where to set the boundaries of the analysis, and the issue of scale. The characteristics of the technology of a society cannot be inferred from those of its individual constituent technologies. Similarly, the effects of technology as a whole on its contexts cannot be inferred from those of its constituent technologies. Macro and micro levels of analyses therefore have to complement each other to create an adequate map. This overview is important when any of the implications of a particular constituent technology are trivial when viewed in isolation. However, when the effects produced by many constituent technologies within a network of activities are similar, they can be significant in combination. Setting the boundaries of an analysis by choosing the 'system' as the subject to be considered is critical in determining the outcome. This can be illustrated by the example of the continuously running furnace, whose impact on the temperature of the home was compensated for by windows equipped with electric motors controlled by thermostats. If the system's boundary is drawn around the furnace, the system may be deemed more efficient than that of a furnace operated by a properly functioning thermostat since the former system avoids the inefficient combustion that occurs during start-ups and chimney warm-ups. If, however, the compensation system is included in the analysis, the conclusion is entirely different.

The issue of scale brings us to what is commonly known as 'the tragedy of the commons.'[1] Clean air in our cities is a 'commons' in the sense that we all use it. If one person buys a car in that city, this particular commons is not affected in any significant way. However, when many people buy cars, it is spoiled, in varying degrees, for everyone. The scale of usage of a technology relative to the natural source of the raw materials or the natural sinks for the wastes becomes a critical issue in determining the compatibility between the technology and the biosphere. Similarly, minute human or societal implications of a technology duplicated to the extent to which that technology is utilized in a particular region or ecosystem, or the world, and the extent to which these effects are further reinforced by similar effects issuing from other technologies, will help to determine the compatibility between that specific technology and its human or societal contexts. The question of scale must be given a concrete form by asking: Who pays? Who benefits? Who bears the brunt of the negative implications? These are but some of the issues that come up in developing an adequate mapping of the ecology of a technology for the purpose of improving its context compatibility.

To sum up, mapping the ecology of an individual technology provides insight into where the technology is on the road, where it is heading, and where it should be according to the values and aspirations of society. It is the basis for adjusting the engineering, management, and regulation of modern technology in order to avoid bumping into human life, society, and the biosphere. Hence, the ability to map the ecology of a technology and the effective use of such a map is an essential skill for prevention-oriented practice. It amounts to intellectually looking beyond the confines of a particular specialty to see what others know about the ecology of a technology in order that we be able to use that technology preventively.

### 2.3 Awareness of Professional Blind Spots

The second driving skill required for preventive approaches is an ability to take into account the existence of professional blind spots that stem from the fact that virtually all of us involved in the engineering, management, and regulation of modern technology are specialists by training and experience. Whatever we decide and do will have consequences that fall outside of our domain of expertise. Hence, effective and responsible practice in general and prevention-oriented practice, in particular, require that we transcend our domain of expertise. This is not a question of becoming an expert in many other domains, if this were indeed possible, nor of amassing additional knowledge, but of fundamentally changing our professional vantage point.

Recall the example of the hunger problem in the Colombian valley observed by many different experts. Their diagnoses could not be added together to form a more comprehensive understanding in a manner analogous to putting together a jigsaw puzzle. Each discipline constitutes a unique intellectual vantage point accompanied by a certain prejudgment of the world derived from professional training and experience. It is generally accepted that professional training helps people to see particular situations differently from those who do not have that training. Under the assumption of objectivity, however, it is rarely acknowledged that this implies an intellectual vantage point and that, if these vantage points vary from discipline to discipline, any synthesis will have to take on a dialogical character, which, in turn, implies transcending the intellectual vantage point of one's specialty. Such conversations must also be encouraged in the public life of a nation, as exemplified by the round-table approach used by the Premier's Council of Ontario, reported in chapter 1.

How can one transcend the intellectual vantage point of one's discipline? What new vantage point must be adopted? Answers to these questions must come from a 'conversation' among all disciplines that can help shed light on the ecology of the technology of a society. In a separate work,[2] I have attempted such a conversation between disciplines to arrive at an interpretation of the ecology of modern technology. The present work anticipates this macro-level analysis, and occasionally draws upon the findings to complement the understanding of the ecologies of individual technologies associated with the development of preventive approaches. Briefly put, the conversation is structured around an examination of the emergence of the ecology of modern technology, beginning with the gradual introduction of industrial technology into a traditional society. It includes changes in the social fabric and institutions, and the effects these had on the consciousness of the members of society and their culture as they experienced these changes and internalized them in their minds and culture, and went on to incorporate them into their institutions and way of life. The relationship between people and their technology is thus a reciprocal one; such reciprocity is at the core of the general pattern constructed by the conversation between those disciplines that can shed light on or challenge particular features of it. The general pattern also includes reciprocal interactions with the biosphere as the ultimate source and sink of all matter and energy, habitat, and life support. In the course of this conversation, more comprehensive vantage points emerge, from which the macro-level transformations can be observed. Since such vantage points include our deepest sense of human life and the universe (for instance, what is good or evil, worth giving up one's life for, or just), such a conversation may produce convergence but no single vantage point. It is therefore essential that the conversation be open and democratic. The process can help us become generalists with a new perspective on our specialties that transcends their limitations by placing them in the context of the ecology of the technology of our society. Once again, it is not possible for this mapping of the ecology to be objective, any more than it is possible to take a photograph of a landscape without placing the camera at a specific point in space.

The need to be able to transcend one's own specialty and disciplinary vantage point can be illustrated by means of the following example. Suppose an engineer is given the task of finding out why one of five relatively similar plants her company operates appears to have persistent quality-control problems. Such an assignment is fundamentally different from the ones she received during her undergraduate engineering education. The nature of the problem has not been precisely defined, nor has all the relevant information (not one piece too many or too few) been provided along with the problem statement, as is customary in course assignments.

The first phase of dealing with the problem is interpretation and diagnosis. A variety of root causes of the quality-control problem can be hypothesized. It could be technological in nature and related to the machinery in the plant. It could be caused by the interface between the machinery and the people who operate and maintain it. Alternatively, it might be rooted in the organizational design of the plant used as a blueprint for the way in which people work together, including the supervisory and managerial reporting lines. Yet another possibility might be low worker morale, resulting from a long history of poor labour relations in the plant. The problem might be externally caused by suppliers who have quality-control problems. The root cause could be a combination of two or more of these possibilities, and further exploratory research could turn up still other problems.

Each possible problem statement creates a verbal image of the situation. In a manner analogous to sketching a landscape, some aspects of the operation of the plant are placed in the foreground because they play a central role in creating the quality-control problem; others may be placed in the background because they play a more peripheral role; still others may be omitted because they appear to have no bearing on the problem. Without an 'ecology of technology' perspective, the engineer, by virtue of her training, may tend to put the plant's technology in the centre. Similarly, a human-factors expert invited to analyse the problem might have focused on the human–machine interface as a critical element, whereas a management consultant might have focused on organizational problems. Viewing from a particular perspective is a normal part of human behaviour. For example, on a visit to an art gallery, you will be attracted to some paintings more than to others, but, after talking to a specialist in art history, you may begin to notice things you had not seen before, and this may affect your appreciation of the paintings. Similarly, persons with an expertise related to one aspect of the plant may not fully appreciate other aspects that fall outside their domain of competence. Besides, perfection never being attainable, those from every area of expertise will notice something that could be improved. When the benefits of the improvements are extrapolated beyond the domain of expertise, they can easily be overestimated. They may even be thought to 'solve' the plant's problems.

There is a strong tendency, therefore, for different specialists to interpret the problem at the plant in different ways. Recall the analogy of the painters' different renderings of the same landscape; in this case, the spatial placement of things in the foreground or background reflects what each artist perceived as having particular significance. If you have not seen the landscape yourself, it will be impossible for you to deduce from the different paintings what the original scene looks like. The risk of a similar situation occurring when different kinds of experts examine the same situation is not trivial. In the case of engineers, it is essential that their technical expertise be developed in the context of an ecology of technology. Only then can the risk of devising a technological fix for a fundamentally non-technical problem be avoided.

The best that can be hoped for under conditions of advanced specialization is that each area of expertise deals with one dimension of reality. Under such circumstances, the interpretations of a particular issue made from different intellectual vantage points would be complementary. However, the earlier example of the hunger problem in a Colombian valley shows that it is more likely that the differing diagnoses of a problem situation put forward by different experts are not complementary. Each discipline has its models and beliefs about how a part of reality functions. This intellectual vantage point constitutes a prejudgment of the world that is not common to all areas of expertise. Hence, each area of specialization implies its own intellectual vantage point from which the world is interpreted. Recall the earlier discussion of how introductory university texts depict technology out of proportion with other phenomena examined by a particular discipline.

The first phase – arriving at an interpretation of the situation at the plant – is followed by a second phase – translating this interpretation into a problem statement. At this juncture, it is decided what kind of expertise is required to solve the problem. Is it essentially an engineering problem, or should it be dealt with by people having an expertise in other areas, such as management science or industrial relations? Many real-life problems, however, cannot readily be classified as belonging to the domain of expertise represented by a particular faculty or department in a modern university. In many cases, these particular perspectives shed light on only one or two dimensions of the situation. The term 'dimension' is used here to indicate the multidimensional reality of the kinds of situations engineers and other experts deal with. If this is the case, the engineer may have to bring together a team of experts from a variety of appropriate fields before a meaningful solution can be devised.

The third phase in dealing with the plant quality-control issue involves the formulation of a solution. If an end-of-pipe approach is followed and the problem is technological in nature, will this phase resemble the kinds of problem-solving activities the engineer encountered during her undergraduate engineering education? Most of the problem sets she solved in school are essentially technical statements of a situation. It is important to emphasize once again that in the real world much difficult work would have taken place before such a statement is arrived at. Bringing to bear on the problem the concepts and methods she has learned in her technical courses is relatively straightforward. Next, an assessment of the context implications of the solution would have to be made to discharge the professional responsibility an engineer has towards the client and the public. To do so effectively, an engineer is often forced to leave the domain of his or her expertise because the consequences of a solution tend to fall outside of that domain. Hence, a responsible specialist must also be a generalist. In the case of an end-of-pipe approach, information about context implications is not used in a negative-feedback manner. When the solution process is prevention-oriented, it must be guided throughout by information about its context implications. Performance values and measures are complemented by context-related ones, to ensure that solving one problem will not create others. For example, if the engineer decided that statistical process-control methods could resolve the plant's quality-control problem, consideration must be given to how the information derived from these methods will be integrated into the supervisory and managerial structures of the plant. The information collected by these techniques could be sent to middle management for analysis and decision making, as was typically done in North America at one point, or it could be adapted to allow the people on the shop floor to more effectively control their work by giving them immediate feedback, thus allowing them to take appropriate action when required. Only when this fails would the next level of decision making become involved. This approach was perfected by the Japanese. The North American approach maintains the hand–brain separation and involves long negative-feedback loops, while the Japanese approach attempts to break down the hand–brain separation by shortening the negative-feedback loops and making each aspect of the organization as self-regulating as possible, thereby reducing the number of administrative levels and flattening the organization.[3] This latter approach, however, may be incompatible with the organizational culture of a corporation, and could violate the contract between management and labour by adding a management function to the job descriptions of people working on the shop floor.

In the fourth and final phase, a negative-feedback loop must be established to check the interpretation of the situation at the plant. Suppose that the industrial-relations aspect of the plant's problem had been deemed only marginally relevant to the situation, and thus had been largely neglected in the subsequent problem statement. If it turns out that implementing the solution has significant implications for labour–management relations and the 'culture' of the corporation, it is essential that the situation be re-evaluated to determine whether the original interpretation should perhaps have placed more emphasis on those aspects in the problem statement and to assess how this shift in focus might have affected the proposed solution. It could also happen that a careful examination of the effects of implementing a certain solution reveals that the quality-control systems of certain suppliers must match those systems that are to be implemented in the plant if the solution is to be effective. If this aspect had received little or no consideration in the original problem statement, the effectiveness of the original proposed solution should be reassessed in the light of this understanding.

In other words, if at any time during the process of dealing with the problem it turns out that any aspect of the operation of the plant was erroneously deemed fundamental, peripheral, or irrelevant in the problem statement, it is important to reassess the situation, retracing the steps taken towards a solution and adjusting each phase accordingly. This feedback loop can safeguard against treating symptoms as if they were the real problem and can avoid technological fixes. It also helps to ensure that any solution complies with relevant regulations and laws.

This example shows that a knowledge of the ecology of a technology is essential throughout the engineering process, particularly when context implications are dealt with in a preventive fashion. It helps the practitioner to transcend her specialty and its limitations by interpreting the situation, not from the vantage point of that specialty, but from the more comprehensive vantage point of the ecology of the technology of her society. This vantage reduces the professional blind spots resulting from specialization, as well as the risk of turning to technological fixes. Any intellectual vantage point, whether it be that of a generalist or of a specialist, involves the prejudgment of a situation based on prior experience. No person ever approaches the world as a blank slate; growing up in a particular society and acquiring its culture and subsequent professional education and experience all contribute to this prejudgment. If this were not the case, there would be no difference in the professional practice of junior and senior engineers.

How well does this example represent the kinds of tasks engineers perform? Does the ecology of the technology of society and the accompanying vantage point offer engineers a way of protecting themselves from their professional blind spots? A study of the career paths of graduates from the Faculty of Applied Science and Engineering at the University of Toronto shows that, within a short time, they move into supervisory, managerial, and executive positions. It also shows that many are working in areas other than the one in which they graduated.[4] A similar study of graduates from the Massachusetts Institute of Technology shows that, a decade after graduation, only about one-fifth continued to be employed as engineers in a strictly technical capacity.[5]

One important implication of these findings is once again that being a good engineer requires more than technical competence. Consider some additional examples. A plant manager may encounter many situations similar to the one described above. It is rare that he encounters either a purely technical or an entirely non-technical problem. As we have seen, the plant is a complex entity, with many 'elements,' such as machines, human beings, work groups, organizational structures, occupational health and safety regulations, environmental regulations, and a corporate style or 'culture,' each of which makes a unique and essential contribution to the plant's operation. Since engineers have an expert knowledge of some of the elements of the operation of the plant but not of others, the risk of creating a technological fix is ever-present. Corrective actions based on an incorrect diagnosis are not likely to be successful.

Many similar examples can be drawn from other branches of engineering. An engineer may be charged with eliminating the bugs in a new information system recently installed in his company, or in a client's bank or hospital. In another case, he may be responsible for obtaining the necessary permits to build a refinery, dam, or highway. This may involve the preparation of an environmental-assessment statement showing that the social and ecological consequences of the project are acceptable and conform to the appropriate standards. In yet another case, he may have made what appears to be a significant invention. Before he decides to set up his own business, he will want to know what its social and ecological implications are in order to determine what laws and regulations he must satisfy. In yet another instance, an engineer may find herself working for a government department or a ministry to help set standards for the use of a new technology.

Studies of career paths show that, as engineers advance in their careers, they are likely to have to deal with more and more of these kinds of issues. They will be designing, building, installing, managing, or dismantling systems within which a variety of technologies and technological systems are embedded. The emphasis on knowing how these technologies and systems function 'internally' will increasingly shift to how these functions relate to and are affected by the larger contexts within which they operate. These technologies and systems become a part of the lives of the individuals, groups, and organizations, and, via them, become a part of the way things are done in a society. The engineer's clients may not know how or why certain technologies function, being concerned only with the outcome derived from using them; therefore, it will be increasingly important to understand technologies as integral parts of larger networks, structures, and systems. The introduction of technologies into these networks, structures, or systems will change their very pattern and character to facilitate certain goals, thereby inevitably inhibiting or blocking others. At the same time, a well-designed technology will reflect the network of relations of which it is a part and, via that network, something of the fabric of a society, its way of life and culture. Archaeologists rely on these connections as they seek to learn about earlier societies by studying their artefacts. Engineers must do the reverse, and learn about their society to incorporate this understanding into their practice.

In recognizing how deeply technologies are embedded into and form a part of human lives, the fabric of a society, and the processes of the ecosystem, it becomes apparent that to be effective an engineer must understand that real-life situations are highly interrelated and do not have distinctive technological, economic, social, political, legal, ethical, and religious components. It is only through a process of abstraction and analysis that we break up an interrelated reality into its separate aspects. An understanding of the ecology of the technologies involved in any engineering task is therefore essential well beyond preventive approaches.

Will the trends noted above continue to hold in the future? If we consider the major challenges we will have to face in the next few decades, this is very likely. Engineers will have to deal with a range of issues that have significant technical aspects, but that are not technical in nature. There is little doubt that the public, politicians, and corporations will continue to enquire about present and future social and environmental implications of technology.

This brief and sketchy survey of engineering career paths, including the kinds of tasks performed and some trends changing the world in which engineers practise, leads us to the following conclusions: The situations engineers deal with range from those in which a focus on the 'internal' functioning of a technology is adequate, to those in which a broader 'ecological' focus is required to show how that particular technology depends on and fits into the fabric of society and the biosphere.

As the focus is broadened, the methods of the applied sciences must increasingly be complemented by those of the social sciences and humanities, but the perspective must remain that of engineering. Similarly, performance values and measures will increasingly have to be complemented by context ones as the focus becomes more ecological. Engineers must have a knowledge of the ecology of technology, of how technology is an integral part of larger structures and systems. The latter wholes are more than the sum of their parts, and their properties and characteristics cannot be understood or derived from those parts.

In other words, an engineer must acquire a kind of cognitive zoom lens that enables him or her to tackle a particular problem or issue from both technical and ecological foci, while maintaining the perspective associated with engineering. This development is encouraged through the creation of codes of ethics[6] and environmental management systems.[7] This is no longer 'soft' or marginally relevant to engineering practice. Engineers can lose their licence, executives can go to jail, and corporations may be bankrupted if, in the case of a serious accident, they are found negligent in these matters. On the positive side, corporations that are 'greening' themselves through preventive approaches enjoy many advantages over those that do not.

## 2.4 Awareness of Metaconscious Images of Technology

A third 'driving skill' required for the engineering, management, and regulation of modern technology through preventive approaches is a critical awareness of what we bring to the task as members of a modern society. We do not exactly arrive on the scene as detached observers. Our lives are filled with technology-related experiences that lead to our having certain mental images of technology and how it fits in as a part of our individual and collective existence. Such images may be considered as a kind of informal ecology that we bring to any task and on which we depend to a considerable degree, particularly outside of the domain of our expertise, where we can rely only on our non-professional experiences. A critical awareness of how these metaconscious images of technology may affect the formal ecology of technology can go a long way in helping us to avoid certain kinds of distortions and misunderstandings. This subject is explored here and in the next subsection.

The acquisition of metaconscious images of technology is an outgrowth of living in a society and making significant use of its technology. Recall that the relationship between people and technology is always and necessarily a reciprocal one. As people change technology through countless activities, they change something of the context in which they live. This experience is internalized in the brain by new neural connections that symbolically map and interrelate the experiences of their lives. Elsewhere[8] I have examined how the structure of symbolized experiences constitutes the mind, as babies and children are socialized into a culture. The mind functions as a mental map that permits them to make sense of and live in the world as people of their time, place, and culture. The significance of this fact may be illustrated by comparing our memory to a road map. It allows us to find our way to places we have never been before. These new places are not unlike the ones we already know: they have many features familiar to us, such as homes, roads, parks, schools, farms, bridges, and the like. But, in many respects, they are different as well. Our experience of familiar areas, together with the road map and an understanding of how that map symbolically represents a new area, allows us to find our way.

We might say that all of life involves map-making in a certain sense. Living involves changes in the brain that allow us to remember. These changes are made coherently and systematically, constituting a structure that symbolizes these experiences. The situation may be likened to 'mapping' the data points of an experiment on a graph. The meaning of each data point taken by itself is extremely limited, but together they form a scatter implying a relationship. However, such a relationship can be identified only if we go beyond the factual evidence, interpolating and extrapolating the data by fitting a curve through them. Rather than calling such a curve unscientific, our confidence in the data is strengthened because of a prejudgment corresponding to our earlier experiences of that world, revealing it to be meaningful and orderly as opposed to random and chaotic. In the context of fitting a curve through the data points, each datum takes on a deeper meaning.

A great deal of insight into human life can be derived from this simple analogy. The structure of symbolized experiences emerges in much the same way – like fitting a curve through the experimental data. This structure, although resulting from individual moments of our life, also goes beyond them so that we can experience what such moments mean in the context of our being. Without it, living a life would be impossible; we could have no personality or biography. This is obvious in cases where brain dysfunction makes it impossible for long-term memories to form in the brain. Patients suffering from such a disorder can no longer live a life, as each moment is lived separate from all others. They cannot remember what they said one minute or one hour ago, or where they came from once they turn a corner in the hall, and so on.

Map-making is a useful analogy to help us understand the interconnectedness of human life. Each moment is lived in the context of all others, and it is that context which gives the deeper meaning to each moment. Although we neither are aware of nor have access to the structure of experience that constitutes our mind, it creates a great deal of metaconscious knowledge (a going beyond the explicit consciousness of each moment) that we use every moment of life without realizing it. This point builds on a previously mentioned study of culture; therefore, I offer here only a few examples by way of illustration.

Consider the many experiences we share with our friends. One day we may say to a friend: 'You really surprise me. I didn't think you would do that!' Why are we surprised? Do we review every memory of our friend to identify patterns in order to determine his or her personality type and then evaluate the behaviour in question in terms of whether or not it fits that type? Of course not. Our surprise is an indication that we must have a mental image of our friend in terms of his or her behaviour patterns from which that particular experience deviates; we respond to that deviation with surprise. In fact, the mental images we have of one another are essential for sustaining social relations. Many examples are found in our daily lives. We may walk into a friend's room one day and remark that something has changed without immediately being able to say what it is. Looking around the room more carefully, we may realize that she has bought new curtains. In the same vein, we acquire mental images of the social roles of others: doctors, teachers, parents, taxi drivers, and so on. These images tell us how we should deal with others, what we can expect from them, and what constitutes normal and surprising behaviour.

On a larger scale, we acquire mental images of the way of life of our society and all the activities that help constitute it. Among these are mental images of modern technology, such as what it normally does, what we can and cannot expect from it, and the role it plays in our way of life and culture, including its significance or value relative to other spheres of human activity. In other words, technology, like any other human creation, finds its place on our mental map according to our experience.

The development of this mental map includes the experiences of elementary, high-school, and professional education in the case of those involved in the engineering, management, and regulation of modern technology. This becomes evident when we consider what young people take with them when they graduate from an engineering school, for example. Is it merely a matter of what they can remember from the many courses they have taken: some formulas, theories, approaches for solving certain kinds of problems, and the like? Is it simply the residue of all the material they crammed before exams? If this were the case, young engineers just out of school would flounder because, as we saw in the previous section, they have to deal with situations that rarely resemble the problems they worked on in school. Yet many of them cope rather successfully. Thus, professional education is more than the acquisition of information. However, the experiences of technology acquired in the classroom are fundamentally different in kind from the daily-life experiences of technology. The former do not deal with the technology of society, and frequently not even with individual technologies, but with sets of processes common to various technologies, represented in an abstract mathematical fashion. Daily-life experiences of technology are limited in another way. Nowhere do we experience the technology of our society. All we experience are individual technologies. These contribute to mental images of technology that are quite different in scope and quality from the mental images built up from the experiences of looking at technology through the lens of a particular specialty. Outside of the domain of specialization, engineers or other technology-related professionals have no choice but to rely on the mental images of technology built up from the experiences of daily life. The two kinds of mental images correspond to two very different modes of experiencing and knowing technology. Nevertheless, together they help constitute the prejudgment with which any practitioner approaches a situation. A formally constructed model of the ecology of a technology of a society and the metaconscious images of technology also correspond to different modes of experiencing and knowing; given the strengths, weaknesses, and blind spots of these different modes, the ecology of technology and the metaconscious images of technology are bound to differ significantly and may even contradict one another.

This may be illustrated by examining two kinds of mental images of technology common in modern societies. In discussing these examples, it is essential that we keep in mind that what we are trying to make explicit are mental images implicit in the structures of our experience, to which our thinking has no direct access and that can only be inferred from our behaviour.

The image of technology included in the mental maps of people living in a modern society has two dimensions: one related to the meaning of technology and one related to its value. Both have their origin in symbolically giving technology a place in our individual and collective life, as represented in our structures of experience. Simply put: the meaning of technology is identified in terms of its significance relative to everything else, and its value is identified in terms of its importance relative to everything else. All this occurs metaconsciously in people's minds, providing them with a sense of the proper role of technology in the way of life of their culture; what they can and cannot expect from it; and how it, along with other human creations, sustains that way of life. I first examine the dimension of meaning of the mental images of technology common in a modern society.

After critically reading the material we have presented thus far, some readers may react (in terms of their mental image of technology) as follows: A knowledge of the contexts of technology may well be useful, but when all is said and done it is really society that decides how technology is used, and that usage determines the consequences. You can use a knife to cut bread or you can use it to seriously injure someone. The manufacturer of the knife cannot control the use a person makes of it. The same is true for a car: It may be used to get from one point to another or to commit murder by running someone off the road. Thus, it is argued that technology is neutral because its consequences are determined by its use. Engineers cannot and should not control the use society makes of technology; instead, they should stick to being technically competent and produce the most efficient and cheapest technologies.

According to this view, the members of a modern society go about their daily-life activities 'using' many technologies. This usage is neutral in the sense that, when a technology becomes available to facilitate a particular activity, it is used and then 'put down' when a task is finishe, without in any way fundamentally altering the activity, the life of the user, his or her relations with others and, via them, the larger fabric of society or the ecosystem. The technology is neither good nor bad, and acquires a value only through the kind of use we make of it. For example, the automobile has simply made transportation more rapid, comfortable, and easy. A car is neither good nor bad in itself, because its usage determines the consequences. From this perspective, technologies are invented and introduced into society for the purpose of facilitating existing activities or permitting the development of new ones for which a need has been expressed. Technological neutrality thus implies that technology is entirely socially determined. A technology is what a society makes of it.

The limitations of this mental image have their roots in our metaconsciously going beyond the daily-life experiences of individual technologies, beyond professional experiences, and beyond the vantage point from which technology is observed (as represented by a person's socioeconomic status in society). This extrapolation allows the word 'technology' to attach itself to our daily-life experiences of individual technologies – a procedure that, on the conscious level, would be the equivalent of extrapolating the characteristics of the parts to the whole, which is problematic, as our earlier example of determining the effects water has on paper illustrates. In the case of a knife, for example, the manufacturer indeed has no control over its use. A knife potentially has a variety of uses, including prying a nail loose or turning a screw, although it is most commonly used to do what it was designed to. In such cases, it is widely recognized that many of the social and environmental consequences are determined by the design. In the case of cars, it should be remembered that we cannot separate the car from the transportation system, without which it cannot function. Do individuals or groups have the ability to use this transportation system for their own ends? Can the influence of this system on our habitat be neutral? The metaconscious integration of professional experiences of technology further reinforces the mental image derived from daily-life experiences. These experiences are not so much those of individual technologies as they are of a two-stage technological approach to a variety of situations. The approach implies a neutral and objective first stage, and a limited consideration of context implications flowing from usage in the second stage. For both sets of experiences, the reciprocal interaction between people and technology is not noticed. This should come as no surprise because an individual does not readily experience the effects of an individual technology. When this fact is extrapolated to technology as a whole, it can be concluded that no significant effect is apparent, further reinforcing the image of the neutrality of technology. As far as the influence of the vantage point is concerned, the mental image of technology as a neutral means is plausible only in those situations in which the influence people have on the technology of their society is experienced as being far greater than the influence of that technology on them and their way of life. This tends to limit the occurrence of this mental image to those sectors of society who see themselves as exerting an influence on technology, as opposed to those sectors of society who see themselves primarily on the receiving end, as is the case for assembly-line workers. The mental image of technology as neutral is therefore emotionally and morally satisfying for one sector of society because it transfers to society the responsibility for the use of technology and for its consequences. These become political matters, which engineers, managers, and regulators of technology must leave to the democratic process. Their professional responsibilities to society are thus sharply reduced, and preventive approaches are all but removed from the realm of possibility.

An apparently opposing mental image of technology is commonly referred to as 'technological determinism.' According to this perspective, technology throughout the ages develops by its own internal logic and processes, and is not significantly influenced by the human, social, or natural contexts. For example, a turbine is invented when all the prerequisite developments have taken place. These include practical and theoretical concepts, special materials and the technologies necessary to work them, critical subcomponents such as high-speed bearings, and other technologies to which the turbine is connected and on which it depends. The strong interdependence of all the different branches of the technology of a society produces an internal 'logic' of development. After all, each technological advance presupposes earlier advances, and it is only when these prerequisites have come into being that the next step can be taken. When the results of this development are accepted by a society, it is obliged to make adjustments and accommodations without which the technology could not be used. Here technological development behaves as a kind of independent variable that creates potentialities a society must decide either to use or to ignore. It is highly probable, however, that there will be at least one society in the world that will choose not to utilize the development. When this happens in a situation where, through a global economic order, societies fundamentally influence one another, the choice may quickly become untenable.[9] This was particularly evident in the arms race. The generals on the one side, learning of the possibility that the latest technological advances could produce a new weapon, argued that they had to have this weapon before the other side could build them. There was no real choice here – after all, national security was a matter of life and death. The same pattern of events is likely to repeat itself with other developments, such as the electronic information highway, high-resolution television, and biotechnology. In a global economy, can any corporation forgo getting involved in the latest technological developments in its field on the grounds that their negative impacts may become unacceptable with widespread use? Can any political party argue that the current state of the art in television technology is adequate and that resources should be diverted to other pressing social issues? Have the dictates of technology that first manifested themselves in the arms race now spread to the global civilian economy?

The flaws of this mental image of technology are similarly rooted in daily-life experience lived from a different socio-economic vantage point, the one that comes with being mostly on the receiving end of technology. People in the workforce whose jobs are at risk of being automated or eliminated as a result of restructuring share the experience of an unstoppable technological advance against which they are powerless and which can radically change their lives for the worse. The technologies of the workplace are not just any technologies, given the vast consequences they have on people's lives. The experiences of other technologies such as television reinforce these metaconscious patterns, to the extent that they, too, require a passive level of acceptance whereby the sounds and images are allowed to flow over a person and generate no more than a reflex response. This mental image of technology tends to be more common among people in the lower socio-economic strata of society. It reflects the experiences of people who fear for their jobs, who have a sense of having lost control over their lives because they no longer know what may happen to them tomorrow. This frequently leads to a sense of resignation, which may be expressed as 'That's the way the system works – you can't do anything about it' or 'As a nation, we have lost political control over our destiny – if we don't make use of the latest technological developments, other nations will, and we will be left behind.' In a slightly modified form, this mental image of technology is also present – although it is still somewhat rare – in higher socio-economic strata of a modern society. For example, it is sometimes found among people involved in the engineering, management, and regulation of modern technology. It corresponds to a sense of being able to do little but provide society with as many technological options as possible from the resources that have been allocated to technological development. It is then up to society to choose what to accept and incorporate into its way of life and what to reject. It may also be felt by young professionals having difficulty finding work or by those being displaced by rapidly changing frontiers. In moments of candour, senior executives of transnational corporations also admit to a feeling that their decision latitude has been significantly curtailed during the last few decades, possibly as a result of the globalization of technological development. Any invention or innovation anywhere in the world can suddenly threaten their business strategies. The classic example is that of the Swiss watchmaking industry, which suddenly found itself threatened by firms mass-producing electronic watches.

Once again, this mental image of technology constitutes a barrier to the development of preventive approaches for the engineering, management, and regulation of modern technology. It implies that, if society ultimately decides which technological developments to put into practice, even when these choices may be constrained by other developments around the globe, it becomes the social responsibility of engineers and managers to provide as many technological developments as possible. In other words, engineers and managers should preoccupy themselves with the internal relations of technology (i.e., inside the technological black box) and leave the external ones (i.e., between the technological black box and its contexts) to the politicians and society at large. This mental image also implies that the regulators are limited to dealing with the unacceptable influences of technology in an end-of-pipe fashion.

In sum, the two mental images of technology – namely, that of social determinism and that of technological determinism – are in many ways complementary and have very similar effects on individual and collective behaviour. Both images are obstacles to preventive approaches for the engineering, management, and regulation of modern technology. They reinforce the distinction between the 'hard' technical and economic 'core' of design and decision-making processes and the 'soft' aspects dealing with the consequences for human life, society, and the biosphere. Both images encourage a certain passivity: a sense that there is little engineers, managers, and regulators can do about the ecology of technology, other than satisfying regulatory and legal requirements. The rest is up to the politicians and society. Why bother with the ecology of technology and preventive approaches? Since we have no direct access to our mental images of technology, the way we can access past experiences, their influence on our thinking and doing remains hidden until they are critically analysed. However, the cultures of the industrially advanced nations imply that such critical scrutiny is unnecessary, and this brings us to the second dimension of these mental images, having to do with the value attributed to a modern technology.

Not only do we assign a meaning to everything within the horizon of human experience, but we also ascribe a value. Just as meanings reflect the place we metaconsciously give to everything in relation to everything else within our world of experience, so values reflect the relative importance of everything in relation to everything else. There is no doubt that technology is exceedingly important. Our lives and our habitat are full of technologies. Almost nothing gets done without one kind of technology or another. Vast technological systems provide us with all the necessities of life, and their failure poses a fundamental threat to our lives. As technology has an implicit role in the structure of all the experiences of our lives, our mental images confer a very high value on it. How could we possibly be casual about and indifferent towards something without which our lives would be unimaginable, unbearable, and unliveable? We are as dependent on technology as prehistoric peoples were on what we call nature. In both cases, individual and collective consciousness is profoundly affected by a particular phenomenon, which is understandably treated as something very special and has one of the highest assigned values in that culture. Consequently, the value dimension of both mental images of technology (social determinism and technological determinism) reflects the fundamental importance of technology for human life and society. Is there any other phenomenon that plays such a decisive and essential role? All this has far-reaching implications. What a society, through its culture, declares as very important and good or inevitable deflects critical scrutiny. When, in a daily-life situation, we are confronted with a serious negative consequence of an individual technology, our mental images of technology may be threatened, and strong emotions can deter us from carefully considering whether that technology should be abandoned. Reactions may include: 'You cannot stop progress! Do you wish to go back to the Dark Ages?' 'If we do not use all the latest technologies, others will, and we may get left behind!' For example, it is very difficult in a modern society to have a reasonably dispassionate conversation about the benefits and risks resulting from the use of nuclear power.

When a culture assigns technology a very high value, negative feedback required for its regulation becomes almost impossible. The reason is that negative feedback requires a 'critique,' much like a thermostat constantly evaluates the temperature in a room by comparing it with a set point or goal. After all, why would a society critique something it has already metaconsciously declared to be very good? The culture and hierarchy of values of a modern society are biased against it, thus further reinforcing the separation of the economy of technology from the ecology of technology and the end-of-pipe approach to mitigation and regulation. If someone insists things be done otherwise, the very roots of our existence are threatened, and emotions run high. Societies have a hard time answering questions such as: Do we really need a supersonic transport plane, smart highways, a map of our genes, still-faster computers, or high-resolution television; or are there more pressing needs, such as jobs for everyone, rehabilitating local ecosystems, and improving people's health? Are current technologies not adequate, given these needs? It is highly unlikely that we could discuss such matters dispassionately, given the value dimension of the mental images of technology implied in the structures of our experience in our minds.

Intellectuals and specialists are not exempt from these issues, which go far beyond the domain of anyone's competence. Hence, they, too, fall back on their own metaconscious images of technology. Philosophers of technology have spent a great deal of energy articulating the two mental images of technology, almost without exception coming to the conclusion that technological determinism is an impossibility, given that technology is a human creation. Anyone who dared to raise concerns over where technology might be taking us was quickly branded a pessimist or, ironically, a technological determinist. Rather than encouraging a critical attitude towards technology, it lulled people to sleep with ahistorical and asociological arguments. We all have within us these barriers to preventive approaches for the engineering, management, and regulation of modern technology.

In discussing the common mental images of technology implicit in the behaviour of the members of modern societies, I have attempted to make explicit what appears to be implicit in our behaviour, intruding into our consciousness only as feelings, emotions, and intuitions. Many people may not associate these images with technology – at least not consciously. It is a question of feelings about one's life, some vague sense of what appears to be happening deep down, and some awareness of powerful currents that from time to time appear to sweep us along, and against which, at other times, we seem to be able to swim.

Both types of images give us a sense of helplessness, preventing us from struggling to regain control over the most powerful creation of humanity. No one is exempt from these deep currents in the metaconscious of the structures of our experience. However, as in earlier times, a growing critical awareness of these undercurrents is the first step in symbolically distancing ourselves from them. This happened when societies emerged at the dawn of history, interposing themselves between the prehistoric group and nature. The same thing must now happen with respect to our technology-permeated life-milieu. It is a question of what drives technology. Are our aspirations, values, and desires freely arrived at, free from the influence of technology, or are they the result of the influence of technology on our minds and cultures? This leads to the issue of human freedom or alienation.

A critical assessment of the two metaconscious images of technology discussed here leads to the following formal interpretation of the ecology of technology. Just as the interaction between people and technology is a reciprocal one, so, too, is the relationship between a society and its technology. If, during a certain period in the history of a society, the ability of its members to create, use, and control technology is much stronger than the reciprocal interaction by which it influences the society, the situation tends towards the 'social determination' end of the spectrum of possible relations. If, on the other hand, the influence technology has on a society is for a certain historical period much more decisive for its evolution than the ability of that society to use technology according to its own purposes and values, the situation tends towards the 'technological determinism' end of the spectrum. Either tendency cannot be ruled out a priori but must be carefully examined and assessed in each particular historical situation. This view is based on the recognition that, as a society develops its technology, that technology also shapes society. The technology becomes a part of the physical setting of the society. As such, its members experience it, and since these experiences are internalized in their minds they become a part of their lives and culture. It is true that, for prehistory and most of human history, individual technologies did not dominate people's habitat. Hence, the influence technology had on people was relatively insignificant and could be neglected. This led to mental habits and approaches in the social sciences and humanities that had so much inertia that, when the human habitat became permeated by technologies of all kinds, the possibility that such technologies might have a greater influence on people than people had on them simply did not come to mind.

A recognition of this possibility would undoubtedly bring about a scientific revolution in the social sciences and humanities that would put much more pressure on the professions engaged in the engineering, management, and regulation of technology.

The recognition that, as a society develops its technology, it is at the same time changed by it raises critical questions. Depending on the relative strengths of these two interactions, a technology can serve a society or vice versa. In the latter case, a society serves the very technology that was created to serve it. This possibility, though largely unacknowledged by us, is implicit in such everyday expressions as 'You can't beat the system' or 'That's the way the system works.' The fundamental questions of human freedom and the possibility of making genuine political choices can be asked only when the relative strengths of the reciprocal interactions between technology and society have been carefully examined in a particular situation. The influence technology has on its host society can manifest itself in the ecology of individual technologies. For example, machines can be designed to help human beings do their work or to control and dominate them. Either case implies very different work-design principles.

This critical view of the relationship between technology and society cannot rest on daily-life experience because technology is not encountered as a whole in that experience and because the extrapolations of the daily-life experiences of individual technologies, as characteristic of technology as a whole, are flawed. Consequently, this formal ecology of technology goes against the common sense derived from daily-life experience and can emerge only from a critical analysis of the situation. This macro-level analysis can be illustrated by a series of examples.[10]

## 2.5 Some Examples

As a first example, consider the reciprocal interaction between people and cars. People design, build, purchase, use, and discard cars. Their influence on cars is obvious and needs no further elaboration. The reverse interaction receives much less attention. No machine has had a greater influence on cities, which constitute the habitat for a growing number of people as industrialization progresses and which affect human beings in a variety of ways. In prehistory, human beings lived in small groups and relied upon hunting and food-gathering, in a habitat we call 'nature.' That habitat is known to have profoundly affected their way of life, including their customs, institutions, morality, religion, and art. There is little reason to believe that the influence our urban-industrial-information habitat has on our consciousness and cultures is any less profound. Since the car has substantially affected the shape of that habitat, cars likely have as much influence on people as people have on cars. Consequently, cars can hardly be regarded as a neutral means in a modern society.[11]

Next, I propose to focus on some examples related to computer technologies to show how these can (in combination with other technologies) produce synergistic effects, and that the technology of a society as a whole produces effects that cannot be inferred from those of individual technologies. First, consider the most elementary use of computers as word processors. The computer as word processor appears to be a simple and neutral improvement on a typewriter, yet it requires a significant adaptation of the office if it is to be used effectively. In addition, the effect extensive computer use has on human beings appears to influence how they think of themselves, and these changes in their self-image affect their social relations with others.[12] Simply put, they begin to think of their brains as the best computer available, and liken a variety of human functions to those of the computer. This is evident in language, when we use the same words for human and machine memory, vision, communication, and other functions. To some people it may become quite natural to say: 'You are not programmed for that. Let's sit down and talk about how we can improve our interface.' Or 'I can't process information on this level.' The Internet appears to be changing the meaning of words used to describe human groups in so far as networks of people exchanging messages on the Internet are essentially seen as similar to traditional 'neighbourhoods' or 'communities.' However, even the most casual comparison will show that such people have almost nothing in common. The influence of the computer on a human user is not surprising when we realize that many of our experiences associated with it can be recalled from memory. This means that the computer does not simply exist as a part of our environment, but also exists within our memory, which is an integral part of the human mind involved in every activity in our life. There exists considerable evidence to suggest that a great many experiences of computer use internalized in someone's mind affect the very structure of that mind and by extension, the person's activities and life. The use of a computer, therefore, is a reciprocal interaction: our using it for some purpose, and these experiences becoming a part of our life, and thus of who we are.

The introduction of computers into the workplace has not been a neutral process. Once again, careful studies show that organizational adjustments have to be made in the way people work together in a plant or office if the computer is to be used effectively. An institution installing computers and various information systems tailors them to the organization's unique needs and requirements; at the same time, its internal functioning and the way it performs various services and interacts with its contexts are also affected.[13]

On the political level, the computer has been heralded as a technology that will allow the decentralization of large organizations and permit greater citizen participation in political decision making. Some observers have pointed out that these expectations resemble the ones people had of electrical-power technology at the end of the last century. It was thought that this technology would decentralize industrial society because people could now have their own workshops in their basements, so to speak. What this expectation overlooks is the huge bureaucracies required to build, operate, and manage the power stations and the vast electrical-power grid that will make it all possible. Any attempt to link computers on a massive scale and to create totally integrated data banks may lead to information utilities, possibly requiring even larger organizations than the electrical utilities, given the vast and growing amounts of information modern societies require. In that case, the balance of power between the individual and large organizations in democratic societies may well not shift towards the individual, even when he or she can use his or her own computer to access these data banks, because people will have little or no control over how data banks are structured and the kinds of information included in or excluded from them.[14]

Some studies have raised concerns over the way computers can be used to invade human privacy. Electronic traces of so many of our daily-life activities – namely, our educational records, health records, credit-card purchases, bank transactions, telephone communications, travel reservations, insurances, employment records, taxation data, encounters with the police and others could be assembled to constitute a profile of what we do. These could be compared with what the 'average' person in the same income bracket, educational level, and social status does for the purpose of detecting any deviance, the cause of which can then be investigated. When the state is threatened in a time of crisis, it may be unable to resist the temptation to invade people's privacy, to locate and investigate those whose lives show some social deviance. The consequences for democracy and individual privacy are profound.[15]

The computer has also affected our legal arrangements. It has made possible new relationships between individuals, between individuals and groups, between groups and institutions. It has opened up many new possibilities, including a new range of socially destructive behaviours. These new activities and altered existing ones have required new legal arrangements, thus changing the legal framework of society. Whether these transformations on balance are positive or not is difficult to determine, but they are certainly not neutral.[16]

There are many more examples of the reciprocal interactions between computers and society. Some artists and musicians have found the computer to be a wonderful new tool, and once again the interactions have not been a one-way street. As computers are absorbed into a variety of other technologies, their effects will undoubtedly continue to spread, with considerable consequences for human life, society, and the biosphere.

It should be remembered that, as the use of computers diffuses throughout society, other technological developments occur as well. Together they create a multitude of new relationships, alter existing relationships, and displace still others. A consequence of the many new technologies diffusing throughout society is an exponential growth of ignorance about their effects, because time constraints, limited research funding, and other economic pressures prevent serious study of these new developments. As a result, society is confident that, with the exponential growth of scientific knowledge, it will enhance its knowledge of human life, society, and nature so as to better cope, but very few have recognized that there is also an explosion of ignorance. What the net growth of knowledge or ignorance may be is impossible to assess, but it is a possibility that ought to rein in excessively optimistic expectations for our common future.

Synergistic effects between technologies may be positive or negative, and transcend the sum of the individual effects. I have suggested that several governments, as well as individual researchers, have expressed concern over the impact information technology may have on the privacy and freedom of the individual. This possibility could become a reality if nuclear technology turns to plutonium as its primary fuel. Concern has been raised that this highly toxic material, of which only a very small quantity is required to make a nuclear bomb, could fall into the hands of terrorists, organized crime, or anyone else who might want to blackmail society.[17] If such an incident were to occur, the public would probably demand that the government do whatever was necessary to recover any stolen plutonium. In such a case, the surveillance potential created by information technology could prove to be irresistible, even to the most open and democratic societies. The invasion of human privacy, the loss of freedom, and the increased power of the state with regard to the citizen would certainly alter the character of a society. My point here is not to debate the likelihood of such a scenario, but simply to point out that the combined effects of these two technologies is much greater than the individual effect each one would have in isolation from the other.

The simultaneous and reciprocal interaction between technology and society has been studied for almost every individual technology. For example, the choice of private automobiles over public transportation has had an enormous influence on the shape of the modern city and the lifestyles of its inhabitants. In a fascinating study of large sociotechnical systems such as nuclear-power stations, chemical plants, airports, shipping, mines, and dams, Charles Perrow suggests that some of them may require a military-style organization to operate safely.[18] This has implications for the kind of society these systems help create.[19] The large corporation is one of the most important institutions of a modern society. Its internal organization is not based on the democratic principle of one vote for each person, and many would argue that a corporation cannot be run without an authoritarian organizational structure. Others oppose this argument on the grounds that it simply justifies the position of a social elite using these corporations for their private gain, and that democracy does not have to stop at the plant gate or the office door.[20] The mechanization of agriculture, and the rise of agri-business and biotechnology, affect the viability of the family farm and, consequently, the entire rural way of life. [21] (It is clear that a large combine is incompatible with a small family farm.) These developments also have a profound influence on the kind of food being produced, which in turn affects the health and all the food-related habits of a culture. As solar technology develops to the point where it can make a significant contribution to satisfying our energy needs, a decision will have to be made whether to construct large centralized systems using solar farms, partly decentralized systems based on neighbourhood installations, or completely decentralized systems where each building has its own collectors. These choices will help to determine the kind of society we will live in tomorrow.[22] The mechanization of bread-making was not simply a question of improving efficiency and productivity. To allow the use of machines, the very nature of bread had to change and, along with it, consumer tastes.[23] Some technological choices have less to do with gains in efficiency and productivity than with social choices. A classical case is that of the bridges over the parkways on Long Island, New York, which prohibit the passing of vehicles taller than nine feet. This height restriction bars buses from the route. Since, at the time, buses were the predominant mode of transportation for poor and black Americans, they were thus excluded from entering certain recreational areas.[24] The adoption of new production equipment in some cases appears to have more to do with the balance of power between labour and management than with gains in efficiency and productivity.[25]

These examples are but a small sample of how individual technologies are involved in the way we create and re-create our daily-life activities. Individual technologies are rarely neutral because, in helping to shape the contexts in which we exist socially and physically, they require that we adapt to their influences. Thus, the technological inventions of a society help constitute its physical setting, in relation to which that society evolves. It is possible, of course, that under certain circumstances individual technologies are almost neutral, but the technology of a society as a whole never is. Examples used to support the former premise must not be confused as evidence supporting the latter one. The popular belief that technology is neutral denies that the interaction between a society and its technology is a reciprocal one.

The widely and deeply held images of technology based on daily experience render the above examples controversial, and so they should be. They raise questions about the kind of society we have created and helped evolve, including the lives people live within it. Our deepest values, beliefs, and expectations act as a 'set point' to which we refer in evaluating what is happening and in determining whether a development is good or bad, acceptable or unacceptable, personally disappointing but socially desirable, and much more. As we share our feelings with others and they share theirs with us, our lives evolve and we learn to adjust our 'set points.' The ability of a society to evolve and adapt to new circumstances depends greatly on its individual and social diversity. Free and open societies have a considerable advantage here over more totalitarian ones, which suppress this diversity. Through the many daily-life interactions, a free and open society conducts an ongoing 'debate' over the kind of society we wish to have for ourselves and our children. It improves the likelihood of society making the best possible choices and decisions.

In sum, the third driving skill is the ability to transcend our mental images of technology by creating a map of the ecology of the technology of a society that shows how technology fits into, depends on, and interacts with society and the biosphere. Two reciprocal interactions must be included. First, there is the effect a society has on its technology through the many decisions individuals make every day regarding the diverse aspects of individual technologies, such as research and development, production, marketing, use and management, and replacement. Second, these decisions help change the social and physical contexts in which we live, and to which we must adapt in some fashion, with the result that, as a society changes its technology, it is simultaneously changed by that technology. The same is true with regard to nature. Changes in a society's technology alter its way of life, and thus the impact it has on local ecosystems. At the same time, the latter changes affect society.

The metaconscious images of technology acquired by living in a modern society affect design and decision making related to the engineering, management, and regulation of modern technology. These images of technology common in modern societies generally imply that it is either neutral or deterministic. In either case, engineers, managers, or regulators will not be motivated to examine the ecologies of the technologies within their domain of expertise. Why should they? After all, their informal ecologies lead to a sense that this would be quite useless because technology either is essentially socially determined or follows its own course. Usually this perspective is reinforced during their professional education, a subject that is examined in chapter 3.

It is clear that the metaconscious images that engineers, managers, and regulators of technology acquire by virtue of living in a modern society readily coexist with their expertise in a particular area, mostly guided by performance values. Beyond the domain of their expertise, they rely on their metaconscious images of technology. Since they do not live two lives – one within their domain of expertise and the other in their daily lives – there tends to be a compatibility between their ecology of technology and their metaconscious images of technology. A critical awareness of the metaconscious images of technology and a prevention-oriented practice are also compatible, but in a very different way, because they imply a reciprocal relationship between technology and society. In most cases, therefore, a critical awareness of metaconscious images of technology requires a comparable change in the orientation of professional practice.

Professionals involved in technology approach any situation with a mental map. Acquired by growing up and living in society, this map includes metaconscious images of technology. This map, full of mental images, identifies which relations are important and which are not, which relations are likely to occur and which are not, which relations are rendered invisible and unimaginable because of fundamental beliefs and convictions, and the patterns of causality that group relations into networks. It is, therefore, essential that professionals are aware of the informal mental maps they bring to their tasks, and how these maps are likely to enhance or interfere with the building of an adequate ecology of technology for preventive practice.

## 2.6 Human Values

A fourth prerequisite for preventively 'steering' technology is a set of human values and aspirations towards which technology can make a genuine contribution. It is operationalized in many forms, including what is considered normal practice, due diligence, codes of ethics, standards, regulations, laws, and institutional 'cultures' embodying the experiences of many years. Any person involved in the engineering, management, and regulation of modern technology by virtue of professional education and experience acquires a mental image of what are considered acceptable or good 'driving habits,' which performs a function analogous to that of the set point on a thermostat regulating a furnace. In addition, the explicit set of values used to 'steer' technology must be independent from technology, which today is no longer the case, given the dominance of performance values over context values. Once again, the acquisition of this fourth driving skill begins with the development of a critical awareness of some bad driving habits before new ones can take their place.

Attempts during the last few decades to improve the 'driving habits' of those involved in modern technology by reinforcing codes of engineering and business ethics, corporate standards, and regulations have not been very effective because they, too, imply an end-of-pipe approach. Recall that the first stage in the conventional approach is based on a fundamental value judgment that permeates all others – namely, that performance values should take precedence over context values whose consideration is deferred to the second stage. Hence, the efforts mentioned above leave the first stage largely unaffected. The same is true for undergraduate courses in engineering ethics or the ethics exam that must be passed in many jurisdictions before professional registration. All such efforts appear to imply an ethical vacuum or an absence of a professional ethos that must be filled by a rational approach developed externally by specialists in ethics, as opposed to beginning with a critical examination of the values implied in engineering theory and practice. Becoming a member of a professional community is not unlike growing up in a society. Probably the overwhelming majority of the members of a society have never read all the laws and regulations that are on the books. Nevertheless, they are spontaneously obeyed by most of the population. When this ceases to be the case for a particular law, a judge may declare it to be inapplicable for the simple reason that it is impossible to prosecute, convict, and fine or imprison a substantial portion of a population. The reason most laws are spontaneously obeyed is that they embody the fundamental values implied in the behaviour of the members of a society, and in their structures of experience. The same kind of process occurs during 'professional socialization' into a 'professional culture,' over and above whatever practitioners may explicitly and formally learn about ethics and values as they grow up in a society. Professionals spontaneously obey those values implicit in the structure of experience that correspond to their professional training and practice. It is these values, implicit in theory and practice, that must be critically examined as a first step towards change. This inevitably leads to preventive approaches guided by both performance and context values. It is a struggle against the influence technology has on our values, which is manifested in the prominence of performance values, now used as if they were human values.

A macro-level analysis of the ecology of modern technology reveals two issues that stand in the way of bringing fundamental reforms to the values that underlie the engineering, management, and regulation of modern technology. The first is the widespread absence of an awareness of the limits of technology in modern societies. Like any other human creation, it cannot be omnipotent – no human creation ever is. Yet most of us have difficulties imagining spheres of human activities to which technology is either irrelevant or harmful. Does this mean that our mental images of technology hold it to be almost omnipotent? Can high technology save us from most of our problems? Can it protect us from our foes, reduce the pressures on our economy from foreign competition, solve the problems created by the older industrial technologies, reduce the threat of serious diseases, reverse the decline in the livabililty of our cities, revitalize democracy, and empower individuals in the face of large bureaucracies? It would appear from the typical public policies being pursued in many of these areas that the answer is 'yes.' This raises the question whether our expectations of high technology are realistic or whether this human creation is seriously overvalued. If the latter is the case, then Abraham Maslow's epigram might apply: 'If your only tool is a hammer, all your problems look like nails.'[26] In other words, if a culture overvalues one of its significant creations and expects too much from it, it will simply discover that this creation cannot deliver, and the society will find itself no better off, despite having committed massive resources. A society may act as if all its major problems are technological ones, that is, amenable to technological solutions. This view leads to the treatment of symptoms rather than root causes. As a result, we are lost at the technological and economic frontiers, no longer having any clear idea where we are going and why, the only certainty being that we must stay afloat in a globalizing economy. Preventive approaches are therefore an act of human freedom in the face of technological values.

The second issue is related to the fact that a society assigns the highest value to what in its experience is so important, central, and fundamental to its way of life that without it life would be unimaginable, inconceivable, and unliveable. It becomes the value of values, or what cultural anthropologists have called the sacred or central myth of a society.[27] Two attitudes become possible in such a case. The members of a society may decide that this phenomenon is all-important and all-powerful in terms of creating and sustaining their world, and thus is their destiny or fate, and resign themselves to it. Fortunately, this socially destructive attitude has never been adopted by a society. The second response, well known in the sociology of religion, is to recognize that there is no higher value, so that such a phenomenon becomes good in itself. This process of sacralization results in the formation of religious attitudes, with the consequences mentioned above for the case of high technology.

In sum, the fourth driving skill for the development of preventive approaches for the engineering, management, and regulation of technology requires that human values dominate performance values to ensure effective negative feedback. For modern cultures, this, in turn, demands an iconoclastic attitude towards technology in the form of an ongoing enquiry into where it serves human goals, where it cannot serve those goals, and where it is simply irrelevant because an issue is non-technological. Developing as realistic a view of technology as possible is essential. Who among us would hire a contractor so fascinated by a newly acquired tool that he refuses to use any other? I am sure no one would, and we can therefore not expect society to be impressed with technology-related professions that do not have a clear and critical awareness of the limitations of modern technology and who confuse performance values with human values. What we need is a critical awareness of our culture and way of life as symbolically represented in our mental maps. To take up the earlier analogy of the set point on a thermostat, our mental maps include a multidimensional set point in the form of a hierarchy of values through which we regulate our involvement in the world. This multidimensional set point is subject to the same limitations as the informal ecology of technology. A critical awareness of who we are as individuals, professionals, and a society is therefore essential for the development of preventive approaches. This work is therefore inseparable from its companion dealing with the relationship between technology and modern cultures.

## 2.7 Other Professions

We have focused on the engineering profession in this chapter for two reasons. First, it is paradigmatic of the other professions associated with modern technology; second, it is the profession with which I am most familiar and which I have critically studied. The applicability of the analysis to medical technology is obvious from the existence of preventive medicine. Similarly, total-quality-management and environmental-management systems may be regarded as dimensions of preventive management. This will become more evident as we proceed with our enquiry.

## 2.8 Conclusion

In this chapter, I have examined four prerequisites for the development of preventive approaches: (1) a formal ecology of technology; (2) the ability to transcend professional specialization; (3) a critical awareness of our metaconscious images of technology; and (4) human values to act as the equivalent of a set point to guide the development of preventive approaches by means of negative feedback and a critical awareness of the limitations of technology with respect to these human values. In the next chapter, I examine another aspect of the ecology of technology – namely, its role in creating useful ignorance.

# 3

# Collective Prerequisites for Preventive Approaches

## 3.1 The Double Meaning of Ignorance

In chapter 1, I drew a comparison between preventive approaches for the engineering, management, and regulation of modern technology and the safe driving of a car. Drivers need to monitor where they are heading, compare this to their intentions (the 'set point'), and use negative feedback to reduce the gap on an ongoing basis. Chapter 2 examined four 'driving skills' that individuals must possess in order to do their work as preventively as possible. The present chapter is concerned with the fact that not one but many people are collectively 'steering' technology into the future.

The difficulties that face a society can be illustrated by extending our metaphor. Suppose it had been decided that driving a car in an old city with narrow streets crowded with pedestrians, pets, cyclists, motorcyclist, and motorists had become too complex a task for a single person. It was to be split among four 'specialists': one person would operate the vehicle according to the instructions given by the three others who would monitor the road ahead, to the sides, and to the rear, respectively. It is not difficult to imagine the confusion that would arise from any gap or overlap occurring among the tasks of the four individuals. It could even be the case that driving would be less safe in critical moments. However, a society has no choice but to create a division of labour for 'driving' its way of life into the future. If preventive approaches are to become an integral part of this development, the strengths and weaknesses of such a division of labour need to be considered. Gaps or overlaps among the knowledge and skills of prevention-oriented participants can create confusion, 'blind spots' or 'ignorance,' with potentially disastrous consequences. The way a society structures and develops its way of specialization and collaboration among individuals working together in a group, between groups working together within an institution, and between institutions that help constitute its way of life will determine the extent to which individual limitations can be overcome.

It is impossible for every participant to have the complete 'driving skills' outlined in the previous chapter. No individual person can know the ecology of the technology of his or her society. Even if it was knowable, nothing would ever get done if everything had to be considered. There are simply not enough hours in a day and sufficient years in a lifetime, not to mention other resource constraints. Hence an effective design for specialization and collaboration is required. This raises the subject of 'useful' and 'harmful' ignorance.

The inseparability of knowledge and ignorance in human life is not simply a question of human finitude that makes it impossible for anyone to know everything and to behave accordingly. A human life lived in a particular time, place, and culture constitutes a vantage point from which each new moment is approached. Consider a group of people looking at a painted portrait in an art gallery. Since the structures of their retinas provide them with detailed vision only in the centre of their field of vision, the portrait cannot be taken in at one glance. The way they let their eyes wander over the picture may be influenced by factors such as their prior experience and knowledge of art, or their personality and interests. One person may be immediately captivated by a particular detail and spend most of the time admiring its contribution to the whole. Another person may focus on the same detail, but examine it in its own right. Another person may do a little of both. Yet another person may be captivated by a different detail. Others may not be attracted by any particular detail and simply admire the painting, or find it uninteresting for a variety of reasons. Even after the group has spent a considerable time looking at the painting, they are able to have a conversation about it. If they had all visually experienced everything about the painting, there would be nothing to talk about. A conversation is possible precisely because their visual experiences of the painting are different. This is, in no small measure, due to the fact that the way they let their eyes wander across the painting and integrate what they saw is profoundly affected by what they brought to this moment of their life, including their experience with paintings, their values and interests. In other words, a conversation about the painting is possible because each person became visually knowledgeable about some of its qualities and remained ignorant about others. None of them perceived him- or herself to be an expert who had seen all there was to see in the work. Art critics can continue to write creatively about certain paintings made centuries ago because each historical epoch contributes a new vantage point.

Just as our life constitutes a vantage point for living each new moment, so, also, although on a more limited scale, becoming a specialist involves developing a unique vantage point for professional practice in the world. We have already discussed three examples of this fact. Textbooks for first-year university courses provide a description of some aspects of our modern world in which the role of technology is out of proportion. Experts examining the hunger problem in the Colombian Valley arrived at mutually exclusive diagnoses of the situation. An engineer attempting to solve the quality-control problems of a production facility may not be successful if the situation is viewed exclusively from the vantage point of her specialty. All these situations involve two kinds of ignorance. The first is related to the fact that, as specialists, we cannot know everything there is to know. However, even what we do know has embedded in it ignorance of a second kind. We have all grown up with the idea that a specialist knows everything our society has learned about a particular part of our world. We forget that any human knowledge is relative to a vantage point determined by our professional experience, formal education, life experience, convictions, values, and, last but not least, the culture of our society. In other words, expertise in one domain inevitably and simultaneously implies knowledge obtained from one vantage point and ignorance about what could be obtained from other vantage points. This hardly means that any knowledge is as good as any other. The vantage point of any specialty is carefully delimited by accepted practice, standards, and values. No specialty can therefore know all there is to know, thus embedding ignorance of one kind into knowledge of another kind. This ignorance can be turned into a useful asset if its existence is clearly recognized, but it becomes an insurmountable barrier to a 'conversation' when denied. I refer to ignorance as 'useful' if an awareness of its existence has positive consequences, including a sense of the limitations of a specialty and the motivation to engage in 'conversations' wherever possible. I suggest that ignorance is 'harmful' when its consequences are the opposite. Unfortunately, there are not many such conversations between specialists within the university or between professions and society as are required to make all parties more aware of the existence of useful and harmful ignorance. An attitude of humility before, and openness to, a reality that is always much larger than any creature can possibly know requires a critical awareness of one's vantage point. Its absence can lead to an ignorance that closes itself off from reality. Science, technology, and culture each deal with reality in a piecemeal fashion, thus embedding ignorance within knowledge, although they do so in very different ways.

Our uncritical acceptance of an objective and detached observer, and objective and value-free knowledge and facts whose status is unaffected by their intellectual context, has obscured what, from a daily-life perspective, is all too obvious: Professional education affects the way we interpret and act in the world. It provides us with a professional prejudgment of a situation. If we interpret and respond to a situation in the same way after graduation as we would have before we entered professional schools, we have wasted our time and financial resources. What we have learned constitutes a professional prejudgment of a part of reality, which is different from the prejudgment we have derived from the mental maps that result from living in a society. We tend to accept the prejudgment (or subjectivity) in the latter case, but we deny its existence in the former by invoking mythical entities such as detached observers, objective knowledge, and context-free facts. It is quite clear, however, that we interpret and act in the world through our minds, and that, without them, we would not experience anything at all. Detached observers, objective knowledge, and context-free facts cannot exist in a human world. At best, they may be regarded as goals towards which we can strive but which we can never reach. To be a member of a culture is to have a vantage point, and to be a member of a professional community is to have that vantage point developed further with respect to a particular area of our world. To deny such a vantage point is analogous to claiming that it is possible to take a picture of a landscape without the camera having to be in a particular place. In that case, there would be no picture at all and, similarly, without these vantage points there would be no human life as we know it. Our prior experience constitutes the prejudgment through which we live in the world. This is how each instant is experienced and lived as a moment of our life. For the same reasons, there can be no science of the sciences capable of transcending the prejudgments of each scientific discipline. Nor can science replace culture as a basis for human life. In a modern society, there can be no science without culture and no culture without science, no technology without culture and no culture without technology. What is possible is that science and technology so permeate culture as to inhibit its ability to give meaning, purpose, and direction to the lives of members of a society. The overvaluation of science and technology has created some of the most detrimental forms of harmful ignorance in our world. A possible symbiosis among science, technology, and culture is blocked by each one denying the existence of a unique vantage point. Cultures achieve this by making their vantage points universal and eternal through myths. Science accomplishes this by assuming that scientists are detached observers gathering objective knowledge, thus making them ahistorical and acultural beings, separated from invisible colleges, scientific communities, and society. Technology accomplishes this by focusing on its 'economy,' which is believed to make it objective and value-neutral.

Ignorance is therefore not merely an absence of knowledge, but simultaneously a denial of the possibility of additional knowledge. All specialists are ignorant in the sense that they cannot possibly know what all other experts know. However, they must get on with their work, and thus at least temporarily close their minds to the possibility that any other knowledge can radically and fundamentally alter what they are doing. All experts must be confident that they know enough to get on with their work. Doing their work as preventively as possible involves the constant struggle to keep their minds open to the greatest possible extent while getting on with the job. This is the art of the possible. However, it is not easily accomplished. Science, technology, and culture are all human creations unaware of their own limitations. Preventive approaches will require conversations that acknowledge the existence of unique vantage points.

## 3.2 The Limits of Specialization

Ignorance, in the positive sense of the word, is simply a recognition that we are human and cannot know or do everything. In the negative sense, ignorance is related to someone's claim that he or she knows everything a society knows about a particular domain. This knowledge then becomes closed, which in essence means it denies that contributions to knowledge of that same domain can be made from other vantage points. This implicitly assumes something about the nature of reality. Closed domains of knowledge must correspond to 'blocks of reality' that are independent from each other; this violates our growing awareness that, in the real world, everything appears to depend on everything else. If this indeed is the 'nature' of reality, then the view that bodies of expert knowledge are entirely independent of one another is untenable. If, on the other hand, we recognize that these bodies of knowledge are in effect open systems in relation to which all other bodies of specialized knowledge are theoretical externalities, then the current division of labour in science and technology creates a great deal of harmful ignorance, and thus a considerable number of blind spots. When technology took on an applied-science basis around the beginning of the twentieth century, the same kinds of claims about objectivity and neutrality were made, and a similar situation ensued.

On the level of daily-life experience, we are all aware of the problem. If a professor standing in front of a class wanted to know everything there was to know about the physical appearance of the students present, he or she would have to observe them from many different vantage points, since one particular perspective limits what can be seen. Similarly, what is observed from one vantage point is not objective, in the sense that other persons observing from the same vantage point would not see the same things. A person with an interest in fashion may notice things that a person without that background does not. Similarly someone preoccupied with nutrition and skin conditions may observe things that those without that interest do not. I could go on, but the point is that a vantage point is not constituted merely by a position in space relative to what is being observed but also by the background and experience the observer brings to it. Not all observations of the university class mentioned above may be potentially complementary if the backgrounds people bring to those observations clash. For example, one of the observers may have a prejudice against members of a certain ethnic minority present in the class. In the same way, the sciences examine one world, but do so from very different disciplinary vantage points superimposed on a knowledge of that world acquired by living in it as a member of a society and culture according to individually unique convictions. Some current approaches in the social sciences delight in exposing the latter two roots, thus showing that science is ultimately a human activity. This is not very helpful when one is trying to determine how science differs from other human activities in terms of the unique vantage points and relative objectivity it brings to the task of understanding our world.

Understanding and living in a reality in which everything appears to be related to everything else is a prerequisite for making our way of life more sustainable. Natural evolution shows how, in the biosphere, everything evolves in relation to everything else without threatening the whole. A sustainable way of life would have to accomplish much the same thing without threatening human life, society, or the biosphere. Will it be possible to establish a symbiotic relationship among science, technology, and culture to accomplish this?[1] It is useful to briefly review the current ways in which these three human creations deal with 'everything being related to everything else.'

Science has become the model for obtaining knowledge about reality that has been incorporated into modern technology since the beginning of the twentieth century. The central problem is the following: How is it possible to study anything situated in a world in which everything is related to everything else and in which life constantly evolves? As previously noted, the answer that science and, later, technology have given is that anything to be studied must first be abstracted from its context and then be placed in the much simpler and more manageable intellectual context of a scientific or technical specialty. If laboratory studies are undertaken, it is placed in a physical context whose complexity is carefully limited so that the effect of a small number of variables can be studied, preferably one at a time. Normally, the knowledge gained will be in context with respect to the body of knowledge assembled by that specialty, but it may well be out of context with the knowledge base of other specialties. Science is based on a knowledge strategy making minimal use of context, thus creating ignorance about how everything relates to everything else. As a result, it can make only a very modest contribution to the goal of improving the sustainability of our modern way of life.

The technological approach to dealing with reality also makes minimal use of context. It begins by abstracting from its context whatever is to be technologically improved, in order to study it. The findings are used to build some type of model. Such a model divides the relations involved in the ecology of a technology into a hierarchy of three categories. The ones that are important with respect to the purpose to be achieved are incorporated into the model, those of marginal relevance may or may not be, and those deemed of negligible importance are excluded. Such a hierarchy may make perfect sense with respect to the purpose to be achieved, but what is its importance with respect to other goals? The problem is that the actual ecology of technology as it exists in reality does not vary according to human purposes. Next, the model is examined to determine under which conditions the desired results may be obtained. Once this has been determined, the real object is reorganized accordingly. Consequently, the results will be compatible with some aspects of the world and out of context with respect to others. This technological approach to reality again makes minimal use of context, beginning, as it does, with a process of abstraction, and being guided by performance values. In this way, out-of-context knowing is complemented by out-of-context doing. With respect to everything being related to everything else in reality, this amounts to a divide-and-conquer strategy in which the world is intellectually and materially taken apart and manipulated in a piecemeal fashion. Thus, the long tradition of regarding science and technology as rational and objective overlooks the fact that they can never be entirely so with respect to a complex and interrelated world. They are obliged to reduce the unmanageable complexity of the real world to a simpler and more manageable complexity, which gives each of these human activities their non-rational component.

In contrast, culture is a human creation for understanding and living in the world that makes maximum use of context. Giving a name, meaning, and value to everything in relation to everything else in human experience unified individual and collective human life. Everything had its proper place, and the tradition-based approach to anything new allowed everything to evolve as much as possible in the context of everything else. The growing roles of science and technology based on minimal use of context undermined the role of culture and thus its ability to sustain the integrity of human life and society. The dominance of the economy over society, performance values over human values, and the economy of technology over the ecology of technology are but a few manifestations of the influence science and technology are having on culture. Although cultures have always made it possible for societies to live in a reality in which everything depends on everything else, they, too, developed a non-rational component. The unknown placed limits on the extent to which everything could be interrelated. This problem was surmounted by the creation of myths, which gave cultures their non-rational character.

The conventional view that science and technology are rational and objective and that culture is non-rational and subjective is therefore simplistic. There is no human knowledge without ignorance, and no ignorance without knowledge. We all live this dialectical tension in a way that is unique to our time, place, and culture. It is institutionalized in our way of life. To understand the limitations of science, technology, and culture in the face of an interrelated reality is an important step towards creating a more viable and sustainable way of life that includes preventive approaches.

No individual can possibly stay abreast of the scientific and technological frontiers. Hence, specialization, a division of labour, and a supporting institutional framework are essential to keep modern societies viable. The question is: How can we structure this, including the institutional framework, so as to minimize our collective blind spots and the role of harmful ignorance in science and technology? Glimpses of an answer can be seen by undertaking a reality check on undergraduate engineering education. I have chosen it because engineering is to technology what physics is to the sciences – namely, the discipline that constitutes the model for all others.[2]

### 3.3 Professional Education and the Transmission of Ignorance

A comprehensive study of undergraduate engineering education has been undertaken to explore the following theses:

1 The typical North American undergraduate engineering curriculum separates the study of the economy of technology from the study of the ecology of technology.
2 The component examining the economy of technology (the technical core) makes minimal use of the knowledge we have of the ecology of technology.
3 The complementary-studies component of the curriculum (traditionally referred to as the 'non-technical' part) does not provide students with a substantial understanding of the ecology of technology, nor does it help them use this information to adjust design and decision-making processes to ensure the greatest possible compatibility between technology and its contexts.
4 The publications of faculty members make almost no use of context information.
5 The informal or hidden curriculum points to a professional ethos that fully legitimates the conditions outlined in the previous four theses. What this means is that courses examining the ecology of technology to make use of this understanding in a negative-feedback mode in order to prevent or minimize harm run counter to the mainstream of the curriculum.
6 The tenure review process makes it extremely difficult for young

faculty members seeking to develop a more preventive orientation in their work to succeed in their endeavours and thus make a contribution to transforming the professional ethos.

7 The nature of the formal and informal curricula and the professional ethos attracts students who excel in a particular cognitive approach to the world characterized by a minimal use of context. Students who also excel in a complementary mode, making much more extensive use of context information, frequently feel out of place and are more likely to drop out. This ensures a dynamic equilibrium between the formal and hidden curricula; the cognitive styles of students and faculty; the professional ethos and the intellectual division of labour; and the corresponding organizational structure of engineering departments, the engineering faculty, and the place of engineering within the university.

8 A prevention-oriented professional ethos will require a paradigm shift as opposed to some cumulative curriculum changes. There is no continuum from conventional approaches being gradually enriched with social and environmental concerns culminating in a prevention-oriented professional ethos. Only a paradigm shift can fully unlock the full potential of preventive approaches and the tremendous benefits for students, faculty, the profession, corporations, government, all members of society, and our common future.

I admit from the outset that these theses anticipate a very serious situation that has profound consequences. The discussion of the study of undergraduate engineering education that explored the theses noted here[3] will therefore be fairly detailed; and to facilitate the task of the reader, the material has been organized under a number of subheadings.

*The Research Instruments*

Two basic questions were asked in reference to the 1988–9 undergraduate curriculum of the Faculty of Applied Science and Engineering at the University of Toronto: (1) How much do we teach students about the influence technology has on human life, society, and the biosphere? (2) To what extent is this knowledge used in a negative-feedback mode to adjust engineering methods and approaches to achieve a greater compatibility between technology and its contexts? The study established a reference point for the ongoing monitoring of the curriculum.

The two questions inquire to what extent a preventive orientation is present in the curriculum. They were operationalized in the form of two mirror-image research instruments. The first was designed for courses in the basic and applied sciences and engineering. References to context and their influence on the subject matter were measured by means of a continuous scale ranging from 0 to 4. In order to make the scale as unambiguous as possible, and to give coders clearly defined categories, the values were defined ordinally as follows:

0 No reference to context issues.
1 Minor reference to context issues which remain peripheral to the thrust of the course. Usually this amounts to little more than outlining the context in which the problem arises, but once the problem is cast in engineering terms little or no reference to context is made.
2 Some reference to context issues with some consequences for the thrust of the course.
3 Major reference to context issues with substantial consequences for the thrust of the course.
4 Substantial reference to context as in the previous value, but, in addition, an evaluation of consequences in order to adjust or reassess the methods or theories under consideration. This value includes some negative feedback whereas the previous one does not.

Context issues include the consideration of: the implications of technology for human life, society, and the natural ecology; ethical obligations for individual practitioners or the profession as a whole in the context of societal values; non-technical aspects of engineering education and the technical mindset; consequences of the application of engineering theory and practice, including the effects of quantitatively representing qualitative phenomena, particularly in the human and sociocultural domains; and implicit and explicit beliefs, values, and assumptions in engineering theory and practice. It is understood that these specifications do not exclude inherent biases and hypotheses in the research design. For example, in a modern society no one even comes close to being a detached observer of technology. The assessment by coders as to the definition of context issues may be affected by their ideological differences. This was greatly minimized by asking them to do their scoring naïvely, by not assessing the quality of the references in terms of content. For example, references to the problem of worker alienation in a course on production automation are taken at face value and not critically assessed in terms of the latest research findings in the sociology of work. As a result, the naïve scores may be higher than those that would have been obtained if a critical evaluation of the context information had been undertaken. Consequently, these scores represent an upper limit. Coders were encouraged to identify any ideological concerns they had when applying the research instrument to particular context issues so that these parts of the data could be assessed further.

A second research instrument was designed to score the complementary-studies courses in the undergraduate engineering curriculum, which comprised the social science and humanities electives. This component constitutes 12.5 per cent of the curriculum, as required by the Canadian Engineering Accreditation Board. The requirements for this component can be met by taking the equivalent of one semester in the social sciences and humanities. Within this category, students can, in principle, take almost any course offered at the University of Toronto. However, as timetable conflicts commonly occur as a result of a large number of contact hours in a professional faculty, the Faculty of Applied Science and Engineering has made arrangements for a number of courses to be offered during the time slot from noon until two o'clock daily. Most students satisfy their social science and humanities requirements by selecting most or all of their courses from this group. Of these courses, twenty-nine were evaluated using a research instrument that is the mirror image of the first one:

0 No reference to technological issues.
1 Minor reference to technological issues which remain peripheral to the thrust of the course. Usually this amounts to little more than outlining some issues, but once the problem is cast in socioscientific terms, little or no reference to technology is made in describing, for example, the economic, social, or political arrangements of a society.
2 Some reference to technological issues that have some influence on the thrust of the course.
3 Major reference to technological issues with substantial consequences for the thrust of the course.
4 Substantial consideration of technology as in the previous value, with the addition of the evaluation of consequences for the context of technology in order to adjust and reassess methods and theories used in engineering or in the social sciences and humanities.

Once again, the scoring was done naïvely so that the results would represent an upper limit of what can be achieved within the current structure.

The first research instrument examines how well students learn to relate technical considerations to human, social, and natural ones, and their preparedness for exercising professional responsibility and keeping the public interest paramount in a highly interactive setting. It also measures what they learn about the limitations of their engineering 'tools' in a particular context. Low scores would confirm a separation of the study of the economy of technology from the study of the ecology of technology. The second research instrument measures the degree to which the social sciences and humanities electives complement the technical core of the curriculum to give students a better understanding of context issues essential for professional practice. Low scores would confirm that little or no attention is being paid to the ecology of technology and to how technological development meshes with everything else. The higher the score of a complementary-studies course, the greater its contribution towards bridging the gap between the technical core and complementary studies within the curriculum and towards the reintegration of the study of the economy of technology with the study of the ecology of technology. Low scores in both the technical-core and complementary-studies components of the formal curriculum would indicate an almost bimodal distribution in which technology either almost entirely fills the intellectual field of vision or has almost disappeared from it. Attempts to restructure the curriculum around a core of design courses and projects involving the synthesis of everything students have learned would then prove difficult because the 'world' of the technical courses is incommensurate with the 'world' of the complementary-studies subjects. The validity and reliability of the research instruments were extensively tested and confirmed by the multidisciplinary research team according to the most exacting standards of the social sciences.

*Research Design*

The starting point for the study was a detailed report containing several thousand pages of course outlines, curricula vitae, and other information prepared for a review by the Canadian Engineering Accreditation Board.[4] The entire population of courses was scored.[5] A standardized composite course score was adopted. It is the average score computed from three equally weighted scores for the lectures, textbooks, and laboratory or tutorial components.[6] The lecture score is a composite obtained from considering the detailed course outline, student notes, class hand-outs, and student interviews, as required. For approximately half the courses scored, several sets of lecture notes were consulted. Wherever necessary, additional information was collected through interviews with students.

The textbook score reflects only one item. For those courses having multiple textbooks, and/or course notes, and/or assigned readings, each item was scored individually and the highest score was used as the textbook component in computing the score for the course. The tutorial or laboratory score is a composite of assignments, laboratory manuals, reports, hand-outs, and interviews. If there were laboratories and tutorials, the higher score was used to represent this component. Finally, if a coder had any doubts as to the value to be assigned to any item, it was scored by another coder and the higher score was used. Taking the higher score once again ensures that the findings trace the upper limit of what can be achieved with the present curriculum.

The decision to weight equally the three components to calculate an overall course score was based on the assumption that for most courses there would be a considerable overlap between the lecture and the textbook components. Thus, two-thirds of the composite course score and one-third for tutorials and laboratories would, in many cases, approximate the balance of time students devote to these two components. This, of course, implies that the influence any portion of the curriculum has on students is directly proportional to the number of hours they spend on it. This, too, is open to question, but it appeared the most reasonable assumption.

In order to answer the two questions with which we began, the composite course scores were aggregated to obtain a measure of the degree to which students are exposed to context issues throughout their four years of study. Average scores were calculated for each of the four years in each department, for the entire curriculum of each department, and for the faculty as a whole. It should be noted that, in the aggregated faculty data, courses shared between different departments are repeated. This is particularly true of the first and second years, where common courses abound. It is true that the multiple inclusion of these courses lowers the faculty average. Thus, it might be argued that, while it is legitimate to use these courses in each one of the aggregated scores for the curriculum of a department in order to show a complete picture, the same argument cannot be made when calculating population parameters for the entire faculty by aggregating departmental parameters. This results, for example, from low-scoring courses in mathematics and the basic sciences taught in the early years. This procedure would also inflate the size of the population, leading to the conclusion that the faculty offers more courses than it actually does. Our answer to such objections is that it is not the intent of this study to offer a descriptive picture of the faculty as a whole, but to give a picture of the faculty the way *students* are exposed to it. The intent is to present a picture of the paths taken by students through their departments within the faculty in order to better understand their professional development. A subsequent aggregation of this data, therefore, does adequately summarize the pictures obtained of individual departments.

Since research makes a substantial contribution to teaching, it was decided to score professorial publications using the research instrument designed for the technical component of the undergraduate curriculum. The publications reported in the documentation prepared for the Canadian Engineering Accreditation Board were scored if they could be found in the libraries.

### *Research Findings*

Mean context scores for lectures, textbooks, and tutorials (or laboratories), as well as course frequencies and percentages within the scoring range, are presented for a typical department in table 1. Table 2 summarizes the data for the faculty in terms of the technical core of the curriculum, as well as for the technical core and technical electives combined. The mean publication scores for each department of the faculty are also presented here. The data for complementary studies is reported in figure 1. Of the twenty-nine complementary-studies courses evaluated, the average score was 1.3. Figure 2 represents the distribution of the faculty publications among the scoring categories as a percentage. Figure 3 presents the context scores of fourth-year undergraduate Industrial Engineering theses from 1981 to 1992, and figure 4 represents the data from a weighted sample of MASc theses in civil engineering from 1950 to 1990.

The overall picture is not reassuring. The mean score for the technical core of the curriculum is 0.82, which reduces to 0.75 when the technical electives are added. Factoring in the complementary-studies

Table 1: Industrial Engineering Scores

| | Course component | Year 1 | 2 | 3 | 4 | Component mean | |
|---|---|---|---|---|---|---|---|
| Mean scores | Lectures | 0.9 | 0.5 | 1.5 | 1.5 | 1.2 | |
| | Texts | 0.4 | 0.4 | 1.6 | 1.0 | 0.9 | |
| | Tutorial/Lab | 1.2 | 0.5 | 1.3 | 1.3 | 1.0 | |
| | Year Mean | 0.8 | 0.5 | 1.5 | 1.3 | 1.0 | |
| | Scores | | | | | Total | % |
| Course distribution | 0.00–0.99 | 6 | 9 | 2 | 6 | 23 | 53 |
| | 1.00–1.99 | 1 | 1 | 5 | 2 | 9 | 21 |
| | 2.00–2.99 | 0 | 0 | 3 | 3 | 6 | 14 |
| | 3.00–4.00 | 1 | 1 | 1 | 2 | 5 | 12 |

Table 2: Faculty Scores

| | Year | CHE | CIV | ELE | ESC | IND | MEC | MMS | Mean |
|---|---|---|---|---|---|---|---|---|---|
| Core courses | 1 | 0.5 | 0.5 | 0.4 | 0.1 | 0.8 | 0.3 | 0.5 | 0.4 |
| | 2 | 0.3 | 0.7 | 0.3 | 0.6 | 0.5 | 0.2 | 0.3 | 0.4 |
| | 3 | 0.7 | 1.2 | 0.7 | | 1.6 | 0.7 | 0.9 | 1.0 |
| | 4 | 1.8 | 1.4 | | | | 1.3 | | 1.5 |
| | *N* | 30 | 38 | 29 | 18 | 25 | 32 | 26 | |
| Core and technical electives | 1 | 0.5 | 0.4 | 0.4 | 0.3 | 0.8 | 0.3 | 0.6 | 0.5 |
| | 2 | 0.4 | 0.6 | 0.4 | 0.6 | 0.5 | 0.3 | 0.4 | 0.5 |
| | 3 | 0.7 | 1.1 | 0.6 | 0.7 | 1.5 | 0.8 | 0.9 | 0.9 |
| | 4 | 1.5 | 1.5 | 0.8 | 1.1 | 1.3 | 0.9 | 0.5 | 1.1 |
| | *N* | 57 | 51 | 67 | 94 | 43 | 67 | 47 | |
| Publications | Score | 0.2 | 0.6 | 0.1 | 0.2 | 0.6 | 0.2 | 0.3 | 0.3 |
| | *N* | 367 | 237 | 306 | 204 | 121 | 290 | 169 | |

Key to abbreviations: CHE – Chemical Engineering; CIV – Civil Engineering; ELE – Electrical Engineering; ESC – Engineering Science; IND – Industrial Engineering; MEC – Mechanical Engineering; MMS – Metallurgy and Materials Science; *N* – Number of courses scored.

Figure 1: Complementary Studies

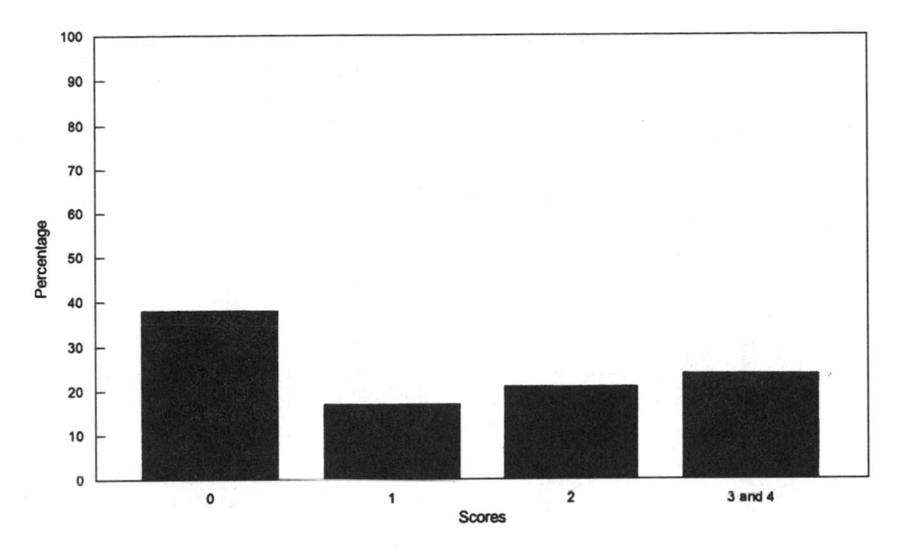


Figure 2: Faculty Publications

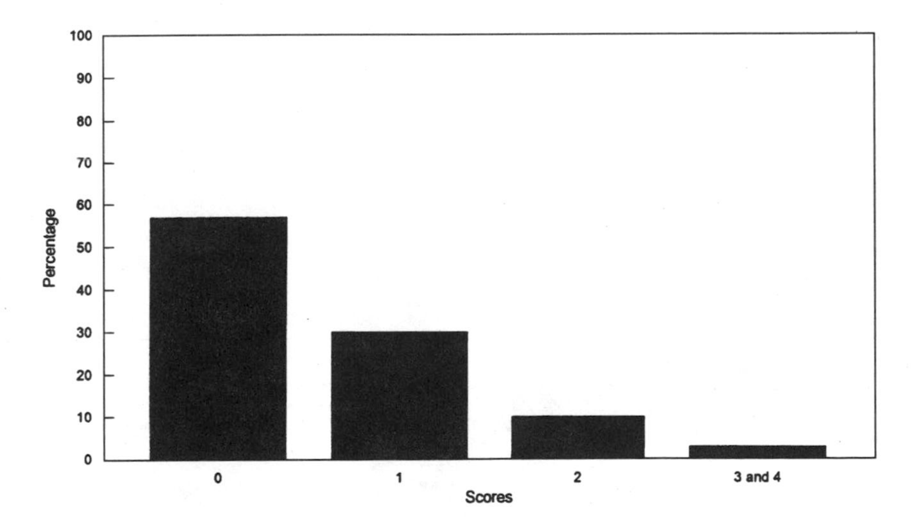

Figure 3: Fourth-Year Industrial Engineering Theses

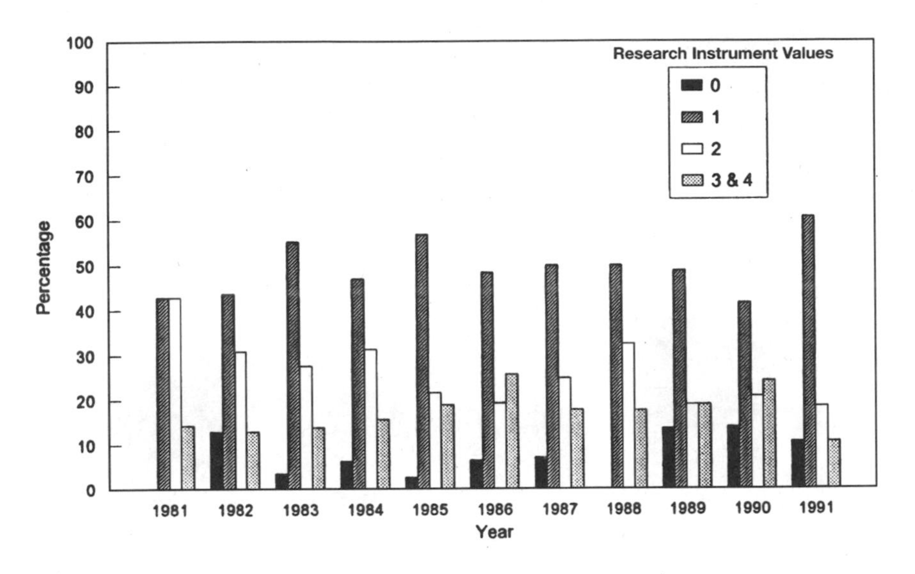


Figure 4: MASc Civil Engineering Theses

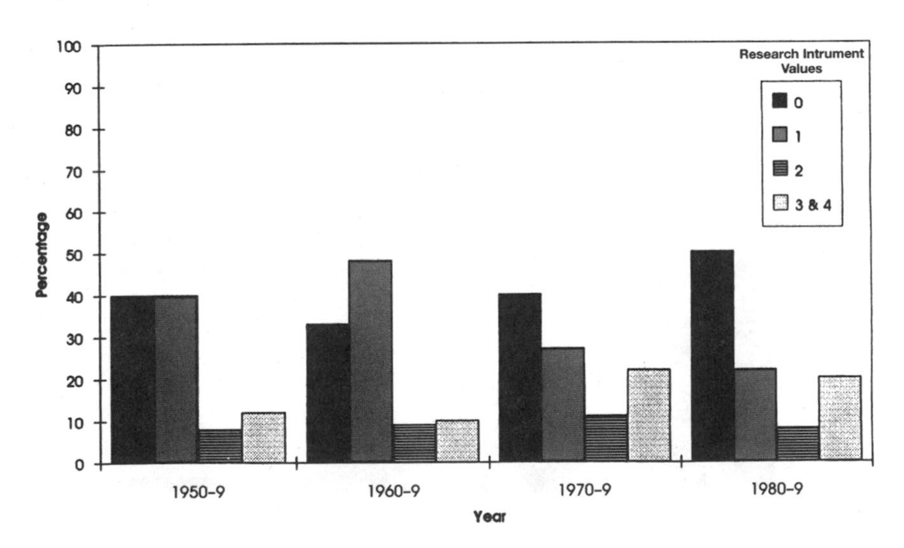

component, weighted as 12.5 per cent of the curriculum, raises the score to 0.8. The mean scores for research publications is 0.3, showing the highly specialized nature and lack of reference to context of the research effort. The scores for all of these components are positively skewed and unimodal, taking on an extreme form of a reverse 'J' curve distribution. This is similar to the so-called reverse 'J' curve of conforming social behaviour, which illustrates the fact that most people adhere to social conventions and laws, and that smaller and smaller numbers show a larger and larger degree of nonconformity. In our case, most of the courses are contextless, with fewer and fewer courses showing more and more context. Moreover, the scores are all tightly clustered around the mean, with the research publications showing the lowest standard deviation, 0.6, as compared with a value of 0.7 for the technical core and electives and 21.2 for the complementary studies. Various association parameters were calculated. A correlation of a professor's publication scores with the courses taught by that professor reveals no significant patterns, especially when controlling for the inherent contextuality of a course. This lends credence to the view that the inherent design of a course contributes much more to its contextuality than do the qualities of the professor teaching that course. Similarly, professors with high publication scores appear unable to raise the poor context scores of courses considered inherently non-contextual in the traditional sense. Frequencies, when associated with the year of study, show a very marginal decline in skewness with the higher years. The same fact is observed when the means and standard deviations are considered. The third and fourth years generally show a progressive rise in the values of the mean and standard deviations, indicating contextuality as well as dispersion. However, the rise in the value of the mean is never more than 1.0 on the scale. The expectation that, after completing the basic and applied sciences in the first two years, the context scores would improve as students sink their teeth into engineering courses is not borne out.

There are thirty-five courses with a mean context score of 2.5 or higher. These courses form the following clusters. Six of them can be classified as environmental engineering, taken in a broad sense; two courses on transportation and transportation planning; five courses dealing with engineering economics, entrepreneurship, professional communication and responsibility, or engineering as a profession; six courses in human factors and systems management; six courses in the aerospace option of engineering science, which has a significant human–machine component; the remaining ten cannot be grouped in any meaningful category. This confirms that those engineering courses scoring highly do so because their subject matter is inherently of a contextual nature.

For a few of these courses, a critical assessment of the quality of the context information being used was undertaken. It suggested that the context information being used was of poor quality compared with the relevant portion of the knowledge frontier. Thus, the scores obtained by the quantitative analysis represent an upper limit below which the results of a qualitative analysis would fall. This was amply confirmed in a follow-up study referred to in chapter 1, where the different methods and approaches used for dealing with social and environmental implications in particular areas of application were scored by means of the research instruments noted above. Preventive approaches were found to make use of the latest knowledge, while conventional approaches, with their focus on the first stage, paid little attention to the research frontiers in other disciplines.

The low mean score of 0.3 for the research publications shows the highly specialized nature and lack of reference to context of faculty research. It is recognized that the editorial policies of many so-called prestigious journals, and the pressure on faculty to publish in them, contribute significantly to explaining such a low score for a profession that accepts the duty of protecting the public interest. It has been said that to be concerned about contextual issues is a luxury young faculty members facing tenure review can ill afford. The data appear to show that, by the time they obtain tenure, their teaching and research patterns are so well established that little attention continues to be paid to context issues.

An evaluation of undergraduate theses in industrial engineering and MASc theses in civil engineering (figures 3 and 4) reveals that there has been no appreciable increase in context scores over the years. Ongoing monitoring of the curriculum shows that, apart from slight gains or losses in specific areas, there is no overall trend towards fundamental change.

How typical are these findings for North American schools as a whole? While the study does not purport to be statistically representative of all engineering faculties, there are many persuasive indications suggesting that it may be taken as such. A comparison of the requirements of the Canadian Engineering Accreditation Board with those of the Accreditation Board of Engineering and Technology reveals that the operating guidelines followed by engineering faculties across North America are very similar. A sample of course descriptions and outlines from the top engineering schools reveals a structure identical to that found at the University of Toronto. A high proportion of the textbooks scored in this study are also in common use at many other schools, and these scores deviated marginally from the course scores. Finally, a comparison of various indicators of performance and academic excellence of departments at the University of Toronto with those of the ten best North American schools[7] leads us to believe that the scores reported for the technical portion of the curriculum are fairly typical for the top North American schools. The scores for the complementary-studies portion are probably above average, since the University of Toronto appears to be leading in introducing context issues into the curriculum. This is based on the ecology-of-technology and preventive engineering approaches offered through a compulsory first-year course and two additional electives leading to a certificate from the Centre for Technology and Social Development. The faculty has also introduced environmental options based on the preventive approach in chemical, civil, and mechanical engineering.

*Implications*

The study and ongoing monitoring of the curriculum amply confirm the theses advanced earlier. The curriculum separates the study of the economy of technology from the study of the ecology of technology. A preventive orientation within the curriculum is almost entirely absent. A gap is thus identified between engineering education and the needs of industries seeking to develop more preventive practices. This has serious implications for the economic competitiveness of the industrially advanced nations, their ability to control spiralling social and health costs, and their success at preventing further deterioration of the health of the ecosystems in which they live. As far as the relationship between the engineering profession and society is concerned, the ability of the former to protect the public interest is greatly limited by professional education, despite an ethics and law exam required for licensing. There is no doubt that, without the efforts of the accreditation boards, the situation would have been far worse, but to ensure more rapid change these boards should probably be made more accountable to society.

On the positive side, the study identifies a largely untapped potential for developing preventive approaches that will benefit all parties. It appears that the engineering profession is at a crossroads. Given current trends, society will increasingly demand that engineers make a significant contribution to the development of a more sustainable way of life. This can be done through preventive approaches, with considerable benefits to the profession. It may help restore the profession's prestige. It may become easier to attract better students, particularly women. In North America, most parents would probably agree that the differences in the way we socialize boys and girls make it likely that girls will develop a greater sensitivity to context. Engineering may appear to be a more attractive career to young people who have a healthy idealism about creating a better future. It can also help reduce the drop-out rate by making the curriculum more relevant and interesting. It appears, therefore, that the profession and society would stand to gain substantially from the development of more preventive engineering practices. Until this happens, however, professional practitioners will remain almost entirely unaware of the preventive potential of engineering practice.

Context scores measured by the two research instruments described above correlate with the goal of sustainable development.[8] In focusing on making our way of life *more* sustainable, we can operationalize the concept of sustainable development in terms of reducing the harm inflicted on human life, society, and the biosphere. From this perspective, the two research instruments measure applied sustainability. Rising scores indicate a move towards more sustainable engineering and managerial practices. Current scores identify an enormous and largely untapped potential for making our way of life more sustainable, with benefits to all stakeholders.

The role of engineering ethics in the undergraduate curriculum and professional education is an uncomfortable one. Since the two-stage approach of conventional engineering delegates social and environmental considerations to the second stage, to be dealt with in an end-of-pipe manner, it compels engineering ethics to operate in much the same way. This converts it into end-of-pipe ethics, which clearly is a contradiction in terms.

Parallel problems occur in the undergraduate and graduate curricula of management science and business administration. These do not acknowledge the fundamental role technology plays in economic and business development and their social and environmental consequences. Significant components of these curricula concentrate on the economy of technology examined by means of conventional economic approaches. A preventive orientation is almost entirely absent, and business ethics, like its engineering counterpart, has an end-of-pipe orientation.

The same kinds of problems occur in medicine. While the social sciences and humanities recognize that human beings are profoundly influenced by their genetic make-up as well as the sociocultural and physical milieus in which they live (the so-called nature/nurture debate), this has thus far had very little influence on medicine. Our conceptions of health reflect this. The curriculum virtually ignores the environmental factors of health, such as the contamination of our food, water, and air, or our exposure to parasites. Environmental medicine is exceedingly underdeveloped and sometimes barely tolerated. The disproportionately large allocation of resources to the genome project reflects an obsolete view of human health. As the definition of the World Health Organization suggests, human health should be interpreted ecologically in terms of the relationship between a human being and his or her environment. Medicine should therefore pay much more attention to how this environment has been transformed by modern science and technology, and what this means for human health. The enormous increase in the body-load resulting from the pollutants in our environment, and the effects of modern work on human health, as noted by socio-epidemiology, are but two examples of how modern medicine creates knowledge of one kind and ignorance of another – a subject to which we return in a later chapter.

The legal profession has suffered a similar fate. Its evolution has increasingly come to be influenced by a desire to better organize society rather than to embody its fundamental values. The connection between law and justice and some of the other values of Western civilization has become severely weakened in the last few decades. Again, the pattern of ever-increasing specialization has led to a parallel decontextualization.

## 3.4 The Hidden Curriculum and the Justification of Ignorance

Although many aspects of modern engineering are transmitted explicitly in the classroom and on the job, many others are only implicit in the behaviour of faculty and practicing engineers with whom the students have contact. It is helpful to think of engineering education as a kind of secondary socialization process into a professional community through the acquisition of a professional 'culture.' Cultural anthropology has understood for a long time that much of a society's way of life is transmitted implicitly, and that what is not said at particular instances is as important as what is explicitly communicated. I am not suggesting, of course, that engineering is entirely separated from the culture of society, but that it constitutes a set of practices that cannot be divorced from profound conceptions about ways of dealing with and existing in the world. Some historians and sociologists of science have persuasively argued, for example, that a scientific community shares much more than what is explicitly transmitted in the classroom, laboratories, and journals.[9]

The hidden curriculum at the Massachusetts Institute of Technology (MIT) has been investigated in some detail by Benson Snyder.[10] His findings imply that the lack of reference to context within the technical portion of the curriculum does not merely create a neutral and uncharted territory in the world of future engineers. The hidden curriculum clearly communicates a message to the effect that the social sciences and humanities subjects are less important than the technical ones, and that any knowledge of reality that cannot be mathematically represented and quantified is less valuable than knowledge which can be represented in that way. This is further affirmed in student newspapers, dress, and in various rituals such as initiation. For example, an advertisement in a recent alumni magazine offered T-shirts for sale with slogans such as 'I am a socially stupid engineer ... and I love it,' 'If you're not an engineer just what good are you?,' and 'Engineering ... the future is in our hands.'[11] The 'Engineering Hymn' also may be read in this context. To dismiss these as jokes is to ignore what sociology and cultural anthropology have discovered about the process of socialization. The regularity with which such sayings are implied in the behaviour of faculty and students at an engineering school is symptomatic. The occasional protests that all this is not professionally responsible, socially acceptable, and, in more extreme cases, legally permissible have had only a minimal effect. The hidden curriculum thus converts uncharted areas related to modern technology in the formal curriculum, about which students should keep an open mind, into areas that are 'soft' and of little value to engineering practice. In this way, the hidden curriculum creates harmful ignorance. It would appear that the hidden curriculum legitimates the blind spots in the technical core to the point that it impairs the ability of engineers to safely 'drive' technology into the future.

The hidden curriculum also implies a hierarchy among the intellectual approaches used in the technical core and complementary-studies components, respectively. It implies that an intellectual approach that is quantitative and mathematical, and that makes minimal use of context is scientifically superior and more objective than one that is largely qualitative and makes a great deal of use of context. It thus creates harmful ignorance of the fact that these two intellectual approaches serve very different purposes. Making sense of the world may be likened to illuminating it with a beam of light. It can be focused narrowly to produce high levels of illumination over a small area in order to accentuate finer details, or it may be beamed very broadly to illuminate a much larger area with low intensity in order to pay more attention to context. There are knowledge strategies based on specialization, or knowing more and more about less and less, and those based primarily preoccupied on context, or knowing less and less about more and more. The former creates a carefully defined and clearly bounded domain of knowledge relatively closed to others. The latter, on the other hand, creates a domain of knowledge that is open to others because it seeks to understand things in terms of their participation in a very broad context. Both strategies have their place, just as a narrowly focused flashlight may be useful for examining a nail lodged in a tire, while a broadly focused one may be more useful for looking for a lost key. In the same vein, the intellectual strategies of the natural and applied sciences are advantageous when examining physical processes or machines that are essentially repetitive by nature, while those of the social sciences and humanities are appropriate for living beings and systems that are essentially non-repetitive. Each of these approaches has unique strengths, weaknesses, and limitations. Both domains of science should be regarded as rigorous, albeit in very different ways.

Some of the most significant differences between the natural and the social sciences may be understood in terms of their having developed different styles of map-making appropriate for examining very different kinds of realities. The natural sciences emerged based on the mechanistic world-view which asserted that the 'nature' of the natural world could be understood in terms of the most perfect machine of the time – namely, the clock. The 'nature' of that world was highly stable over time, and each 'part' of the universe machine existed within its own time and space, independent of all other 'parts.' When attempts were made to extend this style of map-making to the study of the human-cultural world, fundamental differences were gradually acknowledged.

The 'nature' of this world was not stable over time, nor geographically universal. Each civilization and society had a unique 'nature' that constantly changed and evolved since life never repeats itself. The 'parts' of this world were much more dependent on their contexts than those of the natural world. Gradually, other styles of map-making emerged, based on different preadjustments of the 'nature' of that world. Different schools of thought continue to coexist in almost all social sciences, in contrast with the situation in the natural sciences.

The differences in map-making styles also need to be understood in terms of very different prejudgments of the role the observer plays. Originally, the natural sciences regarded the relationship between the observer and the observed as being non-reciprocal, but this view is now being challenged, particularly within physics. In the social sciences, there has been a somewhat greater awareness of this reciprocal interaction. Implicit in the natural and applied sciences is the idea that we can do research *on* nature as detached observers. The findings of Heisenberg and others which suggest that one can only do research *with* nature, since the observer is an integral part of society and nature, have had little influence; and this is particularly problematic when the limits of the underlying methodologies remain largely unexplored. Can these methods and approaches be extended to human reality? Can we do experiments *on* human subjects, or should our attempt to understand the human world be based on a strategy in which the researcher works *with* human subjects?[12] The concept of the 'environment' can illustrate this point. All the other members of our society help to constitute my 'social environment' while I help to constitute theirs. There exists, therefore, no separate 'environment.' The same is true for our 'natural environment.' Humans help to constitute the 'living environment' for all other creatures as they help to constitute ours. Since we are an integral part of the biosphere, the term 'natural environment' appears to justify an approach to the biosphere that treats it as a kind of separate object which we can manipulate for our purposes without this fundamentally affecting us. The environmental crisis is beginning to teach us that, if we impair the health of the biosphere, serious implications for our own health will be the consequence, since we are an integral part of it. I return to this point later in this book.

### 3.5 Cognitive Styles and Ignorance

In daily life, babies and children grow up having to make sense of and relate to both living and non-living entities, which requires different cognitive styles. They rarely develop them to the same extent. This is related to why they may like and do well in certain subjects in school and not others. If they decide to go to university, their cognitive style will likely shape their choice of area of study and specialization. Consequently, the message of the hidden curriculum falls into the fertile soil of the experiences of engineering students. Their intellectual strengths tend to correspond rather closely to those required for success in the natural and applied sciences, and their weaknesses to those of benefit in the social sciences and humanities. Benson Snyder, in a detailed longitudinal study,[13] found that students at MIT could be divided into four groups, based on the relative dominance of their using two kinds of intellectual abilities, one usually associated with the applied and natural sciences and the other with the social sciences and humanities. The former he called 'numeracy' and the latter, 'literacy.'

One group of students started out by predominantly using the numeracy mode, but, after graduating, they experienced its limitations in their private and professional lives and began increasingly to use the literacy mode as well. In a second group, this complementarity of the two modes did not develop. Another group successfully used both modes as students, and kept on doing so after graduation. The last group was strongest in the literacy mode and developed the other mode only to a limited degree. This topology may be regarded as a refinement of a larger topology of personality types. You may prefer to understand something by visually and mentally abstracting it from the world and analysing it apart from the usual context. It is also possible to seek to understand something in terms of how it is a part of, fits into, and depends on larger 'structures' and 'systems' that help constitute the world. Of course, these are extreme tendencies on the continuum of possibilities.

Differences in the four student groups can, in part, be understood in terms of the role that context plays in the intellectual styles and methodological approaches in the applied and natural sciences, on the one hand, and in the social sciences and humanities, on the other. This may be illustrated by the fascinating research of Sheila Tobias,[14] which explores the barriers to a full and equal participation of women in mathematics and the sciences. This is a serious issue for any modern society fundamentally depending on science and technology. In Tobias's study, diaries were kept by a group of social science graduate students and faculty taking a course in university physics or chemistry and by a group of 'hard' science people taking a course in English literature. The experimental design rules out inadequate intellectual ability as a factor to explain barriers to mastering the subject.

Some members of the group learning physics or chemistry found that the lack of an overview of the subject matter presented at the beginning of the course constituted a real barrier to learning the subject. In their own field, such an overview was essential as a basis for integrating the presented information and placing the subject in the context of what they already knew about reality. They found the relatively cumulative development of such a course, where the bigger picture tends to emerge only towards the end, highly problematical. Despite assurances from the teacher that they did not need to know the bigger picture to learn the subject, some people felt that this hindered their ability to put what they were learning into their own words, and hence to fully engage themselves in the subject; others had a sense of being lost, having no idea of where they were and where they were going, and hence had difficulty participating in the subject in their own ways; and still others felt they were treated like children, being spoon-fed small amounts at a time and having to trust that everything would work out. Some people complained that they had difficulty doing the problems because they had no idea of the reason for them, or their significance within the larger scope of the subject.

The instructors saw the situation very differently. Not to expose the students to the full complexity of the subject at the beginning of the course, but to introduce them to it step by step, appeared to be a pedagogically sound, and perhaps the only possible, strategy. The instructors could not understand why students were uncomfortable at not being able to make their own contribution to the problems, and essentially being limited to imitating the instructors' examples. The students' discomfort at what they saw as merely mastering techniques as opposed to struggling with the fundamental concepts and their broader implications, and the fact that they got feedback only on problem sets rather than on the conceptual complexity of the subject, was difficult for the instructors to understand. As a result, their behaviour implied, to the students, a disregard for their questions and feelings. The hidden curriculum appeared to suggest that the intellectual context and history of the subject do not matter.

To the students it seemed as though physics or chemistry had no general structure and no context since only a few relationships were connected in each chapter, and no larger connections appeared necessary. In terms of personal involvement in the subject, this appeared to be limited to mastering the techniques; and no contributing personal opinions or intellectually pushing the limits of the subject were welcome or necessary. As noted above, one member of the group said that when she could not put what she was learning into her own words, she felt excluded from the subject.

In a mirror-image experiment, where 'hard' science people took an English literature course, much the opposite kind of situation resulted. This subject is much more contextual and relational in the sense that it cannot be broken up into relatively self-contained building blocks. It requires a higher tolerance for ambiguity, cannot be reduced to mathematical equations without disfiguring reductionisms, and has no questions with single 'right' answers. Some members of the group complained that there were just too many words, which of course was difficult to understand for the instructor because it is through words that concepts and understanding are built up. Members of the group found the way the subject appeared to meander, the lack of linearity, and concepts that were not stripped of most of their intellectual, historical, and cultural baggage, rather disquieting.

These two experiments confirm in a novel way the presence of two different 'intellectual cultures' in the sciences and humanities. These 'cultures' must be made more accessible to each other. The natural sciences are a human creation and thus have a rich context that could be included within the subject matter. This would help explain the assumptions, beliefs, and values implicit in the concepts, which is crucial for intellectual scrutiny as well as for scientific advances. It would reduce the barriers to learning for people using a contextual intellectual style. For example, Sheila Tobias suggests that mathematics exams could test three aspects of understanding of the subject: getting the right answer, proposing alternative ways of obtaining that answer, and commenting briefly on what makes this problem mathematically interesting. The far from trivial implications of the use society makes of mathematics could constitute another approach, to make the subject more appealing to learners with a contextual intellectual style. The STS (Science, Technology, and Society) movements in the United States and in other countries have been attempting to implement these approaches.[15]

All of this suggests that an engineering curriculum low on context tends to reinforce the intellectual mode in which most students excel, and to inhibit the development of complementary modes that are more context-based. This generates an intellectual and professional culture that is stable precisely because of the relationship between the general dominance of the numeracy mode among most students, the minimal use of context in the curriculum, and the justification of this situation through the hidden curriculum. All of this is easily understood, but the implications for making engineering practice more prevention-oriented are not trivial. The professional culture is likely to reduce the openness of many practitioners to certain domains of reality essential for negative feedback. Like societal cultures, it contributes to the growth of harmful ignorance in a way that blocks its practitioners from more effectively meeting the need to reduce the harm our modern way of life does to its contexts.

The consequences of this ignorance are not limited to professional practice. Snyder's research shows that, after graduation, in both their private and professional lives, people who did not develop their literacy mode had difficulties with interpersonal relations and contextual issues, which grow in importance as people climb the organizational ladder. This can undermine marriages and friendships.

Preventive approaches will necessitate modifications to both the hidden and the formal curricula to nurture both intellectual modes. This task cannot be relegated to the complementary-studies portion of the curriculum, because these subjects do not fill a vacuum but clash with the 'world' built by the technical portion of the curriculum. It follows, then, that the relationship between the two portions of the curriculum can contribute to harmful ignorance, blocking the development of preventive approaches.

The argument above may be summed up as follows: A dynamic equilibrium exists between a lack of reference to the human, societal, and natural contexts of technology in the formal curriculum; the effects of the hidden curriculum on how students learn to value and utilize a contextual understanding of technology within the technical core of that curriculum; and the student's intellectual abilities being either nurtured or held back. This situation has characterized undergraduate engineering education for many decades in North America, and possibly throughout the industrially advanced world. In so far as most of these societies are relatively open and democratic, this situation is surprising, because of the strong and growing concerns of the general public for the negative social and environmental implications of technology and our recognition that we must develop a more sustainable way of life. It is also surprising because, within the engineering profession, some have expressed concern over issues such as the perceived decline in the status of engineering as a profession; the difficulty of attracting excellent students into some, if not all, branches of engineering; growing student complaints about the curriculum being too dry, technical, and lacking in relevance, accompanied by a needlessly high drop-out rate; and the inability of the profession to follow others in attracting substantially more women. Nevertheless, only a modest strengthening of the ethical and professional aspects of undergraduate engineering education has resulted.

From the discussion in this and the previous sections, it follows that undergraduate engineering education, combined with the intellectual resources students bring to it, creates a resistance to some domains of knowledge, including their methods and approaches, essential for unlocking the preventive potential of engineering practice. This harmful ignorance has hardened during many decades and still stamps the ethos of many engineering faculties. It is the outcome of a long historical process in which the values of society interacted with those of the engineering profession in the context of scientific, technical, and economic growth. This ignorance has constrained engineering creativity in a manner that has done incalculable harm to human life, society, and the biosphere.

### 3.6 Specialization in Context

The structure of the engineering curriculum follows what has sometimes been referred to as the 'linear' model. It holds that fundamental science leads to applied science, which in turn drives engineering, which leads to technological and economic growth, which in turn results in higher standards of living and a better quality of life. Although this model has been widely discredited, we cling to it in the way we structure engineering education, as we start down this linear causal chain in first year. We should not be surprised, therefore, when we encounter difficulties in teaching engineering design or developing effective capstone courses in fourth year. We need to recognize that, in real life, engineering is a social process that is value-laden and open-ended. It cannot be compartmentalized according to disciplines whose theories and approaches are disconnected from one another, and from any economic, social, and environmental context. All the highly abstract, closed, and de-contextualized problems in basic science, mathematics, and engineering can just as well be formulated in terms of real-life problems.

I will take a typical problem from mechanics as an example.[16] A person is pushing a heavy roller over a step. All relevant features (not one too many or too few) are specified: the slope of the handle, the radius and weight of the roller, and the height of the step. The force required to push the roller over the slight step must be determined. Students are encouraged to draw a free-body diagram that strips off all contexts. This is very useful for solving the problem, but it omits much of what would happen in a real situation. Suppose an engineer was working for a manufacturer of hospital beds. How large should the rollers on a hospital bed be so that any staff member can move a bed, even with a heavy patient, across a threshold or other minor obstacle that can be encountered in a hospital? What range of diameters should be considered for these rollers? What should the final choice be? The mechanics problem is still the same, but a context has been built into it which fundamentally changes what the students must consider.

The problem is now open-ended. What kind of bumps may be encountered? Does it matter whether these bumps occur while moving the bed in a straight line or around a corner? What are the normal dimensions of a hospital bed? What is a reasonable upper limit of the weight of a patient? Where is the centre of gravity? What force can an attendant exert in moving the bed? I would argue that, in dealing with these kinds of problems, the students will simultaneously learn the principles of mechanics and the contexts that must be taken into account in all real-life engineering situations. It goes without saying that these kinds of context-rich problems have no single right answer and require a dialogue between teacher and students. The last four chapters of this book will show how this can be done in the areas of materials and production, energy, work, and the built habitat.

In more advanced courses, the task of introducing context is even simpler. Problems can be derived from state-of-the-art engineering practice. For example, in a course on heat transfer the new super-windows with radiation filters could be analysed in terms of how they change window characteristics with respect to conduction, convection, and radiation, and the implications this capability has for the design of buildings. These are not trivial since a properly designed building shell equipped with such super-windows can greatly reduce the need for heating and cooling equipment. A thermodynamics course could analyse some of the latest cycles in co-generation, how co-generation could transform the electrical grids by decentralizing electricity production, and the implications this effect could have for the overall impact of such energy systems on the biosphere. I have already referred to the pioneering work of Sheila Tobias in suggesting that mathematics courses can also be opened up to more context. This shift can be extended by examining the implications of the many uses modern societies make of mathematics. In sum, it is not difficult to open up the technical core of the curriculum to the kinds of connections that modern engineers encounter each and every day in their professional lives. I am not suggesting that every problem be put into a real-life context. This would require too much time. What I am suggesting is that enough problems be put into such a context that students become aware that all their problems could take this form, and that the kinds of situations they will encounter after graduation always do.

I have already suggested that the mirror-image situation occurs in complementary studies. Introductory textbooks for subjects such as psychology, sociology, economics, political science, religious studies, and law appear to be able to describe their subject matter on the assumption that technology has little bearing on it. This is entirely incomprehensible to me: Who would we be, how would we live, what would our communities and habitat be like; how would we work, amuse ourselves, and maintain friendships and family ties without all the technologies at our disposal? Since these dominate our life-milieu, they have an enormous influence on human consciousness, culture, and the institutions of society. Yet technology appears to play little or no role in the 'worlds' described in these textbooks. This, too, can be remedied by incorporating into the curriculum contemporary issues directly related to the use we make of technology.

By thus opening up the technical core and the complementary-studies components of undergraduate engineering education, we are essentially introducing, across the curriculum, engineering design, oral and written communication, engineering ethics, and sustainable development. The obvious model for this approach to curriculum change is mathematics, which has so permeated the engineering curriculum that it is difficult for us to imagine that not so long ago it was a separate subject. Today it would be difficult for many of us to teach our subjects without it. The same has happened with computer science; and I am suggesting that these other dimensions of engineering can and must follow suit.

### 3.7 Professional Conversations with Society

Does professional education predispose future engineers towards 'conversations' with society, and with all those on the receiving end of technology? There is an important difference between the values dominant in the natural sciences and engineering, and those dominant in society in general. Almost all the social science and humanities students taking a physics or chemistry course in Sheila Tobias's experiment were struck by the highly competitive atmosphere they encountered in their class. Their fellow students were constantly comparing marks – which creates an atmosphere that is destructive of cooperative learning. The assumption (real or imagined) that the class average would be kept within a very narrow range created a kind of zero-sum game, in which gains made by some students would translate into losses for others. In some sense, every student was another student's enemy. One of the study participants was struck by the contradiction between a learning environment in which everybody behaves like an intellectual warrior and the future working environment, in which, as in many technology-based firms, they are part of a team assigned to a project. The former environment teaches them to fear cooperation; the latter, that only through cooperation will they and the corporation succeed. Of course, you could argue that companies are engaged in fierce competition. An objection could be raised, however, that those modern societies in which governments, corporations, and unions cooperate to the highest extent possible by means of an effective national technology strategy, pre-competitive research, and greater worker democracy tend to do much better than those in countries where this is not the case.

Our value systems also tend to favour quantitative over qualitative knowledge. Quantitative knowledge 'performs': it is precise and objective, and leads to consensus. Qualitative knowledge does not. We again encounter the reasons for the tremendous social success of science over other ways of knowing. The former can build consensus in a way the latter cannot. Recall the profound religious disputes that used to divide Western civilization. For some time, there appeared to be no way of settling these disputes. Science could not convincingly demonstrate that it led to a more accurate understanding of the world, but it certainly excelled at building an account on which people could agree.[17] Map-making as a synthesis of scientific knowledge for the purpose of dealing with real-life problems may lead to a loss of consensus, and can thus be seen as socially destructive. It may also be argued that it will bring more sharply into focus the loss of consensus that now already exists, because the conclusions reached by experts generally diverge very sharply. The pursuit of power values makes it very difficult to build consensus through an open sharing of divergent opinions. Nevertheless, the building of a common future on a finite planet appears to necessitate it.

### 3.8 Recognizing Ignorance in Science

Map-making is an art that includes the sciences, but cannot be reduced to them. This is evident when we examine the structure of the knowledge base modern science offers to society. As mentioned before, there is no scientific way to integrate the findings of individual disciplines into a comprehensive understanding of the world in which we live. I attributed this to the fact that the scientific process produces knowledge either out of context or in a simpler context from unique disciplinary vantage points and prejudgments of the corresponding part of reality examined by a discipline. An analogy from engineering may illustrate the point: When engineers conceptualize some new creation, they sooner or later make a drawing of it. When these drawings are fully developed, they will include front, side, and rear views; cross-sections; and possibly some special projections. Each time a particular view is shown on the drawing, it is carefully labelled so that anyone using the drawing will be able to determine how the different views and cross-sections together represent a three-dimensional whole. These labels indicate the vantage point from which the object is seen.

Now contrast this with modern science: We are confronted with scores of disciplines that each draw an extremely rational and coherent picture of some aspect of our modern world. How this component relates to the whole is not at all clear. It is obvious that sociologists are not describing a part of the societal machine, nor do the economists describe a separate economic subsystem of society. Society is not built up from subsystems, each existing in its own space and time, that interact with one another via external boundaries, the way the components of a mechanical system interact. Human life is not lived in different compartments corresponding to the social, economic, political, legal, moral, and religious categories. Nor are the societal institutions fragmented along these lines. What these different scientific disciplines appear to be doing, therefore, is drawing cross-sections of a society that essentially describe different aspects or dimensions of individual or collective life.

There is an important difference between this situation and the making of an engineering drawing. No discipline informs its students –

and its faculty have only a vague idea – of how their 'cross-section' relates to all the others and to the whole. In other words, what I am suggesting is that, when you look at the scientific knowledge base as a whole, you find extraordinarily rational micro worlds separated by an ocean of ignorance of how they all interpenetrate to constitute the whole.

In science, we do not encounter a situation in which each discipline uses the same base map, which it elaborates in great detail in its area of specialization. The different maps cannot simply be superimposed to form one detailed and coherent picture of the world in which we live. As mentioned previously, when engineering students are sent to another faculty to take economics, sociology, or political science courses, they encounter a world to which they cannot relate, via either their daily-life experience or their disciplinary training. These subjects do not enrich their knowledge of their specialty by offering a further understanding of many aspects of the world they know. They introduce them to rather different worlds in which technology plays, at best, a peripheral role. These subject 'worlds' contradict daily-life experience, the impressions of technology gained from the mass media, and their technical education.

Science contributes to the growth of ignorance in two other fundamental ways. When the scientific knowledge base plays a decisive role in economic development, its fragmented character is transferred into human life, society, and the biosphere, thereby weakening their integrity. Another component in the production of ignorance of our world results from technological development. Any new undertaking introduces something into the world, creating new relationships and new interdependencies that did not exist before and about which we are ignorant until we study them. For example, the introduction of hundreds of new chemicals every year, whose effects on human life and the biosphere can be tested in only limited ways, produces on average more ignorance than understanding of their actual potential effects. The only testing that takes place involves exposing laboratory animals to high dosages for relatively short periods of time. There are no tests for long-term low-dosage exposure and, what is even more serious, there are no tests for the complex positive and negative synergistic effects. Since the overall effect on human beings of the many chemicals present in any particular environment is a non-linear combination of their specific individual effects, nothing scientific can be said about their effect on human health. Similarly, the introduction of computers into many spheres of human activities creates an enormous ignorance of their effects because they help produce new situations, new interactions, and new patterns and social structures that never existed before.

### 3.9 Recognizing Ignorance in Technology

Once again we encounter the kinds of issues referred to in an earlier chapter. For example, what specialists observe when they examine an accident in a large sociotechnical system such as a nuclear power plant depends on their disciplinary vantage point and the extent to which they adjust their 'cognitive zoom lens' to focus on a part, a subsystem, the entire system, or larger structures in which it is embedded.[18] The interpretation varies according to the discipline of the expert interpreting the data. Naturally, experts see what their education and experience predispose them to see. Engineers may explain an accident in terms of the failures of technical components; human-factors specialists focus on problems in the control room; organization theorists and sociologists may point to organizational problems and possibly an ill-suited organizational culture; economists may point to severe competitive pressures; political scientists may point to weaknesses in the regulatory framework; lawyers may focus on a law that artificially protects an installation from full liability in case of an accident (which might make the installation economically unsound if exposed to market forces); while psychologists may point to inadequate testing and screening procedures for potential operators. The problem is that each places certain aspects in the foreground and others in the background in interpreting the event. As a result, these interpretations are, to some extent, mutually exclusive. They are not jigsaw-puzzle pieces that can be fitted together to form a complete description, nor can they be superimposed on one another, because their basic gestalts are different. The question of what caused the accident and what should be done becomes troublesome, especially when the costs of hiring teams of experts and lawyers are so great that only large organizations can afford them, with the result that the public interest is rarely adequately represented.

This discussion forms a basis for mapping what we know about a particular technology and the way it is embedded into, interacts with, and depends on its contexts. The knowledge or ignorance a society has of it could be represented as a four-dimensional space. A first dimension represents the disciplinary or professional vantage point that gives rise to the knowledge. These vantage points comprise the humanities, social sciences, biological and life sciences, natural sciences, applied sciences, engineering, and other professions. Each is further subdivided if competing schools of thought coexist within a certain scientific or professional community. This happens when there are incommensurate beliefs about the nature of the reality studied, leading to different methodological approaches deemed appropriate for the study of such a reality, and hence to different findings. In sociology, functionalism, structuralism, phenomenology, and dialectical materialism are some of the most influential examples.

A second dimension represents the focus employed by the 'cognitive zoom lens' used for the observation. As noted in a previous chapter, these range from an 'internal' focus at the base to broader 'ecological' foci as we move up the axis.

A third dimension is constituted by the level of cooperation and coordination between the disciplines. Several levels may be distinguished, moving up the axis as cooperation and coordination increase. Multidisciplinary cooperation and coordination results from the sharing of concepts and structures emerging from the most influential discipline or profession within the group. To advance further along the road of integration, umbrella concepts and structures are required. These must derive from, but also influence, a larger number of disciplines by creating a meta-level of analysis 'above' them. Its effect will be the blurring of the boundaries between cooperating disciplines as they seek to articulate how they together relate to the same reality and to each other so as to give a more satisfactory account of the integrality of the reality as it is known to them. This constitutes the interdisciplinary level of coordination and cooperation. Still higher levels will be required to draw in other areas of science and the professions through the creation of yet higher-level umbrella concepts and structures to account for the integrality of reality as we know it through the sciences and the professions. Once again, these meta-level analyses must reciprocally interact with the research in individual disciplines and professions, and with lower meta-levels of analysis. This interaction may be referred to as the 'transdisciplinary' level of integration. Map-making seeks to overcome the difficulties encountered in all previous efforts that stem from the fact that each discipline has its unique vantage point and prejudgments with which it observes a part of the world. It is based on a dialogical approach or critical conversation between disciplines in so far as it imposes a critical understanding of the cultural vantage point of a society. It seeks to reinsert the findings of the scientific disciplines into a formal ecology of technology drawn from the vantage point of a culture of a society itself critically examined. It asserts that there can be no science without culture, and no culture without science.

A fourth dimension represents a time scale. It shows the rise and fall of different vantage points, including prejudgments of the world reflecting, in part, the history of a particular discipline or profession, on the one hand, and the fact that the members of the scientific or professional communities involved are people of their time, place, and culture.

The practical value of this representation of the knowledge we have of a particular technology is that it allows us to map our ignorance of it by suggesting additional ways in which knowledge could be gained. This is not to say that every vantage point, focus, and mode of integration will be fruitful and result in useful knowledge. Here is where the art of map-making has to transcend scientific and technical specialization in order to deal with real situations that do not respect those boundaries. The growth of scientific and technical knowledge about our world has been characterized by ever-increasing specialization. Its 'knowing more and more about less and less' must be complemented by the map-making of 'knowing less and less about more and more.' Either approach has its strengths and weaknesses. The latter can be reduced by understanding that there is a complementary relationship between map-making and the sciences, in which the former incorporates but is not reduced to the latter.

## 3.10 Conclusion

Every culture must create a balance between breadth and depth, between performance and context, by encouraging or discouraging high levels of specialization in its institutions, traditions, and values. This is both cause and effect of many important features of a society's way of life. Our civilization, more than any previous one, is based on high levels of specialization, particularly in the way we gain and apply knowledge. This opens up the risk of scientific and technological blind spots in the design and decision-making processes that guide technological development and 'steer' it into the future.

Given high levels of specialization, a greater awareness of the limitations of one's knowledge creates a greater openness to the knowledge of others, and thus to reality itself. We thereby reduce the ignorance that blocks creativity in the area of preventive approaches related to modern technology. By briefly reviewing how ignorance is institutionalized in the intellectual division of labour, the organization of institutional bureaucracies, a professional ethos, and the values of the culture of a society, we have initiated a process to soften one of the components of ignorance, which resists openness to the rich complexities of the ecology of modern technology. A critical self-awareness of who we are as professionals and as members of a culture turns our ignorance into a useful asset by allowing us to see things where we saw nothing before, by valuing relations that had little or no value, and by appreciating the complexity of our relationship with technology. The more realistic we can be as professionals, the greater our ability to make professional practice more preventive. It will, of course, constantly remind us of our limitations, and oblige us to participate with a professional humility in debates over controversial issues related to modern technology.

It might be useful if society complemented such a critical awareness on the part of its professionals with mechanisms to increase negative feedback about the contributions they make to society. A friend once suggested that she was quite convinced architects would not build the same buildings if they were obliged to live in them for a few years; engineers would not design the same assembly lines if they had to work on them for a while; and doctors would modify professional practice and the organization of modern hospitals if they had been patients there for an extended period of time. Without a greater critical self-awareness, however, such experiences might serve to reinforce the status quo, because one might conclude that people on the assembly line would be unable to adapt to a change in routine, for example. We must become aware of our ignorance. Since the above arrangements are hardly practical, the next best thing is obviously to take seriously the experience and knowledge of others who live and work in these buildings, factories, or hospitals. In other words, feedback from all the parties concerned is essential for the development of preventive approaches. We need to learn to have 'conversations.'

Although the examples in this chapter have been principally drawn from engineering, similar situations exist in management, accounting, medicine, social work, and industrial relations. The undergraduate curricula that prepare future practitioners suffer from the same shortcomings and generate harmful ignorance, which prevents more prevention-oriented practices. For example, preventive medicine would no longer be a low-status sideline but would help constitute the core of modern medicine. To develop management approaches equally applicable to a great many different activities is to externalize all specifics, and thus to block the potential for preventive approaches. Green accounting would provide managers with the information they need to practise preventively. Social work, which ignores the fact that modern science and technology have created the very life-milieu for human existence, cannot intervene in serious social issues in a preventive manner. Only a greater self-awareness of who we are and how we have come to be that way can create the greater openness that is required for the reduction of harmful ignorance, necessary for effective negative feedback in all professional activities. A critical awareness of ignorance transforms the very nature of that ignorance to create a valuable personal and professional asset: a greater openness to the world in which we live, including a greater openness to ourselves. Strengthening our numeracy and literacy modes, and learning to use them interdependently, is a necessary step on the road to more prevention-oriented professional practice. After acquiring some basic tools for the task in the next part of this work, we develop the theory of design and decision matrices, before turning to the development of preventive approaches in four areas of application.

# PART TWO

# THE ECOLOGY OF TECHNOLOGY AND MAP-MAKING

# 4

# Tools for Map-Making

## 4.1 Mapping the Ecology of Technology

Mapping the ecology of the technology of a society or of an individual constituent technology is integral to the development of preventive approaches aimed at achieving a greater compatibility between technology and its contexts. To avoid the pitfalls of conventional approaches, such ecologies, and the design and decision making they guide, must keep together what belongs together: technology with its contexts, desired with undesired outputs, and performance values with context values. Our failure to do so has produced, among other things, the environmental crisis that, in turn, has given us a different perception of the integrality of the world, commonly expressed as 'Everything depends on everything else.' We can no longer take this integrality for granted and have had to resort to inventing new concepts such as appropriate technology and sustainable development. Hence, the question that must be faced in the engineering, management, and regulation of modern technology is: How much of this integrality must be taken into account in order to reverse present trends, and can this be done within existing constraints on time and resources? This chapter takes a first step in facing this issue.

In its simplest form, we can represent the ecology of a technology as a set of concentric circles. At the centre, we find technology. It is embedded in and created by a society, constituting the second circle. The society, in turn, is embedded in one or more ecosystems, and thus, by extension, is embedded in the biosphere, these two representing the two outer circles. This 'map' is the simplest way of showing how technology is embedded in, interacts with, and depends on its contexts.

Even this most elementary conceptualization presents us with some difficulties we do not encounter with traditional geographical maps. Technology is simultaneously cultural and natural. It comes into existence through a variety of human activities, but, since these can neither create nor destroy matter and energy, technology is also integral to the biosphere. As a result, it enfolds something of the culture of a society and something of the biosphere. Those who participate in technology are also members of society. A society is simultaneously involved in cultural history and natural evolution. There are no separate individuals and no separate society. Any individual helps constitute a society for others as they help constitute a society for him or her. Similarly, there is no separate biosphere, since we, in concert with all other life forms, are its constituents. For every individual, technology and the biosphere are not simply 'out there'; through experience, they help constitute the mind, thereby connecting individuals internally to these contexts. All relationships between technology, human life, society, and the biosphere are reciprocal. As people change technology, technology simultaneously changes people. As society changes technology, technology changes society. As technology is changed, the biosphere is simultaneously changed. These examples can be multiplied, but the point is becoming evident: what is involved is much more than everything being connected to everything else; what is involved is the interpenetration or enfolding of technology, human life, society, ecosystems, and the biosphere into one another. These kinds of relationships are not encountered when making a conventional geographical map. There, everything exists in its own space, external to everything else. No point on such a geographical map can simultaneously belong to a road, river, railway, and building. How do we map the connectedness of technology to the contexts into which it is simultaneously enfolded? As the four concentric circles at least partially blur into each other, none has clear boundaries.

There is a second major methodological difficulty with this conception. Attempting to mark everything on a map would clutter it to the point of unintelligibility. A map is made for a particular purpose, and it is in relation to this purpose that decisions are made whether or not to represent something on it. What loss of information and what blind spots will such decison making involve? How great an obstacle do these represent in terms of accomplishing the purpose?

Acceptance of the fact that everything depends on everything else and that some things may interpenetrate or be enfolded into each other does not mean that they are all equally significant when mapping the ecology of a technology for the purpose of making technology more compatible with its contexts. Within this larger purpose, some relations depend much more on each other than they do on other relations within their immediate context, thus creating significant variations in the apparent connectedness of reality. This point will be explored by means of the concept of a whole. Within its context, it represents a locus of relatively denser connectedness. Via a lesser density of connectedness, a whole participates in larger wholes, which, in turn, participate in still larger wholes, and so on; in turn, each whole is internally constituted of smaller wholes, themselves constituted of still smaller wholes, and so on. In other words, a whole is not a system in the way that concept is commonly used in engineering. The connectedness it represents is not that of holism, popular in interdisciplinary studies some decades ago. As well, only some of these relationships and causal interactions resemble the ones in what was supposed to be a mechanistic universe, in which each constituent exists in its own space and time, and causally interacts with others via distinct external boundaries. This undoubtedly occurs, but it accounts for a minority of relationships and interactions.

The use of the concept of a whole may be illustrated as follows. If I examine a cell in my body, I am not looking at a distinct 'part' but rather a locus of dense interconnections that make up a tissue that contributes to an organ that functions within a 'subsystem' that, together with others, constitutes my body. Within it, my brain, unlike any other organ, is symbolically affected by the experience of the 'outer' world, connecting me to others and, via them, to social groups, communities, and society. My society is, in turn, connected to others that together constitute a civilization, connected less densely to other civilizations and more densely to the ecosystems in which it exists and through which it is connected to the biosphere, via which it is connected to the planet, our galaxy, and the universe. This construct must not be interpreted as signifying a hierarchy of wholes in which the larger the whole, the greater its importance. Rather, it is a question of making sense of a reality in which everything depends on everything else in terms of variations in the quality and relative density of connectedness. Whatever whole we focus our attention on represents the locus of a high level of connectedness relative to its context, made up of the next immediate whole. It, in turn, is similarly distinguished from its context, representing the next immediate whole. As we proceed, the relations with the originally considered whole become less and less direct. By considering an ever larger scope of connectedness of the original whole, we in essence uncover many layers of interconnectedness, each successive one being less and less direct. In practice, this progression of connectedness may well turn out to be less clear because the next-larger whole may in fact be constituted by several enfolded wholes. Whatever pattern of connectedness emerges is that of the whole being studied, and this pattern would change if another whole were being examined. It is our study of the whole that creates a model of a multilayered connectedness, which is subject to change as a result of new discoveries. A whole is therefore a conceptual tool to make sense of an interconnected reality in which some or most wholes may turn out to be enfolded into one another. Because each whole is a locus of dense interconnectedness relative to its context, the whole may be said to be more than the sum of its parts.

The notion of a whole is therefore a conceptual tool for minimizing the loss of information that occurs through the inevitable process of abstraction. This is of immense theoretical and practical importance for the development of preventive approaches and the avoidance of harmful ignorance and professional blind spots. For example, it can help us avoid the common error of assuming that technology is nothing more than its constituent technologies, or that the effect of technology on a human life or society is nothing more than the sum of the effects of its constituent technologies. A human life is not separable into parts, one spent with computers, another with television, still another with cars, and so on. The impact of technologies in the workplace is not limited to people's working lives, as if these were separate from the remainder of their lives, their families, and their communities. The effects of technology on society cannot be broken down into those on the economy, social structure, political framework, legal institutions, morality, religion, and artistic expressions, as if these were separate parts. The question of the whole being more than the sum of its parts may also be implied in the purpose for which the map of the ecology of a technology is drawn. Attempting to assess how technology has affected the quality of life in a society is largely dependent upon our understanding of the wholes referred to as 'life,' 'society,' and 'technology.' This is the case because the map essentially seeks to clarify how the technology of a society and its constituent technologies are embedded into, interact with, and depend on their contexts. Any relation or whole that cannot be reduced to its constituents without a significant loss of information with respect to the purpose for which the map is drawn must be included in order to minimize harmful ignorance and blind spots. The same is true for any relation or whole whose significance cannot be grasped without mapping it as a part of larger wholes.

This illustrates an important feature of the world as we know it. Everything has a context, and within that context larger wholes may be constituted, which in turn may contribute to still larger wholes, and so on. Each higher-level whole brings something new into the world that cannot be understood in terms of the properties of the smaller wholes within it. This approach to understanding the world continues to broaden the observer's ecological focus so that he or she can grasp something in terms of how it fits into, is a part of, and depends on larger and larger wholes. In this way, the map-making approach to studying the world complements the traditional scientific approach, which abstracts wholes out of their context and, using an 'internal' focus, seeks to discover the fundamental building blocks and relationships out of which the whole is constituted. Map-making includes this approach but seeks to overcome its limitations.

Mapping the ecology of a technology thus uses the ecological foci to 'zoom out' to ever-higher levels of integration and organization, and the internal foci to 'zoom in' on the wholes that help constitute any larger whole, and still others that in turn help constitute them, continuing as far as is meaningful. Map-making looks out from a particular whole and also looks in towards its internal structure. Each focus constitutes a level of analysis requiring the application of the concept of a whole. Map-making is the essential complement to the process of abstraction.

## 4.2 The Concept of a Whole

The technology of a society includes a great many wholes whose characteristics cannot be understood in terms of those of their parts. For example, electronic components can be organized into many different circuits that perform functions that cannot be understood in terms of these components. This is also true for molecules, composite materials, space frames, hydraulic circuits, computers, and almost everything else.

I have suggested that, in many technology-related issues, we often overlook the fact that the effect a technology of a society has on its contexts cannot be understood in terms of the effects of its constituent technologies. On the receiving end, a similar methodological issue is usually ignored as well. Can the effect of technology as a whole on human life or a society be dealt with in a piecemeal fashion, whereby the effects of each constituent technology are examined one at a time? Is the life of a person merely an aggregate of sets of experiences associated with particular technologies? Is a society merely the sum of components such as its industry, economy, and legal system?

The concept of a whole can clarify a variety of technology-related issues. It has already been noted that nothing scientific can be said about the implications of a polluted environment for human health. Similarly, when someone claims that a particular level of a pollutant is safe, he or she is forgetting the simple fact that a threat to human health cannot be determined in a piecemeal fashion. It is the total body-load, and other factors such as lifestyle, diet, and genetic make-up, that determine the short-term and long-term effects over several generations. In other words, what the real long-term health implications are is not known and probably cannot be known scientifically. Some physicians have, in fact, argued that we may well be under as great a threat from the contamination of the biosphere as we are from nuclear weapons.[1]

By requiring technology assessments and/or environmental assessments for major undertakings, many societies seek to ensure that environmental impacts (in most cases these include human, social, and natural effects) are reasonable and acceptable. This, of course, is very important. However, let no one believe that this provides us with a handle on how technological developments are affecting the health and well-being of human life, society, and nature. What may appear to be relatively modest and acceptable influences in a great many studies may well accumulate into major and unacceptable consequences. How, for example, do the stress levels of many different daily-life activities interact within a person's life? How do the many stressors in an urban environment (such as social overload, sensory overload, crowding, and pollution) affect the liveability of cities? How is the overall integrity of an ecosystem affected by the many sources of pollution within it? The reverse side of this coin is that, because the overall effect involves the complex synergistic effects of many individual technologies (telephones, televisions, computers, video games, streetcars, and so on), it is very difficult to examine the implications of a single technology such as television.

At this point, it is useful to introduce a definition of a whole. A 'whole' may be defined as an entity comprising a diversity of elements within its unity. These elements are interrelated in such a way that a change in any one element affects at least some others, which in turn may affect still others and potentially the whole itself, while a change in the whole affects at least some constituent elements. The elements interact more directly with each other than with those of their surroundings, so that they behave as a single entity towards the outside world. The properties of this whole cannot be derived from those of the constituent elements. The relationships between the whole and its elements are fundamentally different in living and non-living entities. A whole tends to maintain its integrality in the face of disturbances originating from within or without. The extent to which this occurs varies greatly in living and non-living wholes.

## 4.3 Living and Non-living Wholes

Three closely related differences between living and non-living wholes are discussed here: the way they come into being, the internal organization that makes this possible, and the necessary interactions with their contexts. The first and fundamental difference, which permeates all other differences, is that living wholes are never constituted from separate and independently existing parts, the way non-living wholes are. A living whole comes about by progressive internal differentiation, through which 'parts' are created. Something of the whole is therefore present in each 'part.' A non-living whole, on the other hand, is built out of separate and pre-existing parts. Contrast the way babies and cars come into existence. Through progressive cell differentiation in the embryo, the 'parts' emerge. Each cell has the DNA-encoded plan for the whole enfolded within it. By means of experience, the brains of babies and young children are progressively differentiated to constitute the mind, thus allowing them to become persons as well as members of a society. A car, on the other hand, is built by first producing the separate parts, then assembling them into subsystems and integrating these into the whole. Living wholes have no 'parts'; instead, they have local manifestations of the whole that are differentiated from others. There are no interchangeable 'parts' : each 'part' is sufficiently unique that special precautions must be taken during transplant operations to avoid rejection. Thus, for example, when people emigrate to another society with a different culture, they face adjustments that are much more profound than simply learning another language.

It is impossible to 'assemble' an ecosystem out of its 'parts.' Suppose we could collect samples of all the species of organisms that constitute the 'parts' of a patch of threatened rainforest. Even if we managed to preserve them in zoos, gardens, or laboratory cultures, or deep-freeze samples of the tissues, the ecosystem could never be 'reassembled,' no matter how many thousands of biologists and how much funding were available. E.O. Wilson puts it this way:

> in the forest patch live legions of species: perhaps 300 birds, 500 butterflies, 200 ants, 50,000 beetles, 1,000 trees, 5,000 fungi, tens of thousands of bacteria and so on down a long roster of major groups. Each species occupies a precise niche, demanding a certain place, an exact microclimate, particular nutrients and temperature and humidity cycles with specified timing to trigger phases of the life cycle. Many, perhaps most, of the species are locked in symbioses with other species; they cannot survive and reproduce unless arrayed with their partners in the correct idiosyncratic configurations.
>
> Even if the biologist pulled off the taxonomic equivalent of the Manhattan Project, sorting and preserving cultures of all the species, they could not then put the community back together again. It would be like unscrambling an egg with a pair of spoons. The biology of the microorganisms needed to reanimate the soil would be mostly unknown. The pollinators of most of the flowers and the correct timing of their appearance could only be guessed. The 'assembly rules,' the sequence in which species must be allowed to colonize in order to coexist indefinitely, would remain in the realm of theory.[2]

In other words, the only way to 'reassemble' the ecosystem is to replicate the process of evolution, which is clearly impossible.

Related to the difference of how living and non-living wholes come into being is a second and radical difference in terms of their internal organization. A classical machine can be thought of primarily in terms of its design, which specifies its internal structure and the way its parts function together. This design is relatively static for the duration of the machine. A living whole, on the contrary, is self-organizing. There is room for growth precisely because the internal structure is determined by complex processes and considerable variations on the level of the 'parts.' Complex feedback mechanisms regulate these processes to form a self-organizing dynamic whole. The complexity of self-organization in the human body is readily illustrated by the fact that, except in the brain, all the cells in the body are renewed every few years. Matter and energy flow through the organism from and back into its context.

These 'internal' differences imply a third difference resulting from the ways in which the whole relates to its context. Apart from an energy input, a clock, for example, does not depend on its context in order to function. It is essentially a closed whole with a separate context. Living wholes are open, which means that they depend on a continuous exchange of energy, matter, and information in the relationships with their contexts in order to stay alive. This exchange involves metabolic processes that break down food and drink to grow and to maintain its organization. Also, the human mind requires interaction with the outside world; when this is curtailed through monotony or sensory deprivation, serious consequences ensue. Ongoing self-organization made possible by a high degree of internal non-equilibrium nevertheless maintains a high degree of stability because of these exchanges. An identity is maintained through complex internal and external changes. We have no difficulty recognizing a friend even after a long time, despite the fact that his body cells may have been changed several times and his mind enriched with many new experiences.

There is a close correlation among the three differences identified above that distinguish living from non-living wholes. According to insights derived from thermodynamics, closed wholes run down and wear out. Disorder and noise can only increase, until a complete breakdown occurs. Thermodynamically speaking, the entropy of such wholes can only increase. Living wholes, on the contrary, are open and self-organizing, and thus have the capacity for self-renewal, adaptation, and growth, and, in the higher life forms, produce mental processes. They take in structures from their contexts and, through metabolic processes, break them down, thus creating disorder (entropy) for the purpose of self-renewal through processes of 'repair,' 'maintenance,' and adaptation. Living is indissociable from re-creation and adaptation. Cells are constantly breaking down and being repaired or replaced by the whole. Capra puts it this way:

> cells are breaking down and building up structures, tissues and organs are replacing their cells in continual cycles. Thus the pancreas replaces most of its cells every twenty-four hours, the stomach lining every three days; our white blood cells are renewed in ten days and 98 percent of the protein in the brain is turned over in less than one month. All these processes are regulated in such a way that the overall pattern of the organism

> is preserved, and this remarkable ability of self-maintenance persists under a variety of circumstances, including changing environmental conditions and many kinds of interference. A machine will fail if its parts do not work in the rigorously predetermined manner, but an organism will maintain its functioning in a changing environment, keeping itself in running condition and repairing itself through healing and regeneration.[3]

In addition to having the capacity for self-renewal, living wholes can also reproduce. They can, therefore, interact with their contexts in a variety of ways not within the capabilities of machines. If we attempt to describe their internal state in terms of interdependent variables, it becomes obvious that all of them fluctuate within various ranges, allowing them to adapt to changes in different ways. This ability is usually referred to as 'homeostasis.' This is not to deny that certain functions within living organisms resemble those of machines.[4] Such 'mechanisms' do exist, but they account for only a very small portion of the functioning whole, at least in the higher life forms. I will not enter into the debate as to whether the human brain is the most complex computer known or whether it is qualitatively different. I will simply point out that many decades of artificial-intelligence research, involving many distinguished researchers and vast resources, have produced an appreciation of how much more complexity any human function has than can be simulated through rules, algorithms, or neural networks.[5]

The fact that a machine has a separate environment, but a living whole does not, has been obscured by the mechanistic world-view. Many thinkers have pointed out that the concept of a nature or environment 'out there' is a relatively recent Western invention.[6] Its origins are highly complex, based on but not limited to the Cartesian dualism between mind and world, between subject and object. This view presupposes an isolated self observing the world. Such a starting point for philosophy and science forgets that all human beings are in the world and can only have a consciousness of the world. The Cartesian starting point for philosophy and science required the observer to prove the existence of the world. If, on the other hand, it is recognized from the start that all human beings are inseparable from the world, then the issue is that of meaning. Pure, detached consciousness makes way for consciousness of the world. The open wholes and non-mechanistic view of the 'nature' of reality resembles those of earlier societies and those implied in some of the most recent developments in science.

**4.4 Enfolded Wholes**

Human beings are integral to their world in two ways – the one sociocultural and historical, the other biological and evolutionary. Beginning with the former, it is clear that every member of a society helps to constitute the social context of his or her fellow members. Each person has learned to function in that context after a long period of growing up socioculturally as a person of his or her time, place, or culture from the social womb made up of the people involved in the upbringing. Since the experiences of the sociocultural context are internalized into the mind, from which they can be recalled, it is clear that the social world is not just 'out there.' We internalize the way of life created, adapted, and passed on by earlier generations. In this sense, something of our culture and way of life, which permits millions of people to live together within a society, exists within us. Much as the DNA enfolds something of our biological whole into each of our cells, so living in a society enfolds a kind of cultural DNA within our minds and lives.[7] Human beings are therefore internally connected to their society. Something of that social whole is enfolded into their very being and lives. At the same time, living a life contributes something to the lives of others and, via them, to the whole of social life. How the members are internally connected to their society is evident from the fact that there can be no separate society 'out there.' Each member helps to constitute the society for all other individuals as they, in turn, help to constitute the society for that individual. Each person is, therefore, integral to a society by being internally connected to it, which is a relationship that could not exist in a mechanistic universe.

The same kind of enfolding takes place with respect to the various spheres of activities a society engages in, such as science and technology. These three 'wholes' (science, technology, and society) do not each exist in their own space and time, interacting with each other across external boundaries, as is the case for the parts of a machine. They interpenetrate and are enfolded into each other. For example, these wholes share 'elements,' since scientists and technical experts are at the same time members of society. Because they are specialists, they rely on the culture of their society for images, conceptions, and models of the larger wholes beyond their specialties. Also, as scientific and technical undertakings are applied and enter into people's lives as products or techniques, they are experienced and consequently internalized into people's minds, thus contributing further to the interpenetration of the wholes of science, technology, and society. As a result of this enfolding, the full effects of science and technology will frequently not become apparent until the analysis reaches the level of culture and society, because these are the wholes that give integrity to human life.

In the same vein, the industry, economy, social organization, political institutions, legal system, morality, and religion of a society are not separate 'parts.' Nor are the lives of the members of a society divided into different segments corresponding to these institutions. Both in the university and in daily life, we have become so used to these institutions being abstracted out of the larger wholes into which they are embedded that we forget that we do not experience and live in society in this way. Going to the corner store to buy a quart of milk clearly has an economic dimension, contributing to the demand for milk. However, it also has a social dimension, as we may speak to the owner, whom we have known for a long time. Shopping can also have a political and a moral dimension if we decide to boycott the products of a certain company because we disapprove of its actions in the third world, or because the country in which the company operates refuses to grant factory workers basic rights. Some consumers are concerned about the environmental impacts of certain products, while others prefer to buy locally made goods to help keep their fellow citizens employed. It is this highly enfolded character of human life and society that makes map-making a complex and non-linear exercise.

From a historical perspective, there is enfolding in time as well. Each generation helps to constitute the social womb out of which the next generation is born as people of their time, place, and culture. Nevertheless, each generation is different because it re-creates in its own way what it inherits from previous generations. This permits a society to evolve, adapt to new circumstances, and have a history. The relationship between a person and his or her society is a two-way interaction. Society precedes the person, and by growing up a person in his or her own unique way embodies something of that society and culture. A person is individually unique and yet culturally typical. As society evolves, its members have to adapt, but at the same time they help to create the social 'environment' for others, and thus contribute to that evolution. There can be no individual without a society, nor a society without individuals. When a person dies, society loses much more than a member. In so far as that life was enfolded into the lives of other members, something of those lives also comes to an end. The historical evolution of a society, therefore, involves the constant enfolding of new members and the 'unfolding' of those who die or move away.

The above interpretation leaves no room for a neutral, detached observer. All people, even if they come from another society, quickly become internally connected to what is observed. Observation is therefore never entirely detached. It can be more or less neutral, more or less objective, but it is never totally so. Neutrality and objectivity are almost impossible when something in a society receives a very high value, because of its importance to its members. I have already noted an example of this in modern societies, where discussions about the role of science and technology quickly become emotional. In the same vein, it would be very difficult to act as a detached neutral observer of rituals in another culture that profoundly offend our fundamental values and convictions. In such circumstances, observers tend to distort the situation so as to reduce their anxiety. This has been extensively documented.[8] Since there is no world 'out there' and any observer is internally connected to it, the facts are affected by the presence of the observer, as has been recognized in subatomic physics and many other disciplines.

In the same way, human beings and societies are enfolded and inextricably connected to the biosphere. There is no nature 'out there' as a separate and external entity. This is how it became regarded in Western culture several centuries ago when the landscape was discovered, along with a separate 'inner' world of human individuals.[9] Nature became a kind of separate object and a collection of resources society had to put to good use through its technology. Without this view of nature, no objective science could have developed, nor would modern technology have become what it is. The writing of this book would have been entirely redundant, for Western civilization would have taken a different course. Modern societies continue to behave as though they have little connection to the biosphere. In a sense, Western civilization has made a fundamental error in the way it created 'nature' as a separate object, and the way it included this object on its mapping of reality. The debate over sustainability is beginning to recognize that in many respects what we do to 'nature,' we do to ourselves. Persons and societies are living wholes open to the biosphere. The food chain is but one example of how these exchanges enfold us into the processes and cycles of ecosystems.

Since the mind develops within the brain, the enfoldedness of human life in the biosphere affects the enfoldedness into society, and vice versa. Babies and children learn to live in the world by establishing relationships. Their only pathway to a knowledge of reality is through these relationships, which create their 'world.' This network of relationships defines who they are socially, and characterizes their being much more than do their physical bodies. How other creatures are treated through these relations will affect how people think of themselves as living beings.

A simple analogy may illustrate the point. By staring at someone, we treat him or her 'objectively,' as if we were looking at an object with which we have no human involvement, no commitment or obligation. There is no reciprocity normally encountered in a human relationship. When noticing the objective stare from someone else, most of us become uncomfortable because, in a sense, our humanity is being denied. A non-reciprocal relationship is established that makes us into a thing. The person staring recognizes no obligation towards us that we might expect from other human beings, and this opens us up to manipulation. It is as if the person staring at us is free to do with us as he or she wills. Most people experience this as intensely frustrating, and shame and anger are common responses.[10] In times of war, atrocities are often committed because people learn to see their enemies as subhuman, and thus more 'thing-like.'

### 4.5 Environment as the Breakdown of Enfoldedness

The significance of the appearance of a concept of 'nature' and later on an 'environment' in Western civilization can be understood in terms of our earlier analogy of staring at someone. When persons and societies see themselves as separate from the biosphere and objectify it, the status of fellow life forms is denied to other creatures. It then becomes much easier to manipulate them for one's own purposes, without feeling any obligation towards them, or acknowledging that any manipulation or exploitation involves profound moral questions. It becomes difficult for the members of a society to recognize and reach consensus on limits of the use of the biosphere and natural resources, which inevitably creates what we call the environmental crisis. The situation is somewhat analogous to the one established with other human beings through pornography. The men and women depicted become objects. They have no name, no past, and no lives. There is no way in which they can remind the people staring of their subjectivity. Their inability to establish reciprocity turns the relationship into the kind people have with objects. In the same vein, the reciprocity between human beings and the biosphere, based on a recognition that all human life is integral to that biosphere, has created many of the kinds of issues that are at the top of the agenda of public concerns today.

The concepts of nature and environment manifest a split between the 'inner' world and the 'outer' world, and between subject and object. Probably the first painting illustrating this schism is the *Mona Lisa*. Her mysterious eyes reveal an 'inner' world, while behind her is the first genuine landscape in a painting. The drawing or painting of landscapes helped mark this transition in human consciousness.[11] This changed awareness of the world opened up new spheres of activities, including sight-seeing and later tourism, allowing people to 'stare' at the world.

These cultural attitudes towards the biosphere are reinforced by an evolutionary world-view. The contemporary understanding of the process of evolution shows that Darwin's focus on the struggle of species to survive by adapting to an environment is only half the story. The 'environment' is largely created by a multitude of fellow species interconnected in complex and ever-changing ways.[12] Their common survival depends on a complex pattern of organization in which all life forms are both species and environment. Their evolution depends on their cooperatively helping to maintain the kind of context in which they flourish. Together the millions of adaptations made by numerous species to their contexts participate in creating the biosphere on which all life depends. It is now increasingly recognized that, in the evolution of higher life forms, it appears that microscopic life helped create the kind of biosphere necessary for its evolution, and that this biosphere in turn helps create new forms of life. In this sense, all living wholes *are* the environment. There is thus no such thing as our environment. The fact that human beings have a consciousness of their world superimposed on the genetic processes of evolution creates the adventure of human history, based on language and culture. Culture has all but obscured our roots in evolution. To once again live *with* and *within* the biosphere will require a cultural transformation. For this transformation to be sustainable, it will have to reintegrate the biological and cultural 'natures' of human life. Social Darwinism, like its natural counterpart, entirely overlooks the fact that, even if the future belongs to the fittest, evolution never threatens the viability of an ecosystem or the entire biosphere, and can therefore not be used to justify behaviour that compromises the viability and integrity of society.

The fundamental shift in human consciousness and culture that made, first, nature and, later, the environment a kind of backdrop for the stage of human history also produced the mechanistic world-view. It eliminated any fundamental distinction between living and non-living entities, and thus made invisible the reciprocal and enfolded relationship of the former with their contexts. Each entity now had a separate environment on which it no longer vitally depended and with which it now interacted by simple cause–effect relations.

In the course of centuries, these developments and others opened the road for the two-stage approach for the engineering, management, and regulation of technology. Technology was separated from the environment, making the latter a kind of warehouse stocked with the resources required for development, and functioning as a storage space for wastes. Hence, technological development and its effects on the environment could be dealt with in two distinct stages. Gross efficiency, productivity, and profitability could not be distinguished from their net values because the performance of any whole was no longer seen as dependent on its contexts for essential inputs and outputs. The reciprocal relationship between technology, on the one hand, and human life, society, and the biosphere, on the other, became all but invisible, thus paving the way for the myth of progress. The latter would have been a valid extrapolation of the experiences of the industrial societies of the nineteenth century had it not been for the effect of technological and economic development on human consciousness in general, and on human needs in particular. The moment it is recognized that many entities in the universe function as open wholes that are vitally dependent on their contexts, the development of technology to maximize performance values becomes seen as fraught with risks, including the potential to undermine all life. When the context on which a living whole depends is degraded, its ability to deliver and receive the kinds of flows that sustain the whole is undermined, and thus the vitality of the whole itself is threatened. The patterns of technological and economic development bear this out: spectacular successes are achieved in the domain of performance values, and equally spectacular failures are encountered in the domain of context values.

These kinds of problems are not limited to the two-stage approach for the engineering, management, and regulation of modern technology. The above developments have also opened the road to scientific and technical rationality in many other spheres, as observed by Max Weber at the beginning of this century and by Jacques Ellul in the latter half of this century.[13] The human consciousness and cultures that result from and affect technological and economic development do not recognize that each whole participates in larger wholes and is in turn composed of smaller wholes, so that it simultaneously participates on several levels of reality. When human consciousness and cultures no longer reckon with this multidimensional functioning (associated with multidimensional meanings and values that are frequently dialectically structured), each dimension is reduced to separate aspects that can be manipulated and optimized one at a time without any apparent implications for the others. This works reasonably well for a machine whose parts generally have only one or two functions that interact with others in clear cause–effect relations; but it does not work for a living and enfolded reality. Making human life or society 'better' by optimizing one or several dimensions seen as separate functions undermines their integrity and vitality. The ability of modern cultures to make sense of and live in a multidmensional enfolded reality has been greatly diminished under the influence of rationality. We have all but lost the ability of traditional cultures to deal with reality in a more dialectical way. The earlier example of attempting to deal with the hunger problem in a Colombian valley illustrates this limitation. Each expert essentially situates the problem in his or her own field of expertise, converting all other dimensions into knowledge externalities. Rationality divides what belongs together and deals with things one at a time. Societies industrialized without recognizing that the network of flows of matter and energy involved in this process ultimately depend on the biosphere; and, for their meaning and value, they depend on society and culture. The inevitable reorganization of society necessitated economic, social, and political technologies that proceeded in the same way. As a result, the power of our means were greatly multiplied at the expense of the integrity of human life, society, and the biosphere.

The breakdown of enfoldedness is also apparent in the way the mass media speak of the economy and markets, much like prehistoric medicine men might have spoken about the gods.[14] They are seen as separate forces that shape our lives, as opposed to the outcome of human actions, decisions, and choices.

## 4.6 Map-Making and Its Limitations

Having explained the definition of the concept of a whole, we will now briefly relate it to the map-making approach and bring into focus its limitations. The map-making approach begins with the abstraction of a particular whole for the purpose of analysis. It can then be studied in terms of four different frames of reference, corresponding to different vantage points. The first places the observer outside of the whole, to examine the 'phenomenological' characteristics it displays towards its contexts. The second frame of reference places the observer within the whole to examine its internal structure as it interrelates the constituent wholes. A third frame of reference involves the disorganization of the whole into its constituents, to learn more about the properties they have in isolation from the whole, and to compare these to those they have within it, according to the findings obtained by means of the second frame of reference.[15] This process gives a deeper insight into the organization of the whole – namely, how it makes use of certain properties of its constituent wholes and suppresses others. For example, growing up in a society represses certain instinctual characteristics and creates others typical of that culture. In order to create a diversity within the unity of a whole, there must be an ongoing tension between diversity and unity. Too great a diversity would threaten the whole's integrity and eventually destroy it, while too great a unity would undermine the unique properties of the whole that emerge as a result of the diversity within its unity. In the same way, the organization of the whole must balance the complementarity and antagonisms of the constituent wholes. Any organization thus embodies both order and disorder, which is essential because the former can be maintained only through negative feedback in relation to the latter. In living beings, there exists, therefore, a kind of dialectical tension between order and disorder, between health and sickness, and between life and death.

A fourth frame of reference extends the above analyses into two directions. It examines the whole in its contexts, including wholes into which it may be enfolded and others to which it directly or indirectly contributes as a building block. This indefinitely extends the analysis begun with the first frame of reference to ever larger wholes. The second direction continues the analysis begun with the second and third frames of reference by examining how the constituent wholes of the whole originally studied are themselves made up of smaller wholes, in turn made up of still smaller wholes, and so on. This nesting of wholes into one another may, of course, be a great deal more complex when some are enfolded in others. This frame of reference seeks to minimize the loss of information that may have occurred when the original whole and its constituents were abstracted from their contexts. It does so by attempting to examine the whole in the full context of reality.

We have already noted that, within the whole, the complementarity between the constituent wholes cannot exist without the antagonisms created by repressing some of their properties. Thus the organizational system creates both order and disorder. The latter is frequently used for self-regulation and feedback. It is now clear that all this happens in relation to the dual role of any whole: as a building block or 'part' in the next-larger whole and as a whole itself. Also, there can no be no simple cause-and-effect relations with such a dual role. In some instances, the situation may even be more complex when a particular whole interpenetrates several others.

From the discussion thuis far, it is evident that map-making includes the scientific study of a whole but also goes beyond it in seeking to assess the loss of information that may have occurred as a result of abstracting it from its contexts. It, too, has its limitations, which are immediately apparent from the description of map-making provided above. The first limitation stems from the open-ended character of the analyses performed with the fourth frame of reference, which seeks to examine the original whole in the full context of all of reality. Since this is clearly impossible, there are limits to what can be known about any whole. A second limitation relates to the impossibility of a detached observer, resulting in an inevitable prejudgment or vantage point implied in any analysis of a whole. The last limitation stems from the impossibility of applying the above four frames of reference in general, and the third frame of reference in particular, to living wholes.

As I have just mentioned, the first limitation is related to the fact that the analysis of any particular whole is clearly open-ended. It progresses in two directions, in which each step includes the next larger and smaller wholes. The findings of each of these steps in the analysis are not cumulative, since what is learned about any particular whole is an input into the analysis of the next larger and smaller wholes. Thus, the findings of any step in this process could affect the analyses of previously completed steps. Any process of map-making thus encounters the difficulty of having to transcend the limitations of the intellectual division of labour in modern science and in the professions as well as facing an open-ended process requiring a great deal of time and resources. The more we try to learn about a particular whole, the more we need to know about many other wholes. There exists, therefore, a certain measure of uncertainty about our knowledge of any particular whole. The information that is lost in any process of abstraction can never be fully assessed, because of the way wholes are nested or enfolded into each other. It should be noted that science has traditionally eliminated this limitation to any process of analysis through abstraction by assuming the nature of reality to be mechanistic, making it possible for it to be analysed one constituent at a time by relatively autonomous disciplines. The fact that the knowledge of any whole is open-ended and relative rather than absolute helps to explain the rapid turnover of scientific facts and theories in comparison to the daily-life knowledge of a society, which is much more contextual and stable. In sum: the knowledge of any whole inevitably includes ignorance stemming from our limited ability to examine it in the full context of reality.

A second limitation is closely related to the first. Just as no whole can be abstracted from reality without some loss of information, so also no observer engaged in a process of abstraction can be entirely detached from what is observed. What this means is that any observer approaches the study of a whole with a prejudgment of the world stemming from two possible sources: as a member of a society and as a participant in a scientific or technical community of specialists, thus having a culture-based and a discipline-based knowledge of the world, respectively. Through both forms of knowledge, the observer is internally connected to the world. The assumption of a detached observer runs contrary to what science has discovered about the relationship between human beings and their contexts, and represents at best an ideal towards which one may strive with more or less success. The map-making approach acknowledges limitations to the level of objectivity and neutrality of any knowledge of a whole. All knowledge of a whole is affected by the knowledge a community has of reality as it knows it, which creates the vantage point from which its members make observations.

What are the implications of the existence of a prejudgment of the world? The 'normal' behaviour of scientific and professional communities implies that they do not regard reality as they know it as limited and relative. To explain this, it is helpful to revisit the distinction between reality as it is known by a particular community and the reality that lies beyond, which is the source of an endless flow of new discoveries.[16] The degree to which the former reflects the latter will determine the level of objectivity of the knowledge the community has. This raises the question of what scientifically can be said about the relationship between the two realities.

Thomas Kuhn[17] has shown that nothing can be said about the relationship between what I call reality as it is known by a scientific or professional community, and the reality beyond it. Reality as it is known by a particular community is not cumulative, in the sense that each new discovery simply adds details to the picture it has created of reality. The situation is not one in which reality as it is known by a community is merely incomplete, with the unknown as a source of details that can be added to the picture as they are discovered. Kuhn has shown that, from time to time, discoveries are made that challenge the basic picture a community has of reality as it knows it, thus causing what he calls a 'scientific revolution.' Such a revolution changes the basic gestalt of reality as it is known by a community. This occurred in physics during the transition from a Newtonian to an Einsteinian picture of physical reality. The same kind of thing happens when, in a society, one historical epoch comes to an end and a new one begins. For example, reality as it was known during the medieval period was fundamentally different from the one elaborated during the next historical epoch in Western civilization.

The implication is that reality as it is known by a society or scientific community is assumed to be identical to reality itself, minus some details yet to be discovered. Thus reality as it is known is extrapolated across the unknown to include all of reality. The unknown is assumed to be nothing else but missing details that will be added to the picture later on. This means that reality is implicitly assumed to have the same 'nature' or gestalt as the reality as it is known. Hence, the unknown is no longer threatening and no longer a force for potential disorder, simply a reservoir of missing bits and pieces that can be added to the basic gestalt of reality as it is known. The assumptions that make this possible are never explicit. To a society or scientific community, they are matters that are so self-evident and obvious that it is simply inconceivable that things could be otherwise. After all, all known experience points in that direction – it has been extrapolated to cover any alternative that might exist within reality. The assumptions are what anthropologists call the 'myths' of a society, and similar 'assumptions' appear to occur in science. During some periods, science appears to contribute new myths to the culture of a society, while at other times it builds on such myths. It is only later that observers with hindsight wonder how, during a particular epoch in the history of a society, no one saw through the myths. All human knowledge and behaviour (including what we bring to the study of wholes as members of a society or as scientists and professionals) inevitably imply myths. It is the direct consequence of living our human finitude.

How a relative level of detachment and the accompanying prejudgment of the world can influence scientific observations has been studied, with some fascinating results.[18] If any scientific observer is assumed to be enfolded to varying degrees into the reality he or she observes, and if that reality itself has a highly enfolded 'nature,' the division of labour between the scientific disciplines and the lack of a 'science of the sciences,' as discussed earlier, raises some troubling questions. Of course, all this is also based on a prejudgment of the nature of reality. The practical consequences of the situation have already been mentioned in terms of how different experts may interpret a particular situation in ways that are partially mutually exclusive. One can therefore receive incompatible advice from different experts about the situation. There exist no scientific ways of determining whose advice best reflects that situation.

The third limitation of the map-making approach is also methodological in nature. The third frame of reference cannot be applied to living wholes. Such wholes have no 'parts,' the way non-living wholes do, and it is therefore impossible to compare properties of such 'parts' outside of the whole. I have said that something of the whole is present in each 'part' in a living entity. It is interesting to note that some physicists, such as David Bohm, have suggested that even physical matter may not be made up of separate parts, such as protons, electrons, and neutrons.[19] He and others have argued that the 'nature' of physical reality is quite different, proposing the implicate order as a fundamental reality – an indivisible whole from which the implicit order of our observations is derived. In the implicate order, each 'part' is internally connected to all the others and to the whole. It is only in the explicate order that we see them as distinct elements. He suggests that some of the current difficulties in physics may be overcome if the hypothesis of a mechanistic world-view, in which each part of the 'universe machine' exists in its own space and interacts with others across distinct boundaries, is abandoned.

What the map-making approach does, therefore, is not eliminate the limitations of the scientific method but include them so that they can be addressed in the best way possible, fully recognizing that they can never be eliminated. It helps restore science and scientific knowledge to their proper places as a human activity and a human creation, which, like any others, are not omnipotent or absolute but relative to a discipline and to a stage in its development. Science is not the road to truth, but it is a knowledge-gaining strategy that, more than any other alternative, has been able to build consensus within and across societies and civilizations. To a large measure, this consensus is possible because of the classical assumptions of a mechanistic world-view, and that of a detached observer helping to build objective knowledge uncontaminated by subjective influences, including the vantage-point of a human community, whether it is a society or a scientific or professional one. The recognition of the limitations of the map-making approach is not an abandonment of the goals of science, but attempts to produce knowledge that is as objective and detached as possible through a rigorous critical examination of the presuppositions and assumptions of the practitioners and their communities. This includes distinguishing between at least two modes of knowing. The first derives from frontier research of the kind customarily found in any modern scientific or technical discipline. This approach produces an ever-greater level of specialization, trading off breadth for depth. Questions of context and broader interrelationships play, at best, a minor role. As a result, this frontier research must be complemented by contextualizing research where breadth is emphasized over depth, including the integration of the findings of frontier research by contextualizing them in relation to each other and their human, social, and environmental significance. In so doing, other aspects, implications, and significance will be unveiled that may complement, negate, or challenge some of the findings of frontier research. Hence, the two levels of analysis are in constant tension with each other. Each one has consequences and implications for the other. The map-making approach seeks to maintain these two levels of analysis, including the creative tension between them. Within the university, this approach is beginning to be institutionalized as theme schools or centres focusing on specific issues in their full context, such as the finding of a more sustainable way of life, the development of more context-appropriate science and technology policies, the creation of healthy cities, or the striving for genuine development for third-world nations.

## 4.7 Applications of Map-Making

The way of life of a society weaves together and enfolds, to varying degrees, many wholes. The manner in which this can be accomplished varies greatly, as is evident from the diversity of the ways of life and cultures created by human societies. The process of industrialization is one way of weaving and evolving the fabric of a society. Such far-reaching implications may be illustrated with the following simple analogy. The components of an electronic circuit can be structured into many different circuits that perform the functions of devices such as amplifiers, oscillators, computers, and tuners. The structure of the circuit determines its positive and negative characteristics, as well as how these are inextricably linked together. For example, an amplifier produces both a signal and noise, and the two are indissociable. In the same vein, societies have 'wired' their technologies into their structures in very different ways, with very different results.

The way the industrially advanced societies have 'wired' their technologies has, as we have previously noted, produced spectacular successes in the domain of performance values and serious problems in the domain of context values. This produces the environmental crisis as much as it produces the many goods and services. Similarly, technology transfer has impelled the 'rewiring' of the structures of many nations, simultaneously producing the well-known benefits and the population explosion by interfering with traditional ways of ensuring a dynamic equilibrium between population levels, productive capacity, lifestyle needs, and the functions provided by local ecosystems. In the same vein, spiralling social and health costs in the industrially advanced nations are as much a consequence of our modern way of life as is the total output of goods and services as measured by the GDP. Thus, public and private policies for developing modern technology must take into account the current and potential structures through which it is 'wired' into society and, via it, into the biosphere. This clearly imposes limits on what can be accomplished through preventive approaches for the engineering, management, and regulation of modern technology, in so far as these deal with specific technologies and not with the entire structure. However, these approaches can have an catalytic function in transforming these structures into sustainable ways of life if these broader consequences are clearly mapped and taken into account.

The 'wiring' of individual technologies is examined through a variety of methods, including energy analysis, life-cycle analysis, environmental assessment, and technology assessment. These examine the ecologies of individual technologies in terms of their implications for human life, society, and the biosphere. Suppose engineers wish to examine the compatibility between a product (such as an automobile) and the biosphere. This is accomplished by what is called a 'life-cycle analysis,' which traces all the materials and energy involved in the fabrication, use, and disposal of the vehicle from their ultimate sources to their ultimate sinks. Such a map shows how the metals in the body-panels are mined, refined, transported, processed, and manufactured into sheets, for example. The panels are stamped out of these sheets and incorporated into an automobile, which is used and eventually disposed of. All these methods run into considerable methodological difficulties because of the enfolded nature of the human activities involved. In the case of the body-panels, for example, should the energy required to produce the machinery involved in the activities from the ultimate source to the ultimate sink be included? What about the energy embodied in the machinery that produced this machinery? What about the energy embodied in the buildings in which these machines are housed? How far should this process be continued?[20] In the case of energy analysis, moving to the next level of analysis generally makes a smaller and smaller contribution, but in the case of life-cycle analysis this is far less straightforward. The enfolded character of the human activities involved makes it very difficult to definitively establish a boundary for the analysis. Consequently, the ecological footprint of any product or human activity varies significantly according to where the boundary is drawn. No standards for doing a life-cycle analysis have succeeded thus far in resolving this problem. The more prevention-oriented approaches such as design for environment have, as we shall see in a later chapter, the substantial advantage of settling for approximate answers that are adequate for the engineering, management, and regulation of modern technology.

Each technology involved in any of the stages of the life-cycle falls into one of three possible categories according to its effect on its natural context. The most common category is made up of technologies that pollute nature, that is, *dirty* technologies designed *against* nature. A second category is made up of those that are integrated into the processes of the biosphere, that is, *green* technologies designed *with* nature. The third category is made up of those that are essentially neutral with regard to the processes of the biosphere, that is, *clean* technologies essentially *neutral* to nature. The map produced by analysing the life cycle of the car thus points to areas where compatibility with the biosphere could be improved. If the technically and economically feasible improvements are not sufficient, alternative technologies may have to be considered.

A similar classification can be made in terms of the negative, positive, or neutral effects on the human and societal contexts. To assess this, it is essential to bear in mind two things. First, before a technology is introduced into a particular context, life obviously goes on without it. There is no 'hole' or vacant 'space' into which the technology can be inserted while all other relations in the fabric constituting that context remain unaltered. Adaptation of the context is inevitably required, some aspects of which may be positive while others are bound to be negative, and still others may be neutral. Together these effects can strengthen or weaken this sphere of life, and they typically create new possibilities while blocking others. This becomes visible only when the context is carefully studied before and after the introduction of the new technology. Such a comparison makes it very difficult to claim a net gain or net loss. Also, the broader the ecological focus used to examine the local context, the more it becomes obvious that there are indissociable links between both positive and negative effects on the fabric of activities and relationships that help constitute the context. Despite these difficulties, an overall evaluation frequently has to be made whenever public decisions about a particular technology need to be taken as to whether, on balance, it 'pollutes,' is symbiotic, or is more or less neutral with respect to its societal context. The values being used can vary. These can be either the values of the people involved in that context, and thus of the culture of their society, or those of the observer, who then may impose something from another society on the situation. Whether the latter is appropriate again depends on the values used.

In the past, belief in progress resulted in very few careful studies being done on the introduction of new technologies into society. Popular thinking about new technologies is still profoundly affected by this belief, which essentially tells society to go ahead because everything will somehow work out for the better. The findings of careful studies sometimes appear quite counter-intuitive. For example, when many North American households began to acquire washing machines, dishwashers, and microwave ovens, it appeared totally obvious to everyone that these appliances greatly reduced the time spent on housework, done mostly by women. After all, is the time required to do a load of wash with the traditional tub and washboard not far greater than that employed in using a washing machine? Historical studies have shown, however, that the time saved as a result of these appliances, taken for granted by almost everyone, does not exist, and may even be negative. For example, district laundries, which picked up, washed, dried, and delivered the household's laundry for a modest fee, were common before the introduction of washing machines. The decentralization of these facilities in the household washing machine was supposed to make laundry so effortless that one could change one's clothes every day. The result was that each household washed a much greater quantity, thus creating more work, mostly for mothers.[21]

Although our belief in progress has been severely shaken, nothing has taken its place. Each new technology continues to be greeted with enormous expectations. This again is because conventional thinking separates the economy from the ecology of technology. People look at what a technology can deliver but do not examine how that potential is modified in a particular context. It was once thought that the new electrical grid would decentralize society, and that educational television would bring to everyone's doorstep, including those of poor nations, the best teachers from the best universities to teach the best courses. We now believe that the computer will revitalize education and that the Internet will create new neighbourhoods and communities, and improve the liveability of cities and the sustainability of our way of life by permitting people to work at home, thus greatly reducing the need for transportation, and so on. The mistake is always the same, and the final results of introducing a new technology are almost always far short of the original expectations. It is essential to remember that, as people change technology, technology simultaneously changes people. Otherwise, the present cultural cynicism[22] will continue to grow.

Because of the highly enfolded character of a society, it may be expected that the examination of the ecological footprint of a part or all of a way of life of a society using input–output methods would vary greatly from the results obtained if the analysis had been undertaken, one activity at a time, using life-cycle-analysis, energy-analysis, or environmental-assessment methodologies. In fact, comparisons that have been made have uncovered huge discrepancies.[23] Similar problems can be expected when the human and societal consequences of technology are examined one constituent technology at a time. As noted previously, what may appear to be negligibly small effects on the level of an individual constituent technology may be replicated over and over again in the fabric of a society, thus constituting significant phenomena. There is also the possibility of synergistic effects. Consequently, carrying out macro-level analyses related to questions about the quality of human life, the viability of a society, or the sustainability of its way of life will require very different methodologies.

In the same vein, the environmental impact of a particular project or activity may appear relatively harmless, but, combined with countless similar or other activities, it may produce what are often erroneously called 'natural disasters.' For example, the practice of clear-cutting forests can cause downstream flooding because of the rapid rise of rivers, since any heavy rainfalls are no longer retained by the forests. Global warming is likely to contribute to a variety of natural disasters. Hence, to a considerable extent these disasters, too, are an 'output' of our modern way of life. Once again, preventive approaches can help reduce the likelihood and scope of such natural disasters. This illustrates the complex and enfolded relationships between a society and its local ecosystems and, via them, with the biosphere.

**4.8 Conclusion**

Before the emergence of the mechanistic world-view, the individual was seen as integral to groups and society, individual technologies as integral to cultures and a way of life, and society as integral to the biosphere, in ways that varied considerably from culture to culture. The mechanistic world-view gradually hid from human consciousness the possibility of open wholes, their having a reciprocal relationship with their contexts, and the necessity of this interaction for their vitality. Everything was regarded as a 'part' and no longer seen as integral to, enfolded in, and dependent on its contexts. The latter were now seen as separate environments. Hence, such concepts as efficiency became thinkable, and the way was open to technological and economic growth becoming guided by performance values. When, centuries later, this road was taken, it became both cause and effect of sociocultural techniques, which continue to separate what belongs together. At the same time some significant limitations became apparent, necessitating concepts such as open wholes, enfolding, appropriate technology, and sustainable development. This is beginning to make room for economic development based on preventive approaches. The latter reconnect technology to its contexts on the conscious level and may eventually lead to its reconnection on the metaconscious and cultural levels.

Until recently, modern science and technology were thinkable only in the context of a mechanistic world-view. Map-making has been proposed as a knowledge-acquiring strategy that includes science but is not limited to it. Its principal implication is that our knowledge of any whole is fundamentally indeterminate as opposed to being merely uncertain (i.e., awaiting cumulative additions from future discoveries). This indeterminacy can threaten our knowledge anytime, for two reasons. First, a whole is never studied in its full context but in a specific intellectual and/or physical context, and the acquired knowledge is indeterminate because of that difference. Second, that knowledge is gathered within the context of what T.S. Kuhn has called a 'disciplinary matrix'[24] (defined as what is implicitly and explicitly shared by a community of experts), which implies myths and tacit knowledge.[25] These amount to assumptions that no one recognizes as being made. All this means that human knowledge is always open to what Kuhn has called 'scientific revolutions,' constituting non-cumulative developments. The map-making approach to the acquisition of knowledge also shows that, in many cases, it will be virtually impossible to determine clear cause–effect relationships, particularly when open and enfolded wholes are involved.

This anticipates two important limitations to any attempts at making our modern ways of life more sustainable. First, such attempts must recognize the fundamental limitations of risk analysis, given the inherent indeterminacy implied in all human knowledge. The implications of this fact have been examined by Brian Wynne.[26] Second, the distinction between the indeterminacy and the uncertainty of human knowledge anticipates the need for the precautionary principle or the no-regrets approach to environmental policy making – a subject to which I return later in this work.

5

# Context Values for Map-Making

## 5.1 The Role of Values

In the previous chapter we showed that technology, society, ecosystems, and the biosphere are open wholes that depend on exchanges with their contexts. Such exchanges establish a reciprocal dependency that is not necessarily symmetrical but must be sustained if they are to be viable in the long term. It is essential to assess the interactions between a whole and its contexts to ensure this viability and prevent harm either to the whole or to its contexts. Performance values cannot help us in this task.

We now turn our attention to the evaluation of such reciprocal relations between a whole and its contexts in terms of their contribution to freedom or alienation, health or illness, sustainability or a lack of it, and integrality or fragmentation. The evaluation is guided by the recognition that any open whole puts demands on the resources of its contexts, while at the same time these contexts put demands on the resources of the whole. The reciprocal exchange is viable if in each direction the demands roughly match the resources available to deal with them and will not be viable if they overtax them. If a whole is damaged by its contexts, they themselves may also be negatively affected because the whole may no longer be able to sustain the exchanges. If, on the other hand, a whole damages the contexts on which it depends, it inevitably brings harm to itself. On this basis, value judgments can be made about how good a reciprocal relationship between an open whole and its contexts is. Context values guide this assessment.

It must be borne in mind that the focus is on preventive approaches, which operate on the level of individual technologies. An analysis of how technology as a whole affects human life, society, and the biosphere is not our concern here. What we are concerned with is how particular technologies strengthen or undermine freedom, health, sustainability, and integrality in a particular sphere of human life and society, or with respect to specific functions of ecosystems or the biosphere. As mentioned earlier in this work, there need be no agreement on the meaning of sustainability but only on the fact that we are imposing too great a load on the biosphere and that efforts must be made to reduce it. Sustainability can then be operationalized as steadily reducing that load, activity by activity and year by year. If this is done, we will gradually learn what a more sustainable way of life looks like. Similarly, there need be no consensus about the meaning of a healthy society as long as there is consensus that the social fabric is under too much pressure. A society will then be seen as becoming healthier if, activity by activity and year by year, needless pressures can be eliminated through the application of a precautionary or no-regrets principle. The assumption is that there exists a sufficiently broad consensus to proceed. I am not in the least suggesting that political and moral debates about issues such as freedom or sustainability are not important. What I am suggesting is that these take place on a different but not altogether independent level of analysis so that only a minimal consensus on these essentially contested concepts[1] is required. The present analysis is inseparable from the preceding one examining the dependence of human life and society on culture,[2] and anticipates the analysis of the effects of technology as a whole.[3]

In the case of freedom, the evaluation of the relationship between people and a specific technology amounts to locating the situation on the spectrum contained between the two extremes of social determinism (or technological neutrality) and technological determinism. The relationship between people and their workplace is a reciprocal one. To the degree that people can exercise control over the work process beyond the demands it puts on them, there is a measure of freedom. In the same vein, the relationship between people and their physical and sociocultural contexts is healthy when their physical and mental resources are challenged but not overwhelmed by the demands placed on them by these contexts. When the opposite is the case, they may become ill. The environmental crisis is the result of our modern way of life overtaxing the resources of the biosphere, thus negatively affecting its role as the ultimate source and sink of all matter and energy, as ultimate habitat and as life support. The integrality of a whole is preserved when the relations with its contexts are viable.

It is also essential to remind ourselves that we approach any evaluation with a prejudgment derived from having grown up in and living in a particular society (or societies). Learning to make sense of and to live in the world involve mapping our experiences, which I have previously likened to plotting the data in a scientific experiment.[4] By fitting a curve through the data, and thus going beyond the experimental evidence by interpolating and extrapolating the data, it becomes clear which are the 'good' data and which are the 'bad.' In fact, the temptation may be to simply reject the experimental points that fall far from what appears to be the relationship between the variables. In the same vein (but on a metaconscious level), young children learning to eat properly may learn, by relating all eating experiences relative to each other, what appropriate behaviour is. The 'data' that fall far from the norm typically have a strong emotional colouring when, for example, the child is messing with the food and experiences disapproval from others.

The informal maps constituted by the structures of experience of human beings therefore colour their lives according to the values of their culture. Implicitly, and later, explicitly, we learn what is good or bad, useful or useless, beautiful or ugly, responsible or irresponsible, and so on. Since each value emerges in the context of all others, a hierarchy of values is implied in our structure of experience. This hierarchy functions somewhat like the set point on the thermostat, making negative feedback in human behaviour possible. The concept of a whole as a map-making tool must be guided by a purpose. For our situation, this purpose is to ensure the best possible compatibility between individual technologies and their human, societal, and natural contexts, or between the technology of a society and its contexts. These attempts must be guided by context values, which express various aspects of this compatibility.

## 5.2 Map-Making and Freedom

The three pillars of Western civilization – namely, Greek philosophy, Roman law, and the Judaeo-Christian tradition – all contributed to the assertion of the worth of the individual versus the collectivity in a manner that was unprecedented in earlier civilizations. Individual freedom is a value that is spreading throughout the world, along with Western science, technology, and modes of economic development.

This is not to deny that this value has been and continues to be abused to justify the exact opposite.

Human freedom or its absence may be evaluated by analysing the reciprocal relationship between human beings and their societal and biospherical contexts. Modern science provides us with a picture of human nature as being extraordinarily determined by external influences coming from its contexts. In so far as the natural, biological, and social sciences as well as the humanities seek to determine 'laws' or regularly occurring patterns in the lives of individuals and societies, they examine the absence of human freedom. This can readily be illustrated by one of the debates that raged in the nineteenth century over the views of the economists of that time – namely, the physiocrats.[5] They believed that the economic regularities they had observed and documented were like the laws of nature. These regularities, like the law of gravity, were simply there; whether one liked them or not did not make any difference. Some scholars argued that these were not laws at all, but ultimately a human creation resulting from a particular way of life. Since, to some, these patterns were unacceptable, immoral, and an offence in terms of what they felt human life should be, they urged a restructuring of the institutions of society. Hence, if one could speak of laws at all, these regularities were temporary and changeable.

The conflict between the picture science creates of the human condition and the value of human freedom cannot be resolved scientifically because there is no science of the sciences. The usual response is an escape, such as progress or some other saving phenomenon.

The view of the human condition implicit in the modern sciences is not dissimilar to that of the Judaeo-Christian tradition, which profoundly influenced Western civilization. It argued that individual and collective human life were characterized by sin, always explained in terms of the model of slavery: a condition of being possessed by someone or something so that you can no longer be yourself. To truly love, a person must be free to give him- or herself. Slavery thus jeopardizes the fundamental relationship between God and his people, and the relationships among persons. It rules out the possibility of full responsibility, commitment, or friendship, making it impossible to have meaningful groups, societies, or civilizations. The condition of sin was manifested by every society that created a sacred, making idols of what was most valuable and essential in the experience of a people, beyond which nothing more important could be lived or imagined. Idolatry overvalues elements of human experience, with dire consequences. At a time when there was little or no evidence of how this could be true, money (Mammon) and the city were regarded as spiritual powers precisely because they had the perceived capability of enslaving human life.[6] The creator did not want his creatures to become slaves and constantly intervened to liberate them.

We recall this teaching about slavery because it inspired a modern secular equivalent – namely, the concept of alienation fundamental in the social sciences. To be alienated is to be possessed by something or someone to the point that one can no longer be free.[7] How can this be? We have noted that a workplace can be alienating when the influence it exercises over people is much greater than the influence they can exercise over it. On a traditional assembly line, for example, workers do not need to exercise any judgment over their work; their decision latitude is non-existent; and the monotony of endlessly repeated, trivial motions make these workers human precursors of industrial robots. In the same vein, the modern industrial, urban, and information life-milieu can be alienating as a result of sensory or social overload and the 'lonely crowd.'[8] When a television interviewer asks passers-by their opinions on any subject, they always seem to have one. To be honest with ourselves, we should ask: what is my opinion based on? Is it based on direct experience, discussions with witnesses, or the reading of books or articles? Research on public opinion in mass societies shows that, in a great many cases, it is none of these.[9] Public opinion is a fundamentally different phenomenon from traditional opinion, which arose out of a person's life. The former demonstrates the influence a mass society has on the opinions of its members. It is but one manifestation of how such a society may 'possess' its members.[10]

Freedom and alienation are dialectically related because of the reciprocal and contradictory relationship people have with their societal context. The same may be argued for the relationship between a society and its ecosystem. If a society bases its legal institutions on a concept of natural law, it makes a fundamental error that greatly delimits its freedom from that life-milieu, as was the case early in human history. To be sure, the extent to which 'nature' determined the cultures of these societies was very great indeed, particularly at the dawn of history, when human beings began to live in societies. To recognize such determinisms for what they are and to struggle against them is an expression of freedom. The decision to imitate 'nature' can also be an expression of human freedom. However, to behave as though nature has commanded a society to follow it is to submit to natural determinisms.

Despite its extensive use in the social sciences, the concept of alienation can never be entirely scientific because it presupposes a norm of what it is to be human. Ultimately, alienation is a moral category since it presupposes that human life was not meant to be a web of determinisms. From whatever source we may derive our conceptions of human life, no person can live without them. Modern societies embody a pluralism of such conceptions. Despite this fact, there is sufficient consensus of what it is to be human that a concept of alienation remains useful in assessing the consequences of modern technology for human life. The heritage of the Jewish and Christian traditions, the value of the individual vis-à-vis the collectivity as affirmed by Western civilization from its beginning, the legacy of the French and American Revolutions, and the abolition of slavery all attest to the fact that Western cultures, with some exceptions, consider the condition of slavery to be unacceptable. Our 'set point' in this matter measures the gap between human freedom and determinism by means of the concept of alienation as a lack of freedom. As noted previously, some freedom is essential if love, commitment, friendship, trust, and responsibility are to have any depth. The influence of technology or individual technologies on human freedom constitutes an important dimension of the negative-feedback processes that guide the development of preventive approaches.

In a secular society, freedom and alienation are expressions of the compatibility or lack thereof between the resources embodied in a person's life and being, and the demands placed on them by society and the biosphere. Because human beings are enfolded into their world, there can be no absolute freedom; nor can there be absolute alienation, unless the human spirit is entirely crushed and human life as we know it no longer exists. Freedom and alienation therefore express the dialectical aspect of the fit between human beings and their contexts. Human values in general, and context values in particular, can make additional contributions to the assessment of the relationship between human beings and their contexts.

## 5.3 Map-Making and Health

What do we mean in our culture when people say that a person or a society is healthy? Conceptions of personal health imply a norm of the 'nature' of human life itself. For centuries, Western culture has essentially regarded the human body as a kind of machine, constituted of many parts.[11] Viewing the brain as the most sophisticated computer known to date has informatized this image. Disease occurs when the biological-information machine malfunctions. In the same vein, health is essentially regarded as the absence of disease, and medicine becomes disease care. This conception entirely ignores what we discussed about the human body in the previous chapter – namely, that, except for the brain, it constantly replaces all its cells. Hence, our relationship with the biosphere is radically different from that of a machine. As pollutants move up the food chain, they become a major threat to human health since they interfere with this rebuilding. We are also beginning to discover how psychosocial stress resulting from certain kinds of relationships with the sociocultural context affects not only the mind, but also the brain and, via it, the human body. In fact, socio-epidemiology shows that the kind of work people do, their socio-economic status, and their lifestyle are fundamental determinants of their health. These and other developments have led many observers to interpret health from a more ecological perspective as a reciprocal interaction between a living whole and its contexts.

To the extent that the mechanistic world-view is beginning to be displaced by conceptions of the nature of reality that include the recognition that human beings are open wholes, internally and externally related to their physical, social, and ecological contexts, the new conception of health is gradually gaining ground. The World Health Organization defines health as a state of complete physical, mental, and social *well-being*, and not merely as the absence of disease or infirmity.[12] A person's health can no longer be determined simply by the typical medical exams and a few lab tests. Conditions at home and in the workplace, a person's lifestyle and physical surroundings, as well as his or her biological make-up, must all be considered. Human health therefore includes the fit between a person and his or her physical, social, and ecological contexts. As an open whole, exchanges with these contexts are vital. A complete picture of someone's health, therefore, requires both internal and ecological foci. It is ironic that the medical profession is putting so many of its resources into the genome project, considering what this entails about human health.[13]

The conventional and more ecologically oriented views of human health have divergent consequences for modern medical practice. Each 'paradigm' may lead to very different diagnoses and treatments for the same problem. A number of examples will illustrate the new conception of health and its practical implications. A person suffering from retinitis pigmentosa was gradually losing his vision, because of a deterioration of the retinas in a manner that is not well understood. At a certain stage in the development of the disease, he read books by bending over a desk, almost literally having his nose on the paper. As a doctoral student, he had to spend many hours in this position. When chest pains occurred, he assumed that they resulted from this situation and not from heart problems. To make sure, however, he had a medical check-up, during which he carefully explained the situation to the doctor. The doctor decided to prescribe Valium. Upset by what this prescription implied in terms of a diagnosis, the patient decided on other remedial intervention. He built a book-stand that permitted him to read with a better posture, and a week later the pains had disappeared, never to reoccur.

The doctor and the patient had very different views as to what constituted health in this situation. The doctor's prescription implied that he thought the symptoms resulted from depression caused by losing one's sight, and that this had some physiological manifestations. The patient, on the other hand, was convinced that the problem was most likely caused by the way his relationships with the life-milieu were altered by his failing vision, particularly when reading. The doctor looked for causes within the individual, while the patient had a more ecological view of health. The result was two dramatically different diagnoses and treatments. This is a description of my own experience.

The World Health Organization's definition of health includes mental and social components. What this means can be illustrated by the emerging concept of stress. It is increasingly recognized that in a modern society stress is a major health problem capable of producing a wide spectrum of symptoms and ailments. A healthy level of stress in a person's life cannot be exclusively defined biologically ('internal' to the individual, as it were), nor can it be defined purely socially (that is, in terms of the fabric of relations in which the individual is engaged). It must be understood in terms of the biological, emotional, and mental resources a person brings to a situation, and how that situation taxes those resources. The balance between these two can range from situations in which a person receives little stimulation, with considerable harmful effects, through situations where the demands placed on a person can comfortably be met, to situations where the demands are so extensive that a person becomes stressed out, causing a diversity of possible pathologies (including depression, nervous breakdown, mental illness, or suicide). Both extremes are unhealthy, although the former, where a person's resources are essentially idle, is less well recognized. The development of babies who have to spend extensive periods in a monotonous hospital environment is much slower than normal. Similarly, persons who have monotonous repetitive work over which they have little control can expect their physical and mental health to be negatively affected. Included in stress is the body-load, resulting in part from the contaminants to which we are exposed in a particular life-milieu. Again, depending on our genetic make-up and our emotional and mental state, such a load could lead to illness of many different kinds, including allergies and environmental hypersensitivity.

Because human lives are open wholes depending on constant exchanges with their contexts, a 'physical' disease can lead to 'mental' effects and vice versa. For example, children exposed to high lead levels become hyperactive, which leads to a variety of behaviour problems at home and in school. The way others react to them will fundamentally affect the social relations of the child and, through them, his or her self-image. This example illustrates how the physical component of health can affect the mental and social ones.

Because of sociocultural attitudes, the physical conditions of blindness and deafness frequently have a negative effect on the mental and social components of a person's health. Terms such as 'the blind' or 'the deaf' suggest that what you need to know about such people, first and foremost, is that they have poorly or non-functioning eyes or ears, and that this permeates their whole lives, including their personalities, social relations, careers, and aspirations. This, of course, was largely true when, around the time of the Industrial Revolution, these people were segregated from society in special institutions.[14] As a result, they were denied the kind of socialization everyone else in society depends on to be able to function effectively, and thus remained critically dependent on these special institutions. Terms such as 'the disabled' and 'the handicapped' refer to this condition, in which people are not able to do anything at all and are thus obliged to beg with 'cap in hand.' Mainstreaming is beginning to undo this harm to some extent, but the sociocultural component of blindness and deafness is, for many people, far more destructive than the actual problem of not being able to see or hear normally. The socially constructed attitudes and beliefs about blindness and deafness lead to these people being treated in a certain way, which has a very destructive effect on their self-image. This, in turn, affects their social relations with others, creating a very unhealthy situation. Traditionally, social workers and special organizations to help these people may have done more harm than good because they disempower them by treating them as 'the blind' or 'the deaf.'[15]

These examples show how 'physical' illnesses can lead to psychosocial impairments. The reverse interaction can also occur. Studies of the effects of hazardous-waste dumps on those who live close to them reveal that, in addition to the health effects resulting from direct exposure to pollutants, there are also significant psychosocial effects. Living with a higher risk of incurring serious diseases adds considerable stress to the lives of people who usually are already in a vulnerable social position, otherwise they would not live near these dumps.[16] Once again, there exists both a physiological and a psychosocial component of these health risks. In public hearings, the latter effects on people's lives are all too often trivialized as hysterical and scientifically unfounded reactions.

The modern view of health is therefore ecological in nature. It focuses on the individual as a living whole existing within the larger living wholes of society and the biosphere. Each helps to constitute the other's context, and their mutual health requires a variety of exchanges, including communication, matter, and energy. Health must therefore be understood in terms of an ecology of reciprocal relationships between a living whole and the larger living context of which it is an integral part. This point will be further illustrated by several examples.

For an individual to be healthy, society must also be healthy. Among other things, this requires a tension between individual diversity and social unity. This is already evident when two people communicate. If two individuals were entirely identical, each one would know everything the other does and thinks, and there would be nothing to communicate. Without individual diversity, the relationship would be impoverished to the point that it is not likely to last. Now suppose two individuals were totally different. Whatever one person tried to communicate would have nothing even remotely resembling it in the 'world' of the other person. The communication could therefore have no meaning. They might as well have come from two different planets.

In a healthy relationship, there must be a tension between differences and similarities. The two people must be different enough that communication brings something new, and thus enriches each of their lives. For example, if one person uses the word 'mother,' none of his experiences of mothers in that society may be shared by the other person. Nevertheless, they both share what that society generally expects from mothers and what constitutes good or bad mothers, and so on, because of the many experiences they each have of mothers and how mothers fit into the way of life of that society. Because of these differences and similarities, these people can use the word 'mother' as a symbol for sharing and enriching one another's experience and life. When this happens on an ongoing basis, and the richness of this sharing is strengthened through it, deep and lasting relations may be formed.

To keep such relationships dynamic and healthy requires that communication continues to be a source of meaning and enrichment to the lives and the parties involved. However, the very process through which people intimately get to know one other and share a great deal of their lives together transforms the relationship and basis for communication. When they know a great deal about how the other person thinks and feels, communication may become much more routine and contribute less in terms of enriching each other's lives. In other words, the relationship tends to run down over time unless the two people learn to constantly re-create and enrich the situation. This is difficult, and when certain kinds of relationships are fundamental to the structure and organization of a society, as was the extended family in pre-industrial societies, a society may step in and artificially keep relationships together through external pressures, such as by making divorce difficult or even impossible. This brief illustration also shows that communications between human beings and 'communication' between machines have nothing in common.

What is true for relationships between two people also holds for a social group. A group remains healthy when its members contribute something to it that enriches the lives of the other members, while they contribute in return. The group thus has a certain vitality and health. When the processes of constantly re-creating the group through ongoing communications and sharing begin to run down, some members may sense a loss of reciprocity in their belonging to the group, and drop out. This can lead to its disintegration. On a larger scale, similar processes happen within a society. Through communication the members of a society simultaneously create individual diversity and delimit that diversity through a cultural unity, as expressed in their way of life. Its maintenance depends on a tension between being different enough to share or disagree about things, but not so different that the members of a society sense they have come from totally different social 'worlds.' A society evolves as its members respond to internal and external changes, as other members interact with these changes, as groups and institutions change as a result, and as the whole within which all members live also changes. In all these changes, there is no complementarity without antagonisms, and thus a society creates both order and disorder. The latter plays an important role in many self-regulating processes through positive and negative feedback. Disorder contributes to the maintenance of order. Of course, we must remind ourselves that within a society order for one individual or group may be disorder for another. For example, the extraordinary diversity of relations that human beings are capable of are a source of disorder on the assembly line, and at the same time account for everything we would call human. It is impossible, therefore, to view unity, diversity, order, disorder, complementarity, and antagonism as absolutes. This shows that every experience, interpretation, and evaluation is context-dependent. Each is made from a certain vantage point in time, space, and history.

For a society to be healthy, the ecosystem and the larger biosphere must also be healthy. Its members will not be healthy if the food they eat, the water they drink, and the air they breathe are polluted. Allergies and other environmental diseases are on the rise. A further dependence of a society's health on the biosphere has been dramatically illustrated by events during the past few decades, where communities based on agriculture or fishing have found their ways of life and communities threatened because of soil erosion, pollution, and vanishing fish stocks. These events show how the physical, psychological, and social well-being of individuals depends on the health of the community, and how that community can be affected by human activities that undermine the health of the ecosystem. This recognition is beginning to affect the concept of the security of a nation.[17] It cannot have any long-term security if its way of life compromises the ability of local ecosystems to provide sources and sinks of matter and energy, as well as life support. As the environmental crisis deepens in many parts of the world, humanity will, one hopes, recognize that here also armed conflict provides no solutions.

The examples discussed above illustrate how health must be regarded contextually, ecologically, and holistically because of the way the human body enfolds the brain, within which develops the mind, which internally connects people to others and their society, in turn enfolded into local ecosystems via a way of life. Health implies a dynamic balance, which must constantly be re-established in the face of internal and external fluctuations that pose challenges to the web of interdependent physical, psychological, and social wholes. There exists, therefore, a strong relationship between the health of an individual, the health of the groups in which the individual participates, the health of their society, and the health of the ecosystem in which they live. Unemployment, poverty, homelessness, violence, drug and alcohol abuse, environmental hypersensitivity, and suicide all may indicate lack of balance and a loss of health on the level of a society. Many regard a society as 'sick' and in serious difficulties when the gap between rich and poor grows too large; when access to education, social, and health services is limited to privileged groups; and when human and civil rights are violated. Unfortunately, measures against 'social sicknesses' are frequently undertaken only when those in power become affected and they can no longer defend themselves against problems such as violent crimes on the street, or when the legitimation of their power is crumbling because of the widespread recognition that the 'system' is fundamentally unjust.[18]

These examples demonstrate how deeply the lives of the members of a society are enfolded into one another. The experiences two people have of each other become internalized into their minds and lives. When one person moves away or dies, it is as though a part of the life of the other person comes to an end, and a profound sense of loss is inevitably experienced. The same phenomenon occurs in a group, although some of the relationships may be more deeply enfolded into one another than others. Of traditional close-knit communities, it may sometimes be said that, if one person suffers, all others suffer with her; or, if one person is dehumanized, all others are dehumanized with him. In a mass society, we do not experience this as directly as in a traditional one. Nevertheless, how often can we walk past a homeless person and not be affected in our own humanity? How often can we use the same word for designating human and machine functions and not degrade life? Those striving for a more secure and peaceful world often point out that this will require a more just relationship between all peoples, and between the rich and the poor in each and every society. As we shall see, sustainable development as a context value embodies the human value of justice.

From the perspective outlined above, it is possible to speak of healthy workplaces and cities. The latter has been described as follows:

> There are, of course, many different opinions as to what is a healthy city, depending on one's discipline, values and point of view. However, com-

monly shared parameters of a healthy city include a clean, safe, high quality physical environment and a sustainable ecosystem; a strong, supportive and participatory community; provision of basic needs; access to a wide variety of experiences and resources; a diverse, vital and innovative economy; a sense of historical, biological and cultural connectedness; a city form that makes all of these possible and a high health status with appropriate, high quality and accessible public health and sick care services. Assessing the health of a city must take all of these parameters into consideration, and must involve an assessment of both qualitative and quantitative indicators – intelligence as well as data.[19]

The context value of health furnishes one perspective from which the relations between the circles representing technology, society, and the biosphere may be examined and assessed. The concept of health discussed above implies a close relationship between preventive approaches in technology and medicine. The spectacular successes and failures of Western medicine are receiving growing attention as a result of a crisis in health care. The prevailing conception of health care is, in essence, disease care; yet little is done about the problems we as a society create. Such an approach to health is, by its very nature, not preventive. As the array of technological means to intervene in the biological-information machine grows in diversity, sophistication, and cost without a corresponding improvement in human health, it will become necessary to convert our illness-care system to a comprehensive health-care system, if it is not to bankrupt us. Health care will have to become more preventively oriented, with a component of disease care to deal with cases where prevention fails. Such a development would be the exact parallel of what is all too slowly happening to the way society deals with pollution.

### 5.4 Map-Making and Sustainable Development

The World Commission on Environment and Development has stated the following:

> Humanity has the ability to make development sustainable – to ensure that it meets the needs of the present without compromising the ability of future generations to meet their own needs ... Sustainable global development requires that those who are more affluent adopt life-styles within the planet's ecological means – in their use of energy, for example. Fur-

> ther, rapidly growing populations can increase the pressure on resources and slow any rise in living standards; thus sustainable development can only be pursued if population size and growth are in harmony with the changing productive potential of the ecosystem. Yet in the end, sustainable development is not a fixed state of harmony, but rather a process of change in which the exploitation of resources, the direction of investments, the orientation of technological development, and institutional change are made consistent with future as well as present needs.[20]

The concept of sustainable development in an absolute sense can be used only with hindsight. It is only at the point a civilization ceases to exist or when human history ends that anything can be said about the sustainability of a way of life and what it left for subsequent civilizations. Much energy has been devoted to demonstrating that the concept is internally contradictory and lacks the potential to be made operational in practical terms. As previously noted, for our purposes it is pointless to get embroiled in these discussions because they are entirely beside the point. It is obvious to many observers that all living systems are in decline. Hence, the concept of sustainable development can be made operational in relative terms by asserting that it is urgent that, every year, our way of life put less pressure on living systems than in the previous year. This can guide public policy for many decades. It leaves open the question of how far we need to go in this direction and how rapidly we must advance. For example, preventive approaches appear to generally be advantageous from all points of view and for all stakeholders, and it is not until we have fully exploited their potential that we need to worry about such questions. I recognize that preventive approaches may well be a necessary but not sufficient condition to sustain humanity and provide for a humane and common future. Getting preventive approaches launched will encounter plenty of resistance, and in the current political climate, a debate over what more should and could be done is unlikely to occur. When eventually it does, we may have seen enough of what is possible with a technological development guided by both performance and context values that our technological and political imagination will have been sufficiently stimulated.[21] In sum, rather than attempting to define, and make operational, sustainable development, it is much more useful to get on with making our way of life *more sustainable* each and every year. In other words, sustainability with regard to technology means a preventive strategy for its development. Different economic indicators will have to be used since the present ones do not measure what we need to know in order to succeed.[22]

### 5.5 Map-Making and Integrality

Integrity for human beings or integrality for all other living wholes is a quality of completeness and of being whole or healthy. It is a concept that most of us have an intuitive grasp of but that is difficult to define in a world profoundly influenced by modern science and technology. I have pointed out how these activities tend to produce tensions between whatever is made 'better' (in terms of performance values) and the context in which it is situated. The same process also tends to distort the integrality of whatever is improved. Recall the procedure for making things 'better.' Whatever is to be improved is not studied holistically, but is examined for a particular purpose. In relation to that purpose, some aspects will be considered as primary, others as secondary, and still others as peripheral or irrelevant. This will be reflected in the model, with the result that the reorganization will affect the internal integrality of whatever is being improved in addition to its relationships with its contexts. Here we encounter the limits of scientific and technical rationality. The point is not that models ought not to be made, but that, like any other human creation, they are good for certain things, harmful for others, and irrelevant for still others. The problem arises when a society treats scientific and technical rationality as omnipotent, a secular religious practice that is the equivalent of what in the Jewish and Christian traditions was called 'idolatry.'

However, in our everyday existence, we experience the integrality of the lives of others, and, through them, that of our society. Consider the example of writing a biography. Reference will have to be made to other people who played a vital part in this or that subplot of the person's life. Their lives positively or negatively complete some aspects of the biography, to the point that the many experiences involved become a part of the person's mind, and thus become integral to their being and life. As noted, when one of the other people dies, something of the person's life dies too. Something of their society's culture that has become embodied in the lives of these two people also ends. Growing up and living in a society causes something of the larger socio-cultural whole to become an integral part of a person's life, the way DNA enfolds something of the biological whole in each cell of his or her body. The genetic code represented by the DNA and the culture of that society connects that person to previous and future generations in a manner that is both biological and cultural. In this way, the life of the person described in the biography is integral to the lives of many others, and, reciprocally, his or her life is integral to theirs. Each person helps to constitute the social life-milieu of everyone else, so that the integrality of an individual's life is fundamentally interwoven with those of others and that of their society. The concept of integrality therefore reinforces those of health and sustainability.

The integrality of a society can be affected in a variety of ways. For example, if wealth is fairly distributed, keeping the gap between the richest and poorest members to a reasonable level, there will be a sense of integrality as opposed to divisions leading to 'us and them' sentiments.[23] It is widely acknowledged that societies within which wealth is justly distributed in the eyes of most members tend to be more stable and less prone to problems of crime and violence. Costa Rica, for example, is the most stable country in a region well known for high levels of insurrection, revolutionary violence, and government repression. The main characteristic that distinguishes Costa Rica from its more violence-prone neighbours is that income and wealth are more evenly distributed. The more egalitarian social structure of Costa Rica has long historical roots. In the industrialized nations as well, levels of violence and other social problems are closely related to the distribution of wealth and income. Among the OECD countries, the United States has by far the highest level of violent crime. Could the fact that it also has the most skewed distribution of income be a mere coincidence?[24]

Income distribution is greatly affected by technology. This becomes evident when one examines how economists have traditionally defined appropriate technology in the context of developing countries. An appropriate technology is one that makes extensive use of factors of production that are widely available while economizing on factors of production that are scarce in a given setting. In most developing countries labour is readily available at low cost, while capital is in short supply. If within this setting the country imports technology that was developed in an industrialized country where labour is assumed to be scarce, and thus expensive, and capital is readily available, it is likely to be inappropriate. It is inappropriate in that it will increase the country's debt problems by using scarce capital, and exacerbate problems of both unemployment and underemployment. The technology used in the industrially advanced nations also has a bearing on income distribution, and thus social inequality. The downsizing of corporations and government bureaucracies was clearly facilitated by the introduction of new technologies such as computers, and led to high levels of unemployment.

Economic inequality leads to political inequality, further affecting integrality. The political systems of all OECD countries and an increasing number of developing countries are based on the ideal (if not the reality) of the equality of citizens within a democratic system. Economic inequalities undermine this equality by giving those with greater economic resources an enhanced ability to influence the political process. More directly, however, the increasing technological complexity of our society has undermined equality through the greater reliance on expert knowledge that decision making within this context requires. Decisions regarding how industries, such as the nuclear industry, should be regulated require a detailed understanding of how these industries function. Such understanding is beyond the reach of the vast majority of the citizens of a democracy, who can therefore not meaningfully participate in the making of these decisions, many of which affect their lives in crucial ways. When the decisions that rely primarily on expert knowledge include those related to trade policy, fiscal policy, regulatory policy, security and defence, and so on (all of which are directly or indirectly linked to the technology of a society), the political equality of citizens is fundamentally undermined. The Science-Technology-Society (STS) movements in many industrially advanced nations recognize the implications of this situation for a modern democracy and are attempting to improve the scientific and technological literacy of citizens.

Organizations tend to function much better when their integrality is high, that is, when income is fairly distributed, participation in decision making is extensive, and the exercise of power is kept to a minimum. Once again this can be fundamentally affected by the technologies being used. Certain types of technology have security implications that make the use of these technologies impossible outside of a military type of organization that requires, among other things, the restriction of access to information.[25] This again undermines the egalitarian basis of our society. If equality is a fundamental value within our society and not mere rhetoric, and if technology has a profound effect on equality, then clearly the design of the technologies must take into consideration the effect that a particular design will have on the economic, political, and organizational dimensions of equality.[26]

In sum, 'integrality' refers to the ability of a whole to change and adapt in the context of natural evolution and sociocultural development while maintaining much of its essential character. The integrality of living wholes is usually taken for granted until it is threatened. It has therefore not been the subject of scientific investigation until recently, when, under the influence of modern science and technology, the integrality of human life, society, and the biosphere has come under pressure. A weakened integrality usually manifests itself by impairments in the functioning of a whole. 'Integrity' is a term generally reserved for human beings, and refers to their ethical integrality.[27]

Context values such as health, sustainability, and integrality imply a constant coping with internal and external changes of all kinds that are integral to the lives of individuals, the history of societies, and the natural evolution of the biosphere. Integral to these ecologies is a wide diversity of opposing phenomena, of which some may strengthen and others weaken the whole. The integrality of the three ecologies has changed a great deal during natural evolution and human history, and yet we have no difficulty recognizing them at any stage. Integrality is, therefore, closely bound up with processes of self-regulation and change rather than with a fixed 'nature.' Context values are related to the process of living – they are dynamic rather than static. They are essential in guiding our map-making to create a healthy interdependence among technology, society, and the biosphere.[28]

### 5.6 Conclusion

The previous chapter elaborated and commented on the simplest possible map of the ecology of a technology of a society, which was represented as a set of concentric circles. The present chapter has developed four context values for assessing the interdependence between the four open wholes shown on that map – namely technology, society, ecosystems, and the biosphere. When used in conjunction with performance values to guide technological development, they represent the kinds of interdependencies required to ensure that technological advances measured in terms of performance values are not undermined by technology-related problems measured in terms of context values.

This chapter has introduced a certain ambiguity that must now be resolved. In their full meaning, the four context values refer mostly to the macro level of analysis dealing with the relationship of modern technology with its contexts of human life, society, and the biosphere. However, the present work focuses on preventive approaches that operate on the level of individual technologies. By means of the theory of wholes presented in the previous chapter, these two levels of analysis can be seen as being interdependent. The individual and synergistic effects of particular technologies and groups of technologies help to determine the freedom, health, sustainability, and integrality of the larger tissue of wholes in which they are embedded. At the same time, these context values are much more than the contributions made to them by particular technologies. Preventive approaches can contribute to the freedom, health, sustainability, and integrality of human lives and societies, but they can never create them. It is in this sense I sought to 'operationalize' the four context values in this chapter. It is now possible to take the next step.

The remainder of this work will be guided primarily by the context value of sustainable development. It forms the basis for constructing a set of design and decision matrices for preventive approaches. By emphasizing the reciprocal interdependence between any whole and its contexts, the minimum condition that all preventive approaches must satisfy is that each one needs to sustain, maintain, support the life of, prolong, and sanction the wholes that are affected. In so doing, they help create the conditions under which their overall effect on human life, society, and the biosphere will satisfy human values and aspirations. For example, workplaces must be designed in such a way that the skills and resources of the people that work there can grow, and that negative spillover effects into the lives of these people are prevented as much as possible, so that they can fully participate in their family lives, communities, and nations. Similarly, the built habitat has to support a diversity of human activities with minimal negative effects. It is only when preventive approaches sustain the wholes and contexts they touch that communities, nations, and humanity as a whole will have a chance to pursue their goals and aspirations. In sum, on the level of individual technologies sustainable development is all about sustaining the local tissue of wholes. As such, it is a necessary but not sufficient condition for the pursuit of human values for shaping the relationship between modern technology and its contexts so as to ensure our human and common future. The effectiveness of preventive approaches will be enhanced if they are also guided by a clear understanding of the ecology of modern technology as a whole and the way it interacts with and should interact with human life, society, and the biosphere. The remainder of this work continues to anticipate that level of analysis.

# PART THREE

# A GENERIC MAP OF THE ECOLOGY OF TECHNOLOGY AND CONTEXT MATRICES

6

# Differentiating Sustainable Development

## 6.1 General Overview

Implementing the design for a new constituent element of a technology involves a great many human activities within a way of life as well as a change in a society's dependence on the biosphere. In a sense, the constituent is being connected to its human, societal, and biospheric contexts during each phase in its technological cycle. Preventive approaches must therefore intervene during the design phase, when most of these connections and their consequences are being decided either by commission or omission. For example, in specifying the materials to be used, the design in effect determines which activities in the way of life of a society will procure the necessary raw materials from the biosphere, refine and process them, and manufacture and assemble the parts. Similarly, the preliminary design points the way to the required innovation and development, and how the constituents will most likely be manufactured. It affects the packaging required for shipping and distribution; determines the maintenance during its use (including the feasibility and cost-effectiveness of minor and major repairs and how these extend the period of use); and essentially rules on post-use activities, such as the feasibility of disassembly in order to retrieve major components for remanufacturing and to separate materials for reprocessing to be reincorporated into the functioning technology of a society, or, alternatively, on its merely being disposed of in a landfill. The design anticipates expected usage with a varying range of applications, which in turn affects how quickly it may become obsolete. It will depend on certain services rendered by the biosphere and will not require others. These services include the role of ultimate source and sink of all matter and energy. In other words, prevention-oriented design, in contrast with conventional-design approaches, seeks to take into account as much as possible the intended uses and implications, as well as the unintended ones, of a future constituent element of a technology by considering the interactions between that constituent and its contexts during all phases of its technological cycle. It conceptually integrates the constituent element into the fabric of connections that make up human life, society, and the biosphere. To aid the designer in this effort, a generic map of how technology is embedded into its contexts will be developed and translated into a set of context matrices to guide design and decision making.

The connections between any design and its future contexts may be regarded as the subset of a generic map of the ecology of the technology of a society. The broad features of such a map symbolically represent how a technology interacts with its contexts in general, and the connectedness of the activities that will be associated with the technological element being designed in particular. Its generic features derive from three constraints on the fabric of connections. The first is related to the integrality of individual and collective human life rooted in the sacred myth and values of a culture, which makes alternative ways of life unliveable. This constraint is limited to a particular epoch in the history of a society during which this cultural unity remains relatively stable. Hence, activities are integrated into individual lives, and these lives are integrated into the way of life of a society through patterns of connections that are individually unique yet culturally typical. The second and third constraints connect those human activities that involve flows of matter and energy, and may be expressed in the forms of the first and second laws of thermodynamics, respectively. The first law holds that matter and energy can be neither created nor destroyed by a society or the biosphere. They can only be transformed, which means that a society temporarily borrows all matter and energy involved in its technology from the biosphere. Within the biosphere, all matter participates in cycles on which human activities are superimposed. The second law holds that energy cannot flow in cyclic patterns because energy transformations are irreversible. Hence, linear chains of such transformations undertaken by a society originate either with the sun or within the biosphere, and then move through the society, eventually returning to the biosphere, whence it may pass into space. Together, these three constraints delimit how a technology can be embedded into a society and, via it, into the biosphere, which on a map of the ecology of technology translates into limits on the kinds of connections and structures these can form. This chapter focuses primarily on those connections derived from the flows of matter and energy associated with the way of life of a society. Chapter 7 examines the kinds of connections made by a modern way of life through a phenomenon that makes minimal use of context.

A generic map of the ecology of the technology of a society can help a designer determine the biosphere-related consequences of a particular design or decision choice by tracing its implications upstream via a chain of human activities to where the matter and energy embodied in the individual technology under consideration were extracted from the biosphere, and downstream through another chain of human activities to where they are returned. These two chains of activities, in their turn, depend on other activities involving additional flows of matter and energy from and into the biosphere. For example, the mechanical spinning of wool depends on the inputs of raw wool and energy into the spinning machines. Secondarily, it depends on the spinning machines themselves constituted from inputs of materials, energy, and other production machines that helped produce them, which in turn depend on other inputs, and so on. Further interdependencies can be traced in so far as that is meaningful, much like the different levels recognized in energy analysis.[1] Tracing the biosphere-related consequences of a design or decision choice thus involves a considerable portion of a society's way of life because of its highly enfolded character. No matter where the boundaries of the analysis are drawn, it will always contain an open whole, but the exchanges of matter and energy across them should make a negligible contribution to the analysis. A generic map is essential for determining where these boundaries should be drawn. Additional connections between activities involved in the technology result from its participation in the way of life of a society, that is, the use the designer anticipates for it. A prevention-oriented designer will, of course, take into account the entire 'life' cycle of the technology under development.

From the perspective of the first and second laws of thermodynamics, the way of life of a society implies a network of activities interconnected through flows of matter and energy. All input flows into the network are derived from the biosphere and all output flows return to it. Any dissipation of matter and energy within the network must be represented as leakage flows into the biosphere. The structure also reflects the fact that cultures organize their ways of life so as to avoid useless activities in the sense that their outputs have no further purpose other than the production of waste. In a market economy, for example, this is largely assured by the mechanism of supply and demand. The expression 'ashes to ashes and dust to dust' reminds us that human beings are physically included in all of this. It is clear that the sustainability of a way of life with respect to the biosphere depends on the ability of the latter to almost indefinitely sustain the flows into and out of the network. Throughout prehistory and history, cultures have created diverse 'systems' involving phenomena such as hunting and food-gathering, agriculture, craft production, trade, and industrial production. Each of these requires very different flows across the society–biosphere boundary. Their implications have also varied considerably. Although, in the past, local ecosystems were sometimes destroyed, not until the present century did the biosphere come under significant pressure as a consequence of supporting such flows. In other words, the culture of a society evolves a 'design' for its way of life, which determines the kinds of demands it makes on the biosphere. For this design to be sustained by the biosphere requires that its network of flows of matter and energy is supported by the corresponding network representing the biosphere. The former evolves on the basis of culture, and the latter on the basis of natural evolution.

The way a technology is enfolded into its contexts thus imposes limits on how this can be done. Since a society can neither create nor destroy matter and energy, limits are imposed within which cultures are free to create ways of life that can be sustained by the biosphere. The emphasis of modern civilization on performance values, at the expense of context values, makes it virtually impossible for us to recognize these limits. Nature's 'way of life,' as best as we can tell from what we know about natural evolution, appears to be based on the principle that everything must evolve in the context of everything else, and that diversity is essential for self-regulation and adaptation. This is diametrically opposed to our modern way of life, which is based on finding the one best way in terms of performance values in every sphere of human existence.

Just as a way of life depends on the biosphere for matter and energy, it depends on human life for meaning, direction, and purpose. These hold together a cultural design and allow it to evolve. In other words, sustainable development has one dimension related to the biosphere and another to human life. The two are interrelated through the cultural cycle, by which people affect their social and physical setting, and are, in turn, affected by it. It is essential, therefore, to pay particular attention to the role of culture in individual and collective human life and to how that role has been modified by technology. From this perspective, what preventive approaches for the engineering, management, and regulation of modern technology seek to accomplish is to reconnect things to their contexts, as opposed to largely disconnecting them through conventional approaches. This would return the relationship between modern technology and culture closer to the ones observed during prehistory and most of history. It is also clear that preventive approaches flourish or die according to their ability to affect the design stage by reconnecting technology to its contexts. This will not be done in the same way as it was with traditional technologies, which flowed directly out of the minds, and thus the experience and cultures, of the people involved. Design and other technology-related tasks must again make a much greater use of context considerations, which will require a mutation in the education of engineers, managers, and regulators. At present, design courses in the engineering curricula are either very narrowly conceived or hardly viable because both the formal and informal curricula make the process of synthesizing what students learn in the technical core and the complementary-studies component almost impossible. Hence, engineers, managers, and regulators must cooperate to give design a much more prevention-oriented approach. This can be greatly facilitated by a generic map of how modern technology is embedded into its contexts. Sections of such a map can then be worked out in greater detail to support prevention-oriented design and decision making related to specific individual constituent technologies. Its role would be twofold. It would show all the likely connections established by an individual constituent technology as it moves through the phases of the technological cycle. It would reveal harmful connections and thus provide the basis for preventive intervention. As such, a generic map would provide a partial substitute for what in traditional cultures was accomplished by technological know-how and practices that were embedded in experience, in turn embedded in the traditions of a way of life shaped by a culture.

Within this broader perspective, a generic map of the ecology of modern technology will be developed. The present chapter begins this task by focusing on how a way of life may be represented as a network of flows of matter and energy that must be supported by flows across the society–biosphere boundary, representing the services rendered by the biosphere as the ultimate source and sink of all matter and energy, as life support and as habitat. The concept of biosphere-related sustainability can be made operational in terms of the ability of the biosphere to indefinitely sustain these flows across its boundary with society. Its capacity for sustaining such flows is related to its internal functioning, in general, and how matter and energy flows are structured within it, in particular. The flows of matter form cyclical patterns where the wastes of one process become useful inputs into the next, continuing around the cycle until the point of departure is reached. As a result, the biosphere is self-purifying and non-polluting in the long term. Similarly, resources are 'produced' in the course of natural cycles. The time scales of the cycles in which all matter participates vary from what is significant only in geological time to what has significance in human time. All this is in sharp contrast with the largely linear flows through which a society derives matter from the biosphere, transforms it, incorporates it into its way of life, and relatively rapidly discards it back into the biosphere. This means that the technological cycle of any constituent element of the technology of a society has multiple beginnings and endings in the biosphere. These may participate in biospherical cycles, directly or indirectly influence them without participation, or contribute to or subtract from stocks that are relatively inertly stored. There are no energy sources in the biosphere or in society. There are only energy-conversion processes that cannot form cycles because of their irreversibility. Within the biosphere, solar energy powers ecosystems by first being captured by plants and irreversibly moving up the food web, sustaining losses that partially take the form of low-temperature heat. At present, society lives almost entirely from the energy capital stored up by the biosphere. As various forms of this capital are rapidly being depleted, we can either turn to other forms or begin to live increasingly from sustainable-energy income from the sun. The analysis will clarify three important differences between society and the biosphere. The first is that, within the biosphere, transformation processes of matter form cyclic flow patterns, while in society these are largely linear. Second, the biosphere is entirely powered by sustainable-energy income derived from the sun, while modern societies are almost entirely powered from energy capital stored in the biosphere. Third, the driving forces that structure the networks of flows of matter and energy are different. For a society, explanations are sought in the events of history, including the emergence, growth, development, mutation, decline, or collapse of societies and civilizations as cultural wholes; for the biosphere, explanations are sought in ecological processes and events of natural evolution. The analysis culminates with differentiating the concept of sustainable development so as to constitute the columns of the context matrices to be developed.

The development of the ecology of modern technology continues in the next chapter. The interdependencies among technology, society, and the biosphere are there examined on the level of human activities, interconnected by means of a way of life and culture as opposed to the level of the flows of matter and energy implied in them. Chapter 7 differentiates a way of life into individual activities or groups of activities affecting sustainable development. It yields the row structure of context matrices. The row and column structures will then be assembled into the context matrices to support prevention-oriented design and decision making. Particular attention will be paid to how modern ways of life delimit the use of context, and the implications this has for sustainable development.

Two features of this approach for developing a generic map of the ecology of technology are worth highlighting. One particular point in this approach may have caught the attention of the reader: The biosphere clearly is not the ultimate source of solar energy. I included it as a reminder that the biosphere 'prepares' solar energy by filtering it through the upper layers of the atmosphere in general, and the ozone layer in particular, so that, in a sense, it is the ultimate source of the kind of solar energy that powers it.

Second, this approach is the opposite of any attempt to manage the biosphere or planet Earth by means of technical interventions, which would constitute a macro level, end-of-pipe approach to the environmental crisis. After all, the environmental crisis has been barely 200 years in the making. On the time scale of natural evolution, it is as though it started yesterday. The beginning of the environmental crisis correlates with the advent of modern civilization and, consequently, it is our modern way of life that we need to manage, and not the planet. To advocate the need for managing planet Earth or spaceship Earth is to advocate end-of-pipe solutions. Our dependency on the biosphere must be changed by going back to the roots of the problem in our modern way of life, thus taking the pressure off the self-regulating character of the biosphere. The reason why a technical fix for the environmental crisis is so appealing can be understood only in terms of modern cultures.

## 6.2 The Biosphere as a Natural Whole[2]

The biosphere comprises many interacting ecosystems. In some respects, the structure of an ecosystem may be likened to that of a society having a social hierarchy built up from social roles required to maintain and evolve its way of life. It demands a variety of occupations such as farming, baking, teaching, healing, defending, and ruling, each carried out by a certain number of people. In traditional societies, children learn the occupation of their parents, thus following in their footsteps. The structure of an ecosystem is similarly composed of niches representing the 'occupations' required to maintain and evolve it. The variety of niches and the numbers of individuals occupying them are determined by the grand scheme and not the breeding strategy of a species. The fittest offspring will occupy the niches, and the rest will die. The closest analogous situation in a society would be to enable as many persons as possible to graduate from business schools but, since only a certain number of business graduates can be employed each year to keep a certain way of life going, those with the highest grades or best connections will get the positions.

The structure of an ecosystem is constrained by the first and second laws of thermodynamics that govern the flow of matter and energy. That structure may be represented by what is called an 'Eltonian pyramid.' The numbers of animals are plotted along the horizontal axis, and the position in the food web (which generally correlates with animal size) on the vertical axis. The base of the pyramid is constituted by a step representing the smallest animals in the ecosystem. It supports a next step, in which animals are about ten times larger but much fewer in numbers. This step supports another one, representing animals that are ten times larger again but whose numbers are still smaller. This continues until the largest animals have been reached. The pyramid may thus be regarded as representing the biomass in each size category. A tenfold jump in size between each step in the pyramid is explained by the fact that each carnivore feeds itself from the step below and needs to be that much larger to easily overcome its prey.

The substantial reduction in biomass present in each higher step of the pyramid can be explained if we look at the structure in terms of energy analysis. Each carnivore obtains its calorie intake necessary for living from the step below. Each step fixes a certain amount of energy in its biomass, which is transferred up as food to the next higher step. Much of this is burned in the process of living and lost to the food web as low-temperature heat, and it is estimated that only about 10 per cent is passed up as biomass. This explains why the numbers of animals decline sharply in each higher step of the Eltonian pyramid. At the base of the pyramid, we find herbivores, who obtain their energy from plants, which in turn obtain it from the sun through photosynthesis. The distribution of herbivores in an ecosystem is similar, despite the fact that they all feed directly from plants. As the size of herbivores increases, their numbers decline for the simple reason that fewer niches can be supported by an ecosystem.

Plants capture energy from the sun by means of their leaves, which, from an energy perspective, function as transducers. Plants convert energy by producing sugar from carbon dioxide and water, and release oxygen by day and carbon dioxide by night. The average efficiency of this process over the lifetime of the plant is about 2 per cent in the wild or in cultivation. One of the reasons for this low efficiency is that the plant can use only the visible portion of the light spectrum, because it alone can create the pulse of energy required for a significant disturbance of the electrons involved in photosynthesis. The infrared portion of the spectrum warms plants, evaporates water from their surfaces, and thus helps drive their circulatory processes. Consequently, only half of the solar energy is available to plants. Another constraint in the efficiency of energy conversions by plants is the carbon-dioxide levels in the atmosphere. Relatively speaking, in dim light there is plenty of carbon dioxide, but in full sunlight there is not, thus constraining the energy-conversion processes of plants. Thus in energy terms, a tiger hunting a herbivore may obtain only 10 per cent of its biomass, which in turn represents 10 per cent of the energy the herbivore obtained from plants, which in turn captured 2 per cent of incoming sunlight. This means that the tiger received only 0.02 per cent of the energy entering this chain of transformations. The remainder was lost in the form of low-temperature heat given off to the surroundings. Hence, the real limit to the number of plants and animals in an ecosystem is ultimately fixed by the amount of carbon dioxide in the air, because it sets the rate of plant production, which is the energy base for the entire Eltonian pyramid. It should not be concluded, however, that carbon-dioxide emissions are a good thing.

It should be emphasized that the discussion above implies real limits to the food supply for the planet. Science has not been able to improve the efficiency of the energy conversion of plants. The Green Revolution did not represent a victory in this area. What the breeding and genetic engineering of plants accomplished was to have more sugars put in those parts of the plants that human beings consume. This comes at a cost. For example, high-yield grains produce more grain at the expense of the rest of the plant, which negatively affects its ability to defend itself against weeds and pests. This requires a more intense usage of herbicides and pesticides, which has many negative consequences for the local ecosystems. The present high expectations from genetic engineering and biotechnology overlook the fact that these techniques are also driven primarily by performance values, and therefore will repeat the kinds of patterns described earlier.

The structure of the plant world reveals further connections to the biosphere. Since weather constitutes a major influence on vegetation, and since vegetation helps produce soil, five kinds of plant formations can be distinguished: those commonly found in tropical rain forests, hot deserts, temperate deciduous forests, boreal forests, and tundra. These in turn correspond to climatic types A through E, and to major soil differences. Once again, some of the differences between these plant formations can be explained by means of energy analysis, developing a 'heat budget' for plants. For example, the 'solar panels' of plants (leaves or needles) create large surfaces that exchange heat with surrounding air through forced and free convection and radiation. In the tundras, plants stick close to the ground, where they can capture solar energy while minimizing heat loss in the relatively still layer of air close to the ground. The shapes of desert plants are a solution to the opposite problem: too much heat and a shortage of water. A tall stick-like shape, for example, minimizes exposure to the sun and maximizes the possibilities for heat loss. Trees in rain forests do not face this problem because they have enough water to cool themselves with. In temperate latitudes, flat leaves constitute an insurmountable problem of heat loss in the winter. Hence, these trees withdraw as many calories as they can from the leaves in the fall, shed the empty husks, and regrow new energy transducers in the spring. This strategy will not work farther north, because of the briefness of the summer. Here the energy transducers take the shape of dense clusters of needles, which lose less heat in the winter but are able to make use of any brief sunny spells that may occur during the winter to produce an energy income of sugar calories. This overall strategy is effective even though, during the summer, needles are less efficient energy transducers than leaves. It appears, therefore, that these plant formations are the best trade-off between the maximum possible energy production, the availability of water, and the need to balance the heat budget.

The underground part of plant formations reflect the five different general soil types. For example, tropical soils lack the basic nutrients for plants because these have been washed out of the soil, thus making them chemically infertile. Vegetation gets along without these nutrients in the soil by recirculating them within the rain forests. Fallen trees and animal wastes shed nutrients that are collected by extremely fine root systems thanks to the fungi that grow on them. Almost nothing that is of value to the rain forest escapes. The situation is very different in temperate forests, where colder temperatures slow down the rotting of any debris from the forest and animal droppings, resulting in soil humus that slowly leaks nutrients into the soil. Root systems do not have to be as efficient, so that crops can obtain the necessary nutrients. This source of nutrients is complemented by unique clay soils acting as filter-beds of local reaction sites through which the rain water trickles. In contrast to rain forests, a substantial portion of nutrient cycles involves non-living processes of which the plants take advantage. Hence replacing temperate forests with crops does not destroy the essential nutrient cycles, so that agriculture is sustainable provided that the soil is not mined by extracting more than can be replenished by natural cycles.

There are, of course, exceptions to these general patterns. For example, the great rivers in the tropics deposit sediments containing many nutrients lost by upstream ecosystems, so that the land can be worked without spoiling it. This is but one of the reasons why damming these rivers carries with it such substantial negative implications. Also, where soils have been made of young nutrient-rich volcanic rock, agriculture can be very successful as well. Fertile land in tropical areas can also occur in regions where there is much less rainfall than usual, resulting in a different soil chemistry.

The transformations of matter that accompany the linear chain of energy transformations occur in cycles. The water cycle is a well-known example. Water evaporates from the oceans and returns to the land by precipitation. Some of it passes through the local ecosystem or through the soils, or simply runs off into streams and rivers to return to the ocean directly. The oxygen cycle involves the transformation of carbon dioxide into oxygen by means of photosynthesis and the conversion of oxygen into carbon dioxide by oxidizing plant and animal carbon. The carbon cycle shares the process of photosynthesis, which in addition to oxygen produces hydrocarbons. Plants also release some carbon dioxide through the leaves or roots. Some of the carbon fixed by plants is consumed by animals, which release carbon dioxide through respiration. Dead plants and animals decompose and oxidize tissue carbon to form carbon dioxide, which returns to the atmosphere. Although nitrogen is the most abundant gas in the atmosphere, it cannot be used directly by most life forms. The nitrogen cycle is complex and involves several transformations carried out by bacteria before serving essential functions in plant and animal life. The other part of the cycle, involving denitrification, also requires transformations carried out by bacteria, producing nitrogen and thus completing the cycle. These four cycles form the basis for life in the biosphere. However, many other cycles are required. These include those of phosphate, sulphur, potassium, sodium, magnesium, iron, calcium, and many more. Many of these cycles are intertwined, not unlike the oxygen and carbon cycles mentioned above.

This brief sketch of how matter and energy flow through ecosystems points to fundamental differences between society and the biosphere. Apart from the fact that contemporary civilization narrows diversity while the biosphere supports an incredible diversity, the transformation of matter occurs in cyclical patterns in the biosphere and in largely linear patterns in modern societies. Four aspects of our unhealthy relationship with the biosphere are worth highlighting. First, there are limits to what the human economy can appropriate from the base of the Eltonian pyramid produced by photosynthesis. Taking too great a share through activities such as agriculture threatens the food webs of ecosystems, impairing, and eventually endangering, all their functions, including those services rendered to societies. Second, economic activities that lead to species loss must be severely limited. Species loss affects the energy flow through ecosystems, and possible domino effects can undermine their long-term viability. Third, modifying constituents of ecosystems by performance-value–driven biotechnology and genetic engineering will almost certainly undermine the health and integrity established through the co-evolution of these constituents over a very long period of time. The successes and failures of the Green Revolution ought to make us very cautious. Fourth, the magnitude of many flows of materials within society now approaches, or even exceeds, those within the biosphere, fundamentally affecting natural cycles. The growing scale of the human economy within the finite biosphere is a significant issue. We are largely ignorant about the long-term consequences of these four implications. Useful ignorance would dictate a strategy of prevention.

## 6.3 The Society–Biosphere Boundary[3]

Ever since the emergence of cultures, humanity has not lived within the biosphere the way animals do. Every human group in prehistory (and later every society) symbolizes its individual and collective life within the biosphere by means of a culture, which establishes and evolves a niche that is in part a human creation. Consequently, the relationship between humanity and the biosphere is very different from that of animals. A spectrum of relations is possible. On the one extreme, humanity could live in its natural niche and have no more impact on the biosphere than do animals. The environmental crisis represents the other extreme where societies create cultural niches that threaten and undermine the biosphere. The in-between range represents almost universally healthy relations with local ecosystems in prehistory, to significantly less healthy relations during history (where some societies even destroyed local ecosystems but where the biosphere as a whole was never threatened). In prehistory, food-gathering and hunting ways of life trod very lightly on local ecosystems. With the emergence of societies at the dawn of history came the development of agriculture, cities, and commerce that transformed ways of life, including their relationship with the biosphere. Most recently, the industrialization of production and the application of scientific and technical rationality to all other spheres have given rise to ways of life that share science, technology, and economic development at their core, leading to almost universally unhealthy relations with the biosphere.

Since matter and energy can neither be created nor destroyed within a society, its way of life determines the kinds and magnitudes of such flows across its boundaries with the biosphere. These flows reveal two primary functions of the biosphere: it is the ultimate source and sink of all matter and energy transformed in a society. An additional two functions are those of ultimate habitat and life support. The former has been explained in terms of the contribution it makes to a life-milieu in an earlier section. The latter, which will not be discussed here, includes protection from harmful radiation and climate regulation. Returning to the first two functions, it is useful to make a distinction between three different inputs from the biosphere: non-renewable flows, renewable flows, and continuing flows. They reflect three different resource stocks. Non-renewable and renewable resources are distinguished on the basis of the time frame required for the cycles in the biosphere that involve the conversions that 'produce' the resource. If this time scale is in the order of geological time (as is the case for oil, natural gas, and coal, for example), the stocks of these resources are fixed relative to any human time frame. Renewable resources derive from natural cycles whose time scale is short enough that the rate of conversion 'producing' the resource is significant in comparison with human usage. Such renewable resources can be depleted when the rate of human use is greater than the rate of production through a natural cycle. The maximum sustainable yield of such a resource is just under the rate at which the natural cycle replenishes it through conversion processes. The situation is analogous to that of living off one's savings. If one lives off earned interest each year, this cycle can go on forever. However, if each year a certain portion of the capital is withdrawn as well, then sooner or later those savings will be depleted. The same is true for 'natural capital.' If agricultural practices result in a rate of soil erosion greater than the rate of its formation through natural cycles, the soil is being mined and the practices are not sustainable. Similarly, a fish stock will be depleted if the annual catch exceeds its rate of regeneration. Falling ground-water levels are a sign that human consumption exceeds the capacity of natural cycles to replenish it. Renewable resources can also be depleted when local ecosystems are affected in a way that slows down or destroys the natural cycles 'producing' them. The rain forests are a renewable resource in principle. However, since the nutrient supply is almost entirely stored in the plants and trees rather than the soils, the process of clear-cutting is irreversible. This contrasts with northern forests, where most of the nutrients are stored in the humus and the soils. Since clear-cutting these forests does not destroy the nutrient source, reforestation or agriculture is possible.

Non-renewable and renewable resources are distinguished from continuing resources because the latter are not affected by human usage, at least not on the scale we can currently imagine. Solar, wind, tidal, and geothermal energy are the most common examples.

In the debate as to whether humanity faces a resource crisis, the stocks of non-renewable resources are usually divided into two components, depending on whether they are available to us or not. Availability is determined by market prices exceeding extraction and refining costs, which in turn are determined in part by the available technologies. In other words, the component available to humanity fluctuates with technological and economic conditions. Many economists argue that there is no resource crisis by affirming perfect substitutability. According to this view, when a particular non-renewable resource becomes increasingly scarce, its market price will rise and demand will fall. The market provides a considerable incentive for devising cheaper alternatives, which sooner or later will be developed. As a principle, perfect substitutability is in effect a declaration of faith that technological and economic development will always find answers to all our resource problems. Whether this is true or not still overlooks the problem that the intensity of the matter and energy dependence of modern civilization is so great that any substitute will be used at a very high rate, thus constituting only a temporary solution. A more prudent approach would at best assume weak substitutability and recognize that a more permanent solution would be to create a way of life that converts matter in cyclic patterns rather than having a linear throughput. In other words, matter would have many different 'lives' by participating in successive materials that constitute closed cycles as much as possible. Only the deficit would have to be made up from the biosphere.

Since a society cannot destroy matter, all resources eventually become wastes. However, the burden on the biosphere can vary enormously, depending on the linearity or circularity of the economy. In a linear economy, wastes are directly discharged to the biosphere; in a more circular economy, wastes are converted to useful materials and products as many times as possible, thus greatly reducing the dependency on the biosphere as source and sink. When wastes enter into the biosphere, they are first dispersed in the atmosphere, bodies of water, and the soils. Over time, larger and less stable compounds break down into smaller more stable ones. In these processes, some wastes (such as some minerals and organic wastes) begin to participate in natural cycles that transform them into resources. The remainder is stored in the biosphere. Wastes that interfere with natural processes in their surroundings are called 'pollutants.' The biosphere produces no pollutants of its own because all its processes of converting matter are organized in natural cycles.

There is no question that modern civilization faces a crisis in terms of the role of the biosphere as sink. The pollution of our air, water, and soils, and global warming and ozone depletion are a few examples of this crisis. The only solution to this problem is to convert wastes into useful materials and products as often as available technologies and economic conditions permit. Closing the loops by the German automobile industry was the first large-scale attempt to move in this direction. The emerging field of industrial ecology is identifying opportunities for converting more wastes into resources. It should be recognized, however, that a more circular economy is no panacea since recycling matter involves transformation processes requiring energy. Energy conversion may well turn out to be the bottleneck in making modern civilization more sustainable, although much can be done here, as we shall see later. Vastly reducing the dependency of modern civilization on the biosphere as ultimate source and sink will also reduce the negative impact we have on it as our habitat and life support.

The phenomenon of pollution can now be put into context. The exponential growth of the flows of matter and energy necessary to sustain a modern economy have given rise to the phenomenon of pollution occurring today on a scale unprecedented in human history, and one that seriously perturbs the biosphere. As noted earlier, the transformations of matter occur in cyclical patterns in the biosphere. These cycles are self-regulating either in themselves or in conjunction with others. In this way, the biosphere produces no waste and is self-purifying. A cycle can be stressed when its self-regulatory or co-regulatory character is stressed, overloaded, or even destroyed. For example, if the water cycle is overloaded with organic animal wastes, the amount of oxygen required by the material that decomposes them can exceed the oxygen available in the water. When this happens, the bacteria die, halting the entire cycle. The capacity of the water cycle to act as a sink for animal wastes is, therefore, limited. This is the case for all sinks. Human-made materials can affect the operations of a particular cycle by impeding or blocking one of its phases. The health of an ecosystem can be degraded when one or more of its cycles are critically affected. This can spread to adjacent ecosystems and the entire biosphere if exchanges between them are distorted in a manner that cannot be regulated by them.

Pollution can also occur when a cycle is speeded up by human interference. Modern agriculture provides an example. When plants are grown for food, they collect from the soil the materials necessary for their growth and maintenance. Since these materials are neither created nor destroyed within the soil, they must be replenished through natural cycles or by human intervention, when the soil is mined rather than being sustainably farmed. Fertilizers artificially add nitrates to the soil's cycle. When the soil is not sufficiently porous because of insufficient humus content, the resulting inadequate levels of oxygen further reduce the already relatively low rates of nutrient uptake by the roots of the plants. As a result, a considerable portion of the fertilizer ends up in the surface water. This creates an external stress on the water cycle by causing algae overgrowth, which restricts the self-purifying character of the cycle. Pollution can, therefore, occur when the flows constituting any phase in the cycle are increased or decreased to an extent that the self-regulating character of the cycle is unable to deal with them, or by the introduction of foreign substances that indirectly cause the same thing.[4]

The development of the new urban-information-industrial life-milieu has created a very unhealthy relationship with the biosphere in all its three functions necessary to support human life. The use of sources and sinks within the biosphere has quantitatively and qualitatively changed. The new life-milieu, which mediates our relationships with society and the biosphere, has influenced human consciousness and culture in a way that facilitates, and even encourages, this unhealthy relationship. The evolution of environmental regulations reflects this changing consciousness. When belief in progress was strong in the 1960s, pollution was regarded as an unavoidable small price to pay, which was reflected in ineffective pollution regulations. As pollution became more widespread and progress less obvious in the 1970s, governments were compelled to mediate between competing interests. As these trends continued, pollution regulations increasingly shifted the burden of proof of acceptable environmental harm to the entrepreneurs undertaking significant projects. Hard economic times and the need for more jobs are now beginning to have some effect on this trend in pollution regulation. Yet the myths of modern society continue to reinforce end-of-pipe approaches.

In sum, the society–biosphere boundary divides two wholes that have very different designs. Modern civilization dominated by performance values and technique sharply reduces diversity, as witnessed by the universalization of science, technology, economic development, and Western (i.e., technological) culture. The biosphere, on the other hand, is teeming with diversity. The reasons why modern civilization sharply reduces diversity are rooted in the fact that the predominant use of performance values leads to the selection of the 'best' means of any kind and discards any others proven to be less effective on these terms. This delimits adaptation to the great diversity of human beings, societies, and ecosystems, in many cases producing severe context incompatibilities. The difference in time-scales of the evolution of the biosphere and the historical development of society results in considerable differences in the extent to which each constituent element has co-evolved with, and thus adapted to, all the others. This is borne out by the incredible resilience and robustness of the biosphere in the face of the massive flows across its boundary with society during the last century.

## 6.4 Challenges, Responses, and Crises

The enormous increase in both the diversity and the magnitude of flows across the society–biosphere boundary in the twentieth century is causing what is commonly referred to as the 'environmental crisis.' Perhaps what matters most is not the magnitude and scope of the environmental crisis, but how we respond to it. Do we perceive it as a crisis that could call into question everything that has been accomplished during the last two centuries? Does the current behaviour of individuals, professions, universities, corporations, and governments contribute to changes in modern ways of life appropriate to the best available assessment of the situation? Does a gap exist between the interpretation of the situation and the response to it, and, if so, can it be attributed to excessive constraints on individual and collective behaviour, or to a lack of resources?

To answer these kinds of questions, it is essential once again to remind ourselves of some of the basics. A civilization is a whole open to the biosphere, itself composed of smaller wholes such as societies, communities, groups, and individuals. In the absence of a science of the sciences, we have little understanding of these wholes as such because no discipline can account for their integrality. As noted before, each discipline treats the subject matter of all the others as theoretical externalities. Each discipline appears to be studying, not different dimensions of the same reality with the same wholes, but a kind of discipline-defined 'part' out of which these wholes are 'constructed.' Daly and Cobb[5] have termed this problem 'disciplinolatry.' It marks a significant departure from more classical thinkers, such as Durkheim, Marx, Weber, and Toynbee, whose works reflect the sense that the wholes noted above are more than the sum of their parts. Macro-level theories have gone out of fashion, and even their need is now seriously debated, although no one appears to be disputing the commonplace of the irreducibility of wholes.

Incorporated into my general theory of culture is Arnold Toynbee's[6] recognition that human civilizations are cultural creations that can grow, flourish, stagnate, decline, and collapse as a result of their ability to give meaning, direction, and purpose, and materially provide for their members. They attempt to create a measure of order and harmony in a chaotic world. They must constantly adapt and evolve in response to internal and external, positive, and negative perturbations of this order. Toynbee interpreted the evolution of societies and civilizations in terms of challenges to their cultural order and the responses to them. Such challenge–response patterns can vary greatly, and it is useful to briefly survey the spectrum of possibilities. We can then apply it to the interpretation of the environmental crisis.

If a perturbation is judged to be small in terms of its implications for the cultural order, a creative response is readily arrived at. If many such responses occurring on a daily basis in a society creatively extend, reinforce, and enrich the cultural order, the civilization grows through these responses. It is based on a relatively cumulative extension of the cultural order. If, on the other hand, such perturbations cannot be dealt with by creatively extending the cultural resources of a society, the creative responses may collectively undermine and weaken the cultural order, with the result that a society enters a period of stagnation that may lead to either a mutation or a decline which, if it goes unchecked, may bring about a collapse. Under these circumstances, the responses are predominantly non-cumulative with respect to the cultural order. It is also possible for perturbations to be ignored, either because they are deemed so trivial as not to be worth bothering with or because they may not be noticed at all. This does not preclude the possibility that many such perturbations slowly and almost imperceptibly cause a society or civilization to lose touch with reality, making future successful adaptations more difficult.

Next, suppose that a perturbation is seen as substantial in terms of its implications for the cultural order. A creative response is now more difficult to arrive at. As noted previously, a healthy culture maintains a dialectical tension between individual diversity and cultural unity, thus enhancing the chances of such major responses occurring. Once again, such responses can be cumulative or non-cumulative, and contribute to the growth or decline, respectively, of a cultural order. In these situations, it is highly unlikely that challenges perceived to be serious are ignored, but this does not guarantee a successful response. For example, some groups within society may benefit from interpretations of the situation that either amplify or diminish the challenge, and here ideologies can play a significant role. Other groups may have quite different responses but, because of differential political influence, social status, and access to power, a plurality of responses is quickly converted into social tensions and, if the stakes are high enough, social conflicts, with eventually one response gaining the upper hand. In other words, while initially there may be a plurality of responses, the fact that a society is a dialectically structured enfolded whole implies that neither unity nor diversity can increase without limits if the society is to remain healthy. For example, in a society faced with the challenge of immigrants from many different cultures, neither a melting-pot nor a multicultural policy can succeed in the long run.

Finally, suppose that a perturbation is perceived as so vast in terms of its implications that it calls almost everything into question. A creative response of an either cumulative or non-cumulative nature now appears almost impossible. The situation is mind-numbing, and the cultural resources of a society appear to be defeated or crushed by the sheer magnitude of the challenge. The only possible response appears to be to attempt to keep going as best as possible, in the hope that somehow the situation may change. When that hope disappears, the cultural order may be crushed by the challenge. Different groups in society begin to go their own way, and make do as best they can. A period of social upheaval begins in which the diversity increases, enhancing the chances of a successful response emerging. If, however, the diversity completely undermines the cultural unity as the basis for order, decline and collapse may follow. This may or may not be accompanied by attempts to ideologically scale down the implications of the challenge, to make it appear less insurmountable and more manageable than it really is. This may bring hope and courage to face the challenge, and again increase the chances of a response.

This simple topology of challenge–response patterns, which begins with an interpretation of a perturbation, converting it into a challenge of some magnitude, shows the diversity of responses that can result. It can be further enhanced through the concept of negative feedback. We have already noted that this concept, derived from mechanical systems, cannot be directly applied to processes in human life or in a society. In a world ordered by culture, relationships are reciprocal and dialectical, as in the case of human communication. There is feedback in each direction, regulated by a hierarchy of values. These are dialectically linked into a relationship. What constitutes positive and negative feedback also becomes more complex. Negative feedback may have to give way to positive feedback to keep a relationship healthy when it has settled into a state of being taken for granted, thus reducing the dialectical tension. The latter may bring a disturbance, required to get the relationship back on track. If it succeeds, positive feedback has performed the function normally accomplished through negative feedback. On the level of a society, disturbances may be required to confront political corruption or economic injustices. They may also prevent institutions from stagnating. These limitations must be kept in mind when the concept of negative feedback is used to explain an entirely different kind of reality.

Does our current unhealthy relationship with the biosphere constitute a crisis? Given a democratic and pluralistic society, and utilizing the above topology, including the concept of feedback, the answer is both yes and no. It fundamentally depends on where a person thinks society is heading and the implications this has in terms of his or her values. The former requires a vantage point and is, therefore, affected by the person's socio-economic position in society, the ideological position adopted, as well as currents of public opinion in a mass society. The latter depends on how deeply the person's life is rooted in the cultural values of a society, which determines the extent to which he or she is in but not of the cultural order in the deepest spiritual and existential sense of that term. For example, the position taken by some groups in a modern society is that the environment is much more than merely an object to be dominated and used for human purposes. They see the present relationship between humanity and the biosphere as unjust to the part of humanity living in extreme poverty, and immoral because we do not have the right to dominate other life forms. They believe that the present course must be fundamentally disturbed before the present 'system' will change and produce a healthy, just, and sustainable relationship with the biosphere. On the other end of the spectrum, we find groups who share a concern over the unhealthy relationship that currently exists between humanity and the biosphere, but who have complete confidence that society (through science, technology, and free markets) has at hand sufficient means to redress the situation. The assumption here is that, through negative feedback, these human institutions will function as well as possible. Although the differences may appear to be irreconcilable, it should be remembered that all these groups are composed of members who are persons of their time, place, and culture. The commonplace that some may be a little ahead of their time, and others a little behind their time, may appear as relatively trivial distinctions to future historians, who with hindsight may note how much the diversity was delimited by the cultural unity of society. However, we do not need to wait for future historians to get a sense of this ourselves. Nobody can possibly be sure that their interpretation of the present situation is sufficiently accurate and reliable to warrant the dismissal of other points of view. An ongoing critical awareness of the relative character of one's own vantage point and moral position could create the kind of openness that allows us all to benefit from the diversity of perspectives in a free and pluralistic society. Whether we are politically conservative, liberal, socialist, Marxist, or green; whether we are practising Jewish, Christian, Muslim, Hindu, humanist, or New Age; whether we are rich or poor, male or female – all this affects the position we take; and who among us has worked it all out without internal contradictions, doubts, and ambivalences? Tolerance and mutual respect could convert diversity and pluralism into a resource necessary to evolve in response to what we each believe constitutes the environmental crisis, as well as the many other issues that currently confront humanity. A constructive and respectful confrontation over these issues is not only possible but essential. However, this requires an iconoclastic attitude towards the secular religious faith in science, technology, and economic development.

By this we are not suggesting that we all relativize our deepest convictions (which is impossible in any case because they are rooted in the sacred and myths of our culture) but that we remain critically aware of them, including the contradictions they generate in our lives, to help us understand and appreciate why others are equally convinced of their positions. Together we are entering into a new life-milieu. The social sciences and humanities show us the extent to which it or any other life-milieu determines a great deal of human consciousness and culture. Judging from the previous two life-milieus, it takes a long time for a society to develop the cultural means to symbolically distance itself somewhat from these determinisms to impose a cultural order of its own making. In addition, the power of our technical means is such that it tends to suppress what ought to have been a reciprocal relationship. Civilizations that overexploited their ecosystems were eventually weakened, and some even collapsed because of their dependency on such ecosystems.[7] Similarly, those who exploited other societies or their own people created an external or internal threat, which in many cases helped weaken or destroy them. Power relationships mask, distort, and destroy relations of mutual dependence, which, for any open whole, invites disastrous consequences. To establish a relationship of exploitation through power is to utilize one direction of the relationship in a way that is advantageous for the exploiter, but the dependence on the relationship in the other direction remains, with inevitable negative consequences. The way a society creates a relationship with the biosphere, for example, helps create its life-milieu, and this in turn fundamentally influences human consciousness and its culture. If the relationship is unhealthy, the consequences are always negative. It becomes a poison that creeps through a society, impeding the abilities of both wholes to regulate themselves. Through power we can suppress minor irritants, but in so doing also reduce the possibility of negative feedback essential for a purposeful evolution and adaptation.

This section asked the question whether or not there is an environmental crisis. As an idea and a set of beliefs and opinions about the impact we are having on the biosphere, and how this in turn is going to affect us, the answer is an unqualified yes. It has become a commonplace. As a challenge to our society requiring a fundamental rethinking of how and why we do things through negative feedback, the answer is an unqualified no. With a few notable exceptions, the fundamental trends created by our individual and collective behaviour over the last thirty years by and large can be summarized as 'business as usual' or as 'for every step forward, there always appears to be a step backward somewhere else.'[8] Yet this is not for a lack of activity. Societies are passing thousands of pages of legislation, and institutions are making large investments in response to the environmental crisis. The two contradictory aspects of our present situation belong together: the 'system' is responding to the situation by compensating for its negative effects on the biosphere through end-of-pipe approaches, and not by dealing with the situation through negative feedback-based approaches. The excessive reliance on the GDP as an economic indicator has masked our true situation.

What are the implications for preventive approaches? The greatest obstacles standing in the way are neither technological nor economic in nature. The most fundamental problem is probably attitudinal and cultural. This is well illustrated by pollution prevention. Despite the facts that it is increasingly being adopted by governments as the preferred strategy for dealing with pollution, that there is growing evidence of its profitability, that payback periods are very short-term (frequently under two years), that it creates win–win situations for all parties, and many other benefits, there is no all-out effort in this area. Pollution prevention has hardly made it into the curricula of engineering and management schools and, even when it has, it occupies a precarious position as an add-on as opposed to the beginning of a paradigm shift away from conventional approaches. As a society, we simply do not believe it can really work, except in some unusual cases, because of the patterns of our culture in general and our predominant university, corporate and government 'cultures' in particular. Just as performance values have largely displaced human values under the influence of our new life-milieu, so also has come a widespread conviction that preventive approaches are suspect since they run against the grain. It is difficult to admit that for some 200 years we could have, and should have, behaved differently. The examination of our relationship with the biosphere using the first and second laws of thermodynamics implies that ultimately pollution is an incurable disease that can only be prevented.[9]

### 6.5 Differentiating Sustainable Development

In this chapter, I have examined the relationships among technology, society, ecosystems, and the biosphere, from the perspective of the first and second laws of thermodynamics, that show the constraints on the internal structure and exchanges with their contexts of these four open wholes. This aspect of the generic map of modern technology may be used to establish the ecological footprint of a constituent element of a technology.[10] It may also be used to reintegrate the economy and ecology of such an element by contextualizing the transformations of matter and energy involved in its technological cycle. Although the cultural aspect of this generic map will be completed in the next chapter, it is already possible to differentiate the concept of sustainable development into various dimensions on multiple levels. To prepare for this task, five conclusions may be drawn from the analysis presented above.

First, the flow of matter into and out of a society must be sustained by the biosphere. The long-term sustainability of any chain of transformation processes of matter involved in a way of life requires that either the biosphere (relative to human usage) is essentially an almost infinite source and sink for the flows crossing the boundary with society, or it becomes a closed cycle through a complementary chain of transformations within the biosphere. The latter piggyback arrangement works, of course, only if appropriate natural cycles exist and if they can deal with the increased flows. Failing these possibilities, the only remaining option is to, as much as possible, convert the linear chain of transformations to a circular one.

Second, chains of energy transformations within the way of life of a society can be sustained over the long haul only if the biosphere can be regarded as an almost infinite source and sink compared with the flows involved. This is no longer the case, and the deficiency will increasingly have to be made up from solar energy and other renewable energy sources.

Third, the flows of matter and energy necessary to sustain a modern society have now grown to a point where they are significant with respect to the flows within the biosphere, thus necessitating a fundamental restructuring that the World Commission on Environment and Development has called sustainable development.[11]

Fourth, the society–biosphere boundary reveals the four primary functions rendered to any society: the ultimate source and sink for all matter and energy (with the exception of solar energy) and the provision of an ultimate habitat and life support.

Fifth, the design for life established through the culture of a society must be brought in line with the design for life evident in the biosphere. The cultural cycle reveals the reciprocal interactions between these two designs.

Following these conclusions, the strategy for establishing preventive approaches can be developed as follows. The ecology of technology concerned with transformation processes of matter and energy can be built up by contextualizing individual processes in relation to one another in three stages. First, they should be contextualized within the way of life of a society up to its boundary with the biosphere. Second, the services the biosphere must provide at this boundary can then be analysed from the perspective of sustainability. Third, these flows should be studied in the context of the biosphere. The concept of sustainable development can be made operational in terms of the flows of matter and energy that may be sustained across the society–biosphere boundary and their implications for both sides: on one side, the modern ways of life established through cultures, and, on the other side, the design for life present in the biosphere. This will establish the limits within which all life can be sustained and within which cultures must give meaning, purpose, and direction to individual and collective human life.

Implicit in this chapter has been the conceptual operationalization of sustainable development. It can be represented as a tree structure, where the branches on each level may be considered as dimensions or reciprocally interacting entities enfolded in the wholes represented on the level above, and which can be further differentiated, as shown on the level below. Thus, on level 1 we find the umbrella concept of sustainable development. The second level differentiates it into the sustainability of the biosphere, the sustainability of the life-milieu, and the sustainability of human life, with the recognition that these are interdependent. This interdependence also exists on all subsequent levels of differentiation. On level 3, the sustainability of the biosphere is differentiated into the sustainability of the role of the biosphere as source, sink, habitat, and life support. The sustainability of the life-milieu is differentiated into that of the primary (technique), secondary (society), and tertiary (nature) life-milieus.[12] The sustainability of human life is differentiated into the metabolic sustainability of a way of life and its cultural sustainability. All entries on this level interact reciprocally. For example, the metabolic sustainability of a way of life depends on the ability of the biosphere to sustain flows of matter and energy across the society–biosphere boundary. The biosphere as habitat is symbolized by a culture and integrated into one of the life-milieus. For a modern society, the primary life-milieu is constituted by the technique-based urban-industrial-information 'system' via which most of society and nature is experienced.

On level 4, the sustainability of the role of the biosphere as source is differentiated into renewable, non-renewable, and continuing components. The sustainability of the role of sinks is differentiated into sinks for wastes that can be incorporated into the cycles of the biosphere, those that cannot and thus interfere with the cycles, and those that are stored as stocks having no effect on their surroundings. The sustainability of the habitat is differentiated into the sustainability of substantially transformed ecosystems, moderately transformed ecosystems, and largely 'wild' ecosystems. The sustainability of life support is differentiated into the sustainability of food-chains, climate control, protection from solar radiation, and as source of human knowledge, ideas, values, and models. The sustainability of the primary life-milieu of modern societies is differentiated into various kinds of technique-based mediation between human individuals, between individuals and society, and between society and nature. The sustainability of the metabolic basis of a way of life is differentiated into specific inputs such as nutritional ones and others. The cultural sustainability of a way of life is differentiated into those of the dimensions of mediation anchored in the cultural unity of a society.[13]

On the fifth level, further differentiation occurs. For example, the sustainability of the role of the biosphere as renewable, non-renewable, and continuing source is differentiated into flows that correspond to the cycles of the biosphere that 'produce' them. The same is true for the sustainability of the various kinds of sinks. Similarly, the sustainability of the technique-based mediation between human individuals and nature can be differentiated into the city, transformed rural areas that function as the hinterland of cities, and transformed rural areas that do not. Some entries on this level can readily be differentiated on the next-lower level. For example, the sustainability of a city can be differentiated into the sustainability of its core, urban, and suburban regions. At the next-lower level, these can be differentiated into the sustainability of neighbourhoods, which in turn can be differentiated, on the next-lower level, into the sustainability of streets, buildings, and infrastructure. These can be further differentiated on the next-lower level in terms of the contribution their 'components' make to sustainability on the previous level, and so on. The next chapter begins the analysis of the cultural aspects of the generic map of the ecology of modern technology.

7

# Differentiating a Way of Life

## 7.1 The Social Fabric

In chapter 6, I showed how the activities that make up the way of life of a society are connected together and to the biosphere as a result of the constraints expressed as the first and second laws of thermodynamics. These activities are simultaneously linked in one, and only one, way of life left open by the cultural unity of a society (comprising what in cultural anthropology are known as a sacred, myths, and a hierarchy of values). In traditional societies, this found expression in religions with 'gods in the sky' and their relations with the earth. This cultural necessity received a great deal of attention in the Judaic tradition, as well as in Christianity, until the belief in a secular society arose. Modern societies produce their equivalent secular absolutes, with the result that contemporary ways of life may be thought of as being anchored 'between the earth and the sky' according to the three constraints set out earlier. The networks of flows of matter and energy, therefore, do not determine a way of life created by a culture, nor can a culture create a way of life free from thermodynamic constraints.

Completing a generic map of the ecology of the technology of a society involves two further steps. The first differentiates a way of life into groups of activities according to their influence on the networks of flows of matter and energy and, via them, on the biosphere in order to clarify the implications for the biosphere-related dimension of sustainable development. The second maps the influence of these ways of life on individual and collective human life in order to explore the other two dimensions of sustainable development. The first step will be taken in only one of many possible ways. It is clearly very difficult to analyse the environmental implications of each and every activity that is part of a way of life of a modern society. The path that will therefore be followed is to group these activities and to examine the contribution such groups make to the load imposed on the biosphere. One way of doing this is to recognize that each way of life 'produces' and 'sustains' a certain population, a consumption of goods and services associated with unique networks of flows of matter and energy, a productive capacity making that consumption possible, and a certain eco-sensitivity reflecting the health of local ecosystems and the biosphere on which that way of life depends.

Grouping the activities that constitute the way of life of a society in this manner is useful for examining the biosphere-related dimension of sustainable development, but not for studying the other two dimensions. A different approach is adopted here that is too extensive to be explained fully in this volume.[1] The question of how our modern ways of life can sustain human life itself and the present life-milieu will be approached as follows: The way of life of a society may be conceptualized as a network of relationships expanded, contracted, or maintained through a variety of daily-life activities. Such a network, representing individual or collective existence, is established in time, space, and the social domain. It can take on many different forms, reflecting the different ways of human life in prehistory and history.

In a traditional village, human life can form a very compact network of relationships in space and the social domain. People live their whole lives in the same social group and in the same local habitat – the village and its immediate surroundings. Contrast this with human life in a modern mass society where the network of relationships extends over a large geographical area involving more people, thus covering a greater portion of the social fabric of a society and even beyond. This situation qualitatively changes the character of many relationships in the network. There is no longer the same time and energy devoted to most relationships, nor is there the attachment to primary groups, the local community, and the habitat. Such a loose network reflects a way of life with weak family and community structures and little identification with place.

How a culture designs its way of life in time also affects the network of relationships. In a culture dominated by performance values, a great deal of effort goes into improving the efficiency of each activity in order to save time, but time for what purpose? For the next activity? It, too, has been made more efficient. As a result, the density of activities in time increases, frequently reducing the quality of the experiences associated with these activities. This negatively affects the social selves of individuals, their relationships with others, and their attachment to a local habitat.

Regardless of the characteristics of the network, human relationships are lived in the context of a person's entire life so that they are enfolded into each other via a person's being and into the social groups and communities in which he or she participates. The relationships representing individual human lives together weave the fabric of a society. There exists a strong correlation between the characteristics and quality of the tissue of relationships of individual human lives and the fabric of a society representing collective human life. All of this raises the question of how well modern ways of life can sustain human life itself. This analysis, together with the one dealing with the biosphere-related dimension of sustainable development, can shed light on the sustainability of the modern life-milieu.

## 7.2 The Master Equation

To diagnostically explore the contributions of various groups of activities of a way of life to the overall harm done to local ecosystems and the biosphere, it is common to write equations of the following kind:[2]

$$I = PCTE$$

where

$I$ represents the total impact a society has on the ecosystems that help constitute its life-milieu

$P$ is the population size

$C$ represents the consumption or throughput of matter and energy per capita

$T$ is the load imposed on the ecosystem per unit of consumption

$E$ is the ecosensitivity expressed as a ratio of the damage inflicted on an ecosystem per unit load.

Although it is impossible to quantify these variables, the equation reproduced above is nevertheless useful for diagnostic purposes. The variables $P$, $C$, and $T$ should be regarded as a shorthand for a network of relations created by a society. The variable $E$ is a function of the structural characteristics of an ecosystem in its present state of being under considerable pressure from human activities. In other words, it also is in part affected by human activities. In this sense, they are all dependent variables created by reciprocally interacting wholes that help constitute an interrelated enfolded reality.

In this equation, 'population' refers to a cluster of activities significantly affecting the living of human life, such as birth, growing up, marrying, raising a family, and being retired. This also includes the movements of people through migration. Consumption refers to the cluster of daily-life activities involving means of one kind or another, including basic necessities such as food, water, clothing, and shelter. The technology cluster includes all daily-life activities that produce, maintain, and dispose of these means. During the last two centuries, economic growth largely based on industrialization has 'produced' three interdependent transitions from one situation to another that follow the typical rising S-curve: first, a demographic transition from populations with high birth and death rates to populations with low birth and death rates; second, a transition in consumption patterns from a low throughput to satisfy basic human needs related to a way of life and culture, to a high throughput based on meeting many non-material needs of a mass society through material means; third, a transition from high context-compatible, low-impact technologies primarily guided by context values to low context-compatible, high-impact technologies guided by performance values. The latter paved the way for a mode of technological development at arm's length from cultural development. A study of how industrialization transformed the structure of technology and how this in turn necessitated the restructuring of society is reported elsewhere.[3] I will therefore limit myself to the following findings of this analysis. The environmental crisis is not primarily the consequence of a population explosion. If that were the case, preventive approaches could do little to affect the present situation. Similarly, if the consumer revolution is seen as rooted in a materialistic human nature, then the present situation is unique only in so far as humanity has succeeded in creating new, powerful means to satisfy this materialistic instinct. Once again, preventive approaches would be able to accomplish little. If, on the other hand, the transformation of technological and economic growth is primarily associated with the ecology of technology dominating the economy of technology, then preventive approaches can do a great deal in directly changing this growth and through this change significantly affect population and consumption. However, it should be kept in mind that the fundamental changes in population, consumption, and technology during the past two centuries have been 'produced' by a complete transformation of societies and their dependencies on the biosphere.

## 7.3 The Technology Factor

In chapter 1, we advanced the hypothesis that the development of modern technology has been primarily guided by performance values and, to a much lesser extent, by context values. We provided detailed evidence of this in chapter 3 through a study of engineering education, and will now discuss additional evidence of this situation. With respect to the harm a society's technology inflicts on the biosphere, its most important characteristic is the environmental load imposed per unit of consumption. If everything consumed by a society were also produced by it, the factor in the equation (presented earlier in this chapter) would represent the environmental load imposed per unit of production. If the equation had been written for all of humanity instead of a particular society, this identity would be ensured in all cases. It would then represent the most important property of the technology of a society in contributing to the environmental crisis. If, in addition, it could be expressed as a single number, it could be used as a context value to assess and compare different technologies, or the same technology at various stages in its development according to how compatible they are with the biosphere. In the equation, technology is therefore a shorthand for its compatibility with the biosphere, which depends on complex networks of relations that structure the technology and embed it into the way of life of a society.

For the technology of a society, it is, of course, impossible to calculate the load on local ecosystems or the biosphere imposed per unit of production. Flows of pollutants cannot be added together because they are not created equal. Some produce much more harm to the health of ecosystems and the biosphere than do others. The fourth factor in the equation deals with this situation.

Despite these difficulties, it is nevertheless possible to obtain a reasonable picture of what is happening as a society develops its technology, by examining individual technologies one at a time and grouping the results in a meaningful fashion. The equation presented earlier in this chapter also applies to individual technologies. Assuming the identity of production and consumption, a negligibly small impact of changes in pollution on the ecosensitivity of local ecosystems over the time period considered, and that the load imposed on local ecosystems and the biosphere can be represented by the quantities of pollutants produced, the equation shows that variations in the impact correlate with changes in the product of population, consumption, and technology. In many cases, statistics are available from which changes in the quantities of pollutants produced, population size, and production or consumption of a particular good, associated with its given production technology, may be obtained or calculated. Suppose that at the end of the period, population size has increased by a factor of $p$ from its initial level of $P_1$, consumption of a particular good produced by the technology by $c$ from an initial level of $C_1$, and technology by a factor of $t$ from an initial level of $T_1$. The increase in the quantity of a particular pollutant produced by the production technology then equals:

$$POL_1 = (pct\text{-}1)P_1C_1T_1E$$

As a particular production technology undergoes development during the period under investigation, three possible trends may characterize its evolution. The first occurs when the consumption per capita of a good or service produced by the technology under question remains roughly constant over the period. The change in a particular type of pollution associated with the technology would then roughly correspond to changes in the population, had there been no changes in the technology itself. This is often the case for basic necessities such as food, clothing, and major household appliances in a modern society. A second trend results when the per-capita consumption of a good produced by the technology declines either in absolute terms or because it grows less rapidly than the population. The most common reason for the absolute or relative decline is that the good is being displaced by another serving the same purpose. In modern societies, this has been happening to the consumption of wool and cotton used in clothing as a result of their partial displacement by synthetic fibres. A third trend results when the consumption per capita steadily increases, because annual increases in consumption outstrip those of population. This situation can arise when one good displaces another, although their joint consumption may be constant, or when the consumption of a new good grows. Secondary consequences of these changes should also be included. Examples are synthetic fibres partially displacing wool, chlorine used in the fabrication of synthetic fibres, and, more recently, video-game machines. It follows that the technological cycles of two or more production technologies may be linked because the flow of goods associated with one becomes partly or completely displaced by that of another.

The approach outlined above was pioneered by Barry Commoner.[4] He is particularly interested in the contribution technological development is making to the environmental crisis. For the American economy from 1946 to 1968, the component of pollution increases attributable to technological development was, according to Commoner, larger than those attributable to population growth and increases in affluence. The displacement of natural fibres (cottons and wools) by synthetic fibres in clothing, the displacement of lumber by plastic, soap by detergent, steel by aluminum and cement, railway freight by truck freight, harvested acreage by fertilizer, and returnable bottles by non-returnable ones increased the load imposed on the biosphere by the U.S. economy. Other technological developments are derived from these. For example, the production of synthetic fibres involves organic chemicals and chlorine, and because chlorine is produced in a mercury electrolytic cell, the use of mercury increased. In addition, many of the newer technologies are much more energy-intensive, thus greatly adding to the burden imposed on the biosphere. The diffusion of consumer electronics, in particular radios and televisions, represents an increase in affluence. In sum, there was a pronounced shift away from natural materials, power-conservative processes, and reusable containers. Generally speaking, the newer technologies introduced since 1946 impose a greater load on the biosphere than those they displaced; Commoner concludes that this is the chief reason for the environmental crisis. Could it be that much of the new wealth produced during this period was gained at the expense of the health of U.S. society and its ecosystems?

We will briefly summarize a few of the details. In traditional agriculture, everything except the crop itself was returned to the soil, and the deficit was made up by nitrogen fixation. Modern agriculture reduces the humus in the soil, and seeks to replace the lost nitrogen by means of fertilizers. When the soil's ecosystem is accelerated, more food can be produced, but the impact on human health and the health of local ecosystems became a serious problem because an appreciable part of the added nitrogen does not enter the crop and appears elsewhere. Similarly, traditional range feeding of cattle returned much of what was grazed to the soil through manure. The move to feedlots changed all this, producing an additional burden. Whereas the energy required to produce cotton comes from sunlight, synthetic fibres require heat and electricity for their production. The gains in the durability of some of the new artificial materials became a liability at the time of disposal. Originally detergents were rich in phosphates that stimulated the overgrowth of algae, which released organic matter into the water when they died, thus overburdening the water ecosystem. The compression ratio of car engines increased dramatically over the period studied by Commoner. It required the addition of lead to gasoline and greatly increased nitrogen-oxide emissions. This by-product was first detected in Los Angeles in the early 1940s, but did not become a serious problem in most other cities until the early 1960s, when it had become a common urban pollutant. The building of suburbs increased the mileage travelled between home and work, worsening the problem. Pollution also increased when trucks began to haul an ever-greater proportion of freight, partially displacing the much more energy-efficient railways. Had context values played a more significant role in technological development, the picture would have undoubtedly been very different. The success of pollution prevention and similar preventive approaches confirms that a great deal of the pollution produced by a modern economy is unnecessary, inefficient, and unprofitable.

Commoner[5] also concludes that the way in which increased pollution is being dealt with is costly and ineffective. The only significant reductions achieved were those where no end-of-pipe approaches were used, but where the production or use of the material was simply banned. Substantial reductions in lead-poisoning, particularly in children, DDT in body fat, PCBs in body fat, mercury in Great Lakes sediments, strontium-90 in milk, and phosphates in surface waters are some of the better-known examples. In reviewing the evidence, Commoner concludes that pollution is almost always an incurable disease that can only be prevented. It is time we began to recognize that environmental harm done by technologies is as much a part of their design as the useful functions performed. High-compression car engines produce more power and smog. Fertilizers produce a greater agricultural output and more pollution. A trash-burning incinerator produces energy and dioxins. The design, management, and control of modern technology is, therefore, in urgent need of more effective negative feedback so as to reduce or prevent altogether the burden imposed on the health of human beings and the biosphere. The evidence is clear. Pollution has to be attacked at its source. Once it is produced, the chances of successfully controlling it tend to be very slim. Despite this evidence, the main legal instrument continues to be pollution-control laws, which do not deal with its sources. This glaring oversight was obscured by the misconception that increased pollution was essentially the result of growing populations and rising levels of affluence. The technologies of production were associated only with economic growth. As a result, pollution was dealt with in terms of its symptoms, and the results have been pretty much the same as if a doctor attempted to treat German measles by using medication to suppress the red spots on the patient's skin. Pollution prevention is now the official policy of the United States, but professional education, the intellectual division of labour in the university, the organization of corporations and government departments continue to be the cause and effect of end-of-pipe approaches.

Is it possible to generalize the findings of Commoner to other nations and time periods? The answer is negative because an advanced industrial economy is very different from a developing one, and the time period studied by Commoner is one of rapid technological change. This is typical only for the industrially developed and some developing nations. Paul Harrison[6] has drawn a more comprehensive picture by considering different levels of industrial development and a longer time period. As might be expected, such a picture shows other possibilities, including ones where population and consumption contribute the largest component towards the load imposed on the biosphere and those where technological development leads to a reduction of this load. Unfortunately, the presentation of the data is less clear than it might be. For example, when some factors contribute to an upward pressure and others to a downward pressure on the load a society imposes on the biosphere, each factor is scored either out of +100 per cent for the upward pressure or out of –100 per cent for the downward pressure. When the results are tabulated in this manner, their relative contribution cannot be assessed. Harrison acknowledges that the fundamental approach assumes a linearized system with simple causality in which population, consumption, and technology do not affect one another. Despite these limitations, it provides us with some useful approximations.

Some general trends are worth noting. In developing societies, population tends to outweigh the other two factors combined for basic needs such as food, energy, and land, provided the technologies involved change only slowly. In the industrially developed societies, on the contrary, population tends to change much more slowly so that the other two factors tend to be more significant. More specifically, population tends to be the dominant factor or a very significant one in deforestation, species loss, irrigation (including the associated phenomena of soil salination, waterlogging, and methane production), increases in livestock (including the associated phenomena of overgrazing and methane production), and carbon-dioxide emissions. It should also be noted that population and consumption increases can drive technological development. Changes in consumption may lead to changes in population, and vice versa. Increases in pollution and their accompanying effects on human life, local ecosystems, and the biosphere can lead to technological changes. This once again confirms that population, consumption, and technology are not independent variables.

From this brief overview of the technology factor, it becomes evident once again that, particularly for the industrially developed societies, preventive approaches in the engineering, management, and regulation of modern technology are essential for reducing the harm done to the biosphere. Given the enormous demographic pressures on developing societies, they need all the help they can get. The consumption of basic goods has to increase as well, and the technology factor is the one that can help reduce the environmental crisis in the short term.

### 7.4 The Eco-sensitivity Factor

We defined 'eco-sensitivity' as the damage inflicted on an ecosystem or the biosphere per unit load imposed on it by the way of life of a society. It represents the consequences, for local ecosystems and the biosphere, of human population, consumption, and technology. An interpretation of what this means brings us face to face with the limits of our understanding when attempting to answer questions such as: What is an ecosystem? What is the biosphere? How do we know the extent to which human activities have impaired the health of these wholes, and are we heading for a major breakdown or collapse? How much of a load can we impose on local ecosystems or the biosphere without significantly impairing their health? What are the vital signs of health or illness of these wholes? As the self-regulating character of ecosystems and the biosphere becomes impaired as a result of human activities, is it likely that we can compensate by 'managing' these wholes? Given the limitations of the scientific approach and the absence of a science of the sciences, will it ever be possible to have a sufficient understanding of ecosystems and the biosphere to find adequate answers to the questions listed above? If such understanding far exceeds what scientific knowledge can deliver, what constitutes prudent and responsible behaviour? Can our ignorance be put to good use in shaping public policy, especially with respect to scientific, technological, and economic development, to ensure our common future? Once again, we are confronted with the fact that modern science has been extraordinarily successful in generating knowledge of one kind and ignorance of another kind, so that our challenge is to avoid becoming victims of our own scientific blind spots.

On a limited scale, we have coped rather well with ignorance. For example, engineers deal successfully with this problem through safety factors. Responses to the challenge of the environmental crisis should incorporate an ignorance factor when we propose concrete solutions. Such 'global safety factors' would create useful ignorance as the power of our means and the uncertainty of the consequences of their use continue to grow. Modern cultures are not well equipped to deal with the situation because performance values, which are values of power, have become prominent in their value hierarchy, inverting the relative importance of information, knowledge, understanding, and wisdom. There is no question that a modern society has much more knowledge at its disposal than any other society in human history. However, this is hardly reassuring, given our equally unprecedented ability to affect ourselves and the entire planet. The adequacy of our knowledge base must be assessed relative to our way of life, our life-milieu, and the potential implications for the future, and not in relation to an earlier society whose way of life required less knowledge. Furthermore, the quality of our knowledge in relation to our present situation must be considered. Scientifically knowing things in a much simpler context is of limited value in helping humanity create a more sustainable way of life in which everything, including technology, is much more compatible with everything else. This will require context-rich knowledge. The commonplace that 'knowledge is power' is, after all, the flip side of power being contextually impotent.

Our ability to proceed with preventive approaches for the engineering, management, and regulation of modern technology does not depend on having answers to the kinds of questions with which we began this section. However, our motivation to do so will greatly depend on this kind of knowledge. In order to significantly reduce or prevent altogether the damage to the biosphere caused by our modern way of life, we need the introduction of negative feedback controlled by a 'set point' of 'the less harm, the better,' which is the opposite of and complement to performance value–based optimization. Answers to the deeper questions are crucial when assessing whether an all-out preventive effort will be a sufficient response to the environmental crisis, or whether it is simply a necessary but not sufficient condition for dealing with it. In the present work, we are concerned with the art of the possible through preventive approaches, and we will leave the larger questions to a future volume.[7]

In a preventive strategy, the fourth factor in the master equation – namely, eco-sensitivity – above all symbolizes useful ignorance. John Passmore has put it very well:

> No doubt, the modern West has more knowledge at its disposal than has had any previous society. But it has neither the kind nor the degree of knowledge which it now needs. Our knowledge of the ways in which societies work is not at all comparable in extent with our knowledge of the workings, let us say, of the planetary system; we know little about the atmosphere, the seas, the life cycles of the scarcely numerable varieties of plant and animal life. While absolutely speaking we know a great deal, we are proportionally ignorant, proportionally in relation to what we need to know. A farmer tilling his fields by conventional methods and living in a traditional society did not need to know a great deal. But, the task we have now to undertake in attempting to estimate the long-term effects of our actions both on the biosphere and on human societies, is so immense that in relation to it our ignorance is almost total.[8]

## 7.5 Calculating the Ecological Footprint

Equations similar to the one on which this chapter is founded can be written to calculate the actual ecological footprint as the 'quantity of biosphere' currently used by a population with particular consumption habits and technology. This may then be compared to the sustainable ecological footprint of that population, which is the 'quantity of biosphere' required to sustain that society indefinitely. The extent to which the former exceeds the latter represents the level of overexploitation and thus non-sustainability. Despite the complexity of the calculations and the large number of assumptions that have to be made to carry them out, this approach is nevertheless useful in obtaining a qualitative sense of how unsustainable our modern societies are.[9]

### 7.6 Differentiating a Way of Life

Thus far, individual and collective human life have been examined through groups of activities that bear directly on the environmental crisis. These can directly or indirectly benefit from preventive approaches for the engineering, management, and regulation of modern technology in order to make modern ways of life more sustainable. As in the previous chapter, the results may be represented in a tree-like structure that differentiates the living of human life into those dimensions that are most significant for sustainable development. This tree may be described as follows.

On the first level, we encounter the umbrella concept of the living of human life (individual and collective). On the second level, this may be differentiated into four dimensions – namely, population, consumption, technology, and eco-sensitivity. On the third level, each of these clusters of human activities may be differentiated into a common spectrum according to the roles played by: food, drinking water, air, clothing, shelter, habitat, employment, education, health care, recreation, security and protection, social support, cultural support, and religious support services. These activities reciprocally interact in the lives of human individuals and in the way of life of a society, and may be seen from the perspectives of population, consumption, technology, and eco-sensitivity. Each smaller cluster of human activities has a metabolic and a sociocultural dimension, and an effect on the biosphere via local ecosystems.

It must be borne in mind that this differentiation of our modern way of life is designed for considering the biosphere dimension of sustainable development. It is not appropriate when considering the other two dimensions – namely, the sustainability of the life-milieu and that of human life itself. The reason for this is that the emergence of the phenomenon of rationality around the turn of the twentieth century, and its subsequent development into the phenomenon of technique during the last half of that century, has led to performance values dominating context values in all areas of modern life. It is therefore not possible to identify groups of activities that are particularly relevant for examining the dimensions of sustainable development related to our life-milieu and human life itself. Instead, it is possible to point to various categories of context issues to which most activities making up modern ways of life contribute. Many rationalized and technicized activities produce various kinds of undesired effects that synergistically interact within a way of life to produce fundamental changes in individual and collective human life. These can be grouped into categories of context issues. I will shortly show how modern ways of life can be differentiated into several categories of such context issues, to which most rationalized and technicized activities contribute and which are a concern for human sustainability. These profoundly change the life-milieu, which in turn affects human consciousness and cultures.

The damage technological and economic growth continues to do to human life and society is probably as serious as what it does to the biosphere. Both problems stem from the same root: behaving as if technology is connected to little else, so that all society needs to be concerned about are its inputs and outputs. Consequently, when measured in terms of performance values, the technological and economic 'system' at the heart of modern civilization has succeeded beyond anyone's expectations. It is capable of providing the members of some societies with material support in excess of 1 million pounds per person while three-quarters of this is disposed of as waste within half a year. This feat becomes less remarkable when measuring the ratio of waste to useful product, which can exceed 50:1. The system has an equally important flaw with respect to human life. While making resources too cheap, it simultaneously has made human beings too expensive. The impact on each new generation growing up under these circumstances cannot be exaggerated. Many young people realize that they are neither wanted nor needed and react in a variety of ways that societies by and large fail to understand. Rather than hearing the anxiety and despair, many believe that bigger and better jails are the answer. The system is as wasteful of human beings as it is of resources. As a consequence, it is as unsustainable with respect to human life and society as it is with respect to the biosphere. It is to this aspect of sustainable development that I briefly turn within the scope of the present analysis.

## 7.7 Context Issues

Culture represents an approach to the world which attributes a name, meaning, and value to everything a human being experiences in relation to everything else. The meaning and value, respectively, represent the place and significance of something in individual and collective human life. They reinforce the possibility of living human lives, as opposed to living sequences of weakly interrelated moments of existence. Culture thus represents a way of making sense of and living in the world based on making the maximum use of context.

The culture-based approach to the world stands in sharp contrast to the one based on science and technique, representing approaches for knowing and dealing with the world that make minimal use of context. As previously argued, in order to study something, science abstracts it out of the unintelligible complexity of reality to examine it in the more limited and manageable intellectual context of a discipline and, wherever possible, in the much less complex physical context of a laboratory. Hence, science knows things out of their usual context. Technique may be regarded as a phenomenon that evolved out of the rationality observed by Max Weber around the turn of the twentieth century. It is defined by Jacques Ellul as the ensemble of means rationally arrived at to obtain the greatest possible efficiency in all spheres of life.[10] It is a phenomenon that implies a greatly diminished role for culture. The technical means are not arrived at through a tradition shaped by experience and culture, the way traditional technologies were. They are aimed at maximum efficiency, which means obtaining the greatest possible desired output from requisite inputs with little or no consideration of the meaning and value of such inputs and outputs for human life, society and the biosphere. Obtaining the highest possible efficiency or other performance value has nothing in common with the satisfaction of human needs and aspirations arising from a culture, which traditional technologies sought to satisfy. In sum, modern science and technique develop with minimal reference to context. The results are obvious. Traditional technologies were context appropriate, while modern societies need to be concerned with sustainable development.

What has been almost entirely overlooked in discussions about the nature of modern societies and contemporary civilization is the fact that our modern way of life considerably reduces the role of culture, in general, and the use of context in human knowing, doing, and controlling, in particular. Contemporary civilization is preoccupied with making everything 'better' by means of what I call the 'technical' approach to reality. In chapter 3, I began to show that it proceeds as follows. First, an area of human life is examined for the purpose of making it 'better.' The findings of the study are used in the next stage to build some kind of model. In the third stage, this model is examined to determine how it can be improved by changing its parameters. Finally, the results are used to reorganize the area originally studied to achieve the desired improvements. By means of this pattern of events, modern societies seek to improve activities as diverse as the production in a plant, the running of a large office or hospital, the effectiveness of classroom instruction, the performance of a professional athlete or hockey team, the functioning of a group, and even the satisfaction derived from a sexual relationship.

This technical approach has an enormous influence on human life, society, and the biosphere because 'better' is defined primarily in terms of performance values. These input–output ratios are supposed to ensure the effective use of scarce resources, but, on the level of the system as a whole, they have created serious problems. We have seen that the main cause is largely the fact that they are entirely mute on the question of whether any gains in output derived from certain inputs are partially or wholly achieved by degrading the human, societal, or biospherical contexts of a particular technology, process, or undertaking. The predominance of performance values over context values within modern cultures leads to technological and economic growth being valued on their own terms rather than in relation to human life and society. The response to growing incompatibilities between technology and its contexts has been the building of higher regulatory fences around it to avoid the most damaging and unacceptable consequences. However, this after-the-fact and end-of-pipe approach is proving increasingly ineffective and expensive. It is for these and other reasons that the technical approach to life is a form of human behaviour making minimal use of context, so that 'bumping' into human life, society, and the biosphere is an inherent feature of the structure of modern civilization.

The technical approach to life is not culturally neutral because culture as symbolization is based on the process of contextualization. I will explore several implications that give rise to four categories of context issues. From the description provided earlier, it follows that an area of human life is studied not holistically, but for a specific purpose. In relation to that purpose certain aspects of the situation will be considered central, and others as peripheral; still others will be externalized as having little or no relevance to the achievement of the goal. As a result, the integrality of the human activity being studied is represented in a modified and frequently distorted form by the model.

When the findings are used to improve the human activity being studied, its integrality is usually weakened and distorted as well. This creates a first category of context issues associated with a distortion of the internal context of any whole.

Whatever the technical approach to reality is applied to is first abstracted from its human, societal, and biospherical contexts. It is then improved on the basis of performance values, which, as I have shown, make no reference to the way it fitted into and will fit into these contexts. This often gives rise to a second category of context issues related to a reduced compatibility between the whole and the context formed by the larger wholes to which it contributes.

A third implication of the technical approach to life is that it separates knowing from doing, and the knowers from the doers, thus requiring an external control over a technicized activity. Assembly-line work is an obvious case in point. As the diversity and interdependence of technicized activities grow in the cultural fabric of a society, control over the networks of these activities tends to be centralized in large institutions such as the transnationals and the modern state. The latter must also control the many externalities produced by contextless development. The rise of the technical phenomenon based on a contextless scientific and technical rationality creates the need for the separate management and regulation of a growing range of activities that used to be self-regulating. The separation of the regulatory and control functions from the activities themselves makes this regulatory apparatus much less effective than the traditional ones, which were built into these activities. It is also much more costly to run and maintain. This gives rise to a third category of context issues related to a weakening of the self-regulating character of living wholes.

When techniques permeate the way of life of a society, it is clear that the context issues discussed above are multiplied within the fabric of human lives, society, local ecosystems, and the biosphere, effecting fundamental changes. For example, the separation of the knowers from the doers and the externalization of control have many significant consequences. Human life in a technicized society is made up of some activities in which a person is a 'knower,' and a great many other activities in which he or she is a 'doer' largely controlled by others. To some extent this has always been the case, but the rise of the technical phenomenon has tipped the balance against self-regulated activities. The result has been a fundamental insecurity about activities that human beings have successfully engaged in for thousands of years. A person is not to rely on his or her own experience but on expert knowers to a degree detrimental to the human individual. There is an excessive dependence on 'how to' books for the bringing-up of children, for communicating with one another, for making love, and for a host of other human activities. Too many activities are regulated and interconnected by the complex policies of large institutions – policies that are neither intelligible nor accessible to a great many people. The result is an erosion of meaning, the displacement of a traditional with a technical morality, and a sense of no one being morally responsible. When people face problems such as a complex set of allergies, the inability to find work, a vague sense of being depressed most of the time, a drug problem, or a tendency to resort to violence, they often have difficulties finding the source of their problem and the responsible party. The common result is a sense of helplessness, anxiety, frustration, or discontent. The consequences of this reification and fragmentation of human life are immense, undermining, as we will see, the very possibility of a genuine human civilization. Similarly, the undermining of the internal integrality and the context-compatibility of many wholes in a society has transformed traditional societies into mass societies, with profound consequences for individual and collective human life. These may be assessed in terms of context values, including health, quality of life, freedom, and well-being. The transformation of a way of life by technique has also immensely reduced the compatibility with the biosphere, producing the environmental crisis. The fourth category of context issues is therefore related to the sustainability of individual and collective human life, and of the biosphere.

The question of human sustainability arises from the influence technique has on modern cultures as the basis for giving everything in human life a meaning and value. Human and social development guided by performance values, and to a much lesser extent by context values, is ultimately based on the technical approach to life having been declared good in itself by means of absolutization. This leads to a number of widely acknowledged phenomena, such as the absence of stable values, a 'statistical' morality converting the normal into the normative, and weakly integrated personalities. Culture continues to assign a meaning and value to everything according to its place in individual and collective life relative to everything else. However, a decontextualization and fragmentation of wholes resulting from the technical approach to life weakens the relational character of individual and collective life, and thus reduces the depth of meaning as well as the value of everything. This leads to a process of desymbolization. In other words, the process of absolutizing the technical approach to life as objective and rational, declaring it good in itself, continues to convert the unknown into extrapolations of the known, rendering all alternatives fundamentally unthinkable, unimaginable, and, in any case, unliveable. At the same time, the cultural unity that gives meaning, purpose, and direction to the lives of millions of people is weakened through desymbolization. This has been most powerfully portrayed by modern art, which for a time became abstract, signalling the end of the human subject. The process of so-called secularization would be more accurately described as desymbolization and the creation of 'new secular gods.'

This sketch of the technical approach to life shows that attempting to make everything 'better' as measured in terms of performance values constitutes a form of technical rationality that limits the consideration of context to such an extent that it may be likened to tunnel vision. While preventive approaches are designed to increase the 'field of vision' of the engineers, managers, and regulators of modern technology, and of the institutions that employ them, it does not get to the heart of the problem, rooted in the nature of contemporary culture. However, it might trigger other developments, which in turn may trigger other ones, to produce a mutation of the system as a whole.

## 7.8 Economic Context Issues

There is a growing awareness of one particular class of context issues – namely, those resulting from the application of economic techniques. These are both cause and effect of the separation of the economy of technology from the ecology of technology. They have their intellectual roots in the sciences, in general, and economic science, in particular. Other kinds of context issues are discussed in subsequent chapters.

As previously argued, economics, like any other scientific discipline, does not specify the 'boundary conditions' where its subject matter meets those of sociology, political science, cultural anthropology, history, ecology, and others, so that they can jointly account for human reality. The resulting autonomy of scientific disciplines deflects critical attention away from whatever assumptions may be implied in the boundaries of the subject matter. It helps to maintain the labyrinth of technology since much of our understanding of the economy of technology derives from economics. Of particular concern is how modern societies see the market as having almost omnipotent powers. The autonomy of economics has led to wonderfully sophisticated theories and models resting on the most unrealistic assumptions. There are a number of 'economic heretics' who have warned society against the dangers of relying too heavily on economic theory, but their voices are being drowned by a wave of economic fundamentalism that has been sweeping the world, undermining traditional political differences. However, it is not true that the economic theory that guided earlier generations is good enough for us. It is endangering the possibility of a humane and sustainable future. I briefly take aim at this fundamentalism, leaving a detailed critique of contemporary economics to others.

The running of a modern industrial economy depends on the use of economic techniques established according to the patterns described in the previous section. For example, extensive economic statistics are compiled by governments into input–output models that can be used to simulate the consequences of various economic policies to ensure optimal results. As with other techniques, the first three stages of the technical approach create theoretical externalities that are translated into context issues during stage 4. Leontief, a Nobel prize winner in economics, expressed his concern over input–output models as follows:

> Page after page of professional economic journals are filled with mathematics formulas leading the reader from sets of more or less plausible but entirely arbitrary assumptions to precisely stated but irrelevant theoretical conclusions ... econometricians fit algebraic functions of all possible shapes to essentially the same sets of data without being able to advance, in any perceptible way, a systematic understanding of the structure and operations of a real economic system.[11]

Daly and Cobb[12] have systematically examined the inadequacy of the assumptions underlying a great deal of modern economics, given the changes that have taken place in the latter half of the twentieth century. Such efforts have given rise to the birth of an alternative approach to economics called 'ecological economics.'[13]

Economic techniques extract a particular domain from reality by means of a process of abstraction. As a result of this knowledge-gaining strategy, economics and its application through techniques isolates its field of study from those of all other disciplines. Each one is largely out of context with all the others. This is scientifically acceptable if, and only if, the phenomena examined by the other disciplines have a negligibly small effect on the phenomena within the domain of the discipline in question. If this is not the case, then the knowledge developed by at least some disciplines will constitute theoretical externalities for the discipline in question. The limitations of modern economics have had a much greater impact on human life and society than the application of any other social science because of the extensive use modern societies make of economics.

Economic theory and technique are increasingly on shaky ground because globalization is leading to historically unprecedented levels of interdependence within societies, between societies, and between them and the biosphere. This development is also forging a greater awareness of how everything depends on everything else. The consequence for many scientific disciplines is that 'neighbouring' disciplines will examine phenomena that present them with a greater number of theoretical externalities that can no longer be neglected. The risk is that science will become increasingly impotent in producing a coherent body of knowledge that is meaningful, relevant, and helpful to humanity in its efforts to deal with a variety of concerns, including the stresses on human life, society, and the biosphere.

There is no doubt that, in the late nineteenth and early twentieth centuries, the scale of human economies with respect to the biosphere was still small. Hence, economists could neglect the influence of the economy on the biosphere, and vice versa. However, its interaction with the human community was put beyond the horizon of economics by means of implicit and explicit assumptions that, even in those days, did not withstand careful scrutiny. The theoretical externalities in modern economics have now become so vast that their collective significance rivals, and possibly exceeds, those of the phenomena studied by economists.[14]

An example is furnished by free trade. When the mobility of financial capital was largely contained within national boundaries, the doctrine of free trade made sense. Each nation would specialize in producing those goods and services that, for reasons of geography, climate, or culture, it could produce with comparative advantage so that all would benefit from trade. This would ensure the highest quality of goods and services at the best possible price for everyone. Jobs would not be lost because the decline of economic activities in one sphere resulting from free trade would cause investment capital to be reassigned to another sphere within a nation. However, with the growing global mobility of capital, the vitality and quality of life in communities, regions, and entire nations can no longer be treated as negligible theoretical externalities. The failure to incorporate adequate social and environmental standards into free-trade agreements has unleashed a powerful force that is dragging individual and collective human life towards a lower common denominator. Weaker communities, groups, and nations can do little to protect themselves and, from their perspective in particular, free trade really means forced trade. Instead, we could get ahead of our end-of-pipe attempts to deal with our economic problems by creating high social and environmental standards that are realistic, given the potential of preventive approaches.

Externalities in economic theory present significant obstacles to preventive approaches for the engineering, management, and regulation of modern technology. It is therefore useful to briefly examine the origins of some of the major theoretical externalities, including their growth and translation into context issues during the process of economic development.

Some of the foundations of modern economic thinking were laid by Adam Smith.[15] He wrote at a time when Western Europe had begun a fundamental transformation. Societies had always created an order, including a way of life and institutional framework, through their cultures. Adam Smith suggested that a new order was beginning to emerge based on an entirely different principle, putting the economy centre stage. The medieval order had long broken down and this new order was gradually taking its place. Smith suggested that, through the 'invisible hand' of the market, self-interested behaviour related to production and consumption was contributing to a new economic order that could constitute the backbone and foundation of the order of the entire society. The situation observed by Smith was one in which there were relatively few cultural constraints on individual self-interested behaviour because no vital cultural order could be opposed to it. Smith argued that this was not to be feared because out of the apparent chaos would emerge a new order that would ensure the greatest good for the greatest number of people. The trouble with this reassuring vision proposed by Adam Smith was that no community had ever survived without imposing severe restrictions on self-interested behaviour by means of its way of life, institutions, and other elements of its culture. It is true that, during his days, the contribution of these cultural elements to individual and collective life in England continued to be weakened by a growing reliance on the market necessitated by industrialization. To take this as evidence of their non-essential character for the order being established by the invisible hand of the market was to overlook many related problems of that time. It constituted a serious intellectual error, with significant practical consequences. Some of these are apparent in Smith's work. In a later chapter, I examine why Smith recognized that a growing technical division of labour would increase the wealth of nations but, as he acknowledged, would also make human beings as stupid as they could possibly become, in which case one wonders what the wealth was for.

All that this reassuring economic vision retains of human life, society, and culture is *homo oeconomicus*. This places the maximization of utility at the centre of human life, neglecting almost everything else. Wage earners spend their income to obtain the maximum possible utility from their wages, thus creating demands for various goods and services that are transmitted through markets to producers. Producers seek to maximize profits by optimizing the mixture of inputs or factors of production to meet this demand with the highest quality and lowest cost. As long as no producer is large enough to have a significant influence on any markets for goods and services, economic democracy will be ensured. What the consumer desires is produced for the lowest possible price, the highest possible quality, and the maximum possible profit for the producer. Hence, out of an anarchical chaos of unrestricted self-interested behaviour emerges an order creating the greatest good for the greatest number of people, according to this economic vision. An important task for economists is to study the performance of markets and to recommend to society how their efficiency may be kept optimal. For this new economic order to function, everything had to be brought within markets, thus producing what Polanyi[16] has called the 'great transformation.' Traditional societies frequently had markets for the exchange of goods and services, but these tended to be local, small in scale, and never regulated the entire economic order of a society. The market now emerged as the regulating principle for the economic order of society.

The economic view of human life and society implicitly and explicitly emphasizes performance values over context values. This became gradually more explicit as it evolved into a self-fulfilling prophecy, because the process of industrialization increasingly necessitated that economic practice be informed by economic theory. However, the theory could only be verified through accompanying economic practice. This led to the neoclassical theory in economics, and to the socio-economic orders of industrial societies. To this day, differences across the political spectrum can largely be characterized in terms of the extent to which this economic vision should be adhered to.

As the economic vision was progressively translated into society, the previously outlined four categories of context issues began to manifest themselves. The psychological make-up of the individual, the role of the economy within society, and the relationship between society and the biosphere underwent fundamental changes that had a mirror image in the theoretical externalities of economic science. How significant are the implications of neglecting the context of *homo oeconomicus* – namely, people's lives? How significant can the influence of individual market transactions between two parties be for other individuals, communities, societies, or the biosphere? What is the influence of the market as an organizing principle for human life and society? If all of this is of little significance, economic science has provided us with a valuable abstraction for guiding economic practice. It has then made an unintelligible complexity intelligible. If, on the contrary, many of these implications are as important, or even more important, than the phenomena examined by economists, it has done society an enormous disservice.

For the market to function as the organizing principle of a society, it must be possible to exchange anything in the market. On the supply side, everything has to become a commodity separated from its contexts and reduced to a monetary equivalent. Land, before it can be bought and sold, can no longer be seen as an integral part of a functioning ecosystem that helps to support life. It becomes regarded as a piece of terrain that can be described in a deed. Similarly, the totality of a human life is reduced to labour power that can be traded as a human resource like any other resource. The patrimony we inherit from previous generations is reduced to capital and regarded as a source of unearned income. We have become so accustomed to these reductionisms that they appear self-evident. However, before the 'great transformation,'[17] traditional societies would have regarded this as incomprehensible, as is evident, for example, from the reactions of Native people to European settlers in the Americas with respect to land ownership and land claims.

The market cannot regulate public goods (not privately owned) because their use by one party does not exclude use by others. Examples include scientific and technological knowledge, lighthouses, national defence, the air we breathe, and the values we live by. The market does not regulate spill-over effects from any transaction on parties that are not involved. Market externalities are the costs of production and the consumption of goods and services not borne by the two parties involved in the market transaction. They are external in the sense that they represent the consequences of that transaction for other individuals, society, or the biosphere, resulting in social costs not included in the private costs of the transaction. Externalities also occur in the use of common resources and public goods, and transactions affecting future generations. If externalities are minor in terms of their consequences to human life, society, and the biosphere, the regulation of the market is very effective for the distribution of scarce resources. However, when these externalities begin to affect the quality of human life, the viability of communities, or the capacity of the biosphere to support all life, the operations of the market will have to be improved since central planning is not an acceptable alternative in democratic societies. It is obvious that the situation cannot fundamentally change by ensuring that the market functions as efficiently as possible. Except for the ideologues, it was obvious from the beginning that the market could never be entirely free. Governments had to constrain the market by, for example, passing laws against child labour, and imposing occupational health and safety standards for workers, safety and other standards for products, environmental protection, and taxation. Although these place constaints on individual economic behaviour, they essentially maintain a level playing field and therefore do not challenge the operations of the market.

There remains a fundamental obstacle to the market's serving human ends, rather than vice versa, as a result of what Jacobs[18] has called 'market forces.' A 'market force' occurs when the individual decisions of producers and consumers in specific markets are integrated by the system to yield 'collective decisions' that cannot be derived from individual decisions. The consequences can be unintended and may be positive or negative. One example is the 'invisible hand' of Adam Smith that converts individual self-interested behaviour into the greatest possible collective good. Another example is what Jacobs has called the 'invisible elbow,' which, as at the dinner table, can knock things over as the 'invisible hand' is reaching for market efficiency. One such consequence is the environmental crisis. A parallel effect on human life and society is also occurring. The premise of modern economics that the greatest possible prosperity produced by an economy from a mix of inputs occurs when the market functions optimally is highly questionable when the externalities of this theory neglect the fundamental dependence of the economy on human communities and the biosphere. The very scope and significance of these 'theoretical externalities' cry out for a rethinking of modern economics that would internalize these dependencies into the theory so that economic practice, now better informed, would not endlessly heap up market externalities.

On the demand side of the market, as an organizing principle for society, equally serious theoretical externalities occur. There exists a considerable gap between the demands on markets for goods and services, and the meeting of human needs and wants. First of all, what are considered needs have varied enormously from one society to another and from one social stratum to another. Human needs are not insatiable, as assumed in the theory of exchange value – they are socially influenced and constrained by cultural values. Needs are insatiable only if the abstraction of *homo oeconomicus* theoretically externalizes cultural influences on individual behaviour as well as all individual virtues and vices other than the maximization of utility. Thus needs cannot be moderated by concerns for others, one's community, or nature. Neither can needs be shaped by advertising, which becomes reduced to neutral information rather than persuasion. The maximization of utility externalizes virtually everything that has characterized human life prior to the emergence of the economic vision of society. Everyone's life is seen as resembling that of Robinson Crusoe: relations with others occur, but solely for the purpose of engaging in market transactions. Social welfare increases when individual welfare goes up, but all individual utility is the same whether it is gained from purchasing a second vehicle or from obtaining some food when one is starving. It externalizes the reality that human beings derive most of their pleasures through relationships with others.

Whatever the source of human needs, they cannot be equated with economic demands without significant theoretical externalities. Translating a need into a demand requires financial resources. Those who have plenty can therefore easily outbid those who have little or nothing at all. The demands of wealthy nations for scarce resources can be made to count more in international markets than can those of poor nations. In other words, equating needs with demands creates theoretical externalities that include phenomena examined by sociology, political science, law, and religious studies.

On the supply side of the market, all producers are not equal either, except if significant theoretical externalities are imposed. There is a well-recognized tendency in the market for the elimination of competition. Producers that have been very successful in the recent past have more resources than others and are therefore more likely to continue to be winners. Galbraith[19] has shown that modern transnationals, having grown very large with respect to the markets for the goods and services they supply, constitute a different economic system than one made up of smaller firms. These global giants are islands of central planning dominated by a full range of techniques that fundamentally change the structure of the firm, its goals, and its interactions with the market. This point has been explored elsewhere[20] and only a few features are highlighted here. On the input side of these giants, the market is largely bypassed through vertical integration and long-term contracts with suppliers. On the output side, there is a parallel reduction of dependence on markets. Consumers are to a significant extent incorporated into the plans because their needs are influenced by advertising. This challenges the fundamental assumptions of modern economics: to the extent that producers can influence consumers, economic democracy is undermined and the market is less free. The behaviour of producers in the economic sector made up of transnationals is largely based on planning and, to a much lesser extent, on the market. In the economic sector constituted by small and medium-sized producers, traditional economic assumptions are a much better approximation of what is happening.

According to economic theory, the market essentially regulates a circular system that excludes the dependence of the economy on communities, societies, and the biosphere. The outputs of producers (goods, services, and wages paid to their workers) are the inputs to consumers who turn their wages into outputs in the form of payments for goods and services purchased. This simple representation in essence illustrates the problem: the economy is essentially depicted as a perpetual-motion machine in which flows continue unabated without any external inputs from the biosphere and society. Yet these constitute the very basis of an economy. All life merely becomes a theoretical externality in modern economics. Thus, an optimal market allocation of resources is simply an economic optimum. An efficient distribution does not create a sustainable way of life, enhance the quality of life, or create a just order. The value choice that underlies modern economics is that of performance values over context values. As a result, society excels in the domain of the former and fails in the domain of the latter. Modern economics essentially describes the market as an optimization scheme of utility, profit, market efficiency, and the greatest common good for the greatest numbers of people. Future generations are drawn into this theoretical scheme through discounting. Resource depletion is non-existent because the market, thanks to modern technology, will make one form of capital substitutable for others.

In sum, the market is a contextless regulator of society. Everything is reduced to its economic dimension, while economic theory externalizes all the others. This economic view of human life, society, and the biosphere is itself a significant theoretical externality in that it holds that no restrictions should be placed on individual enterprise and markets. This has a corrosive influence on the human community. No traditional culture ever valued or condoned self-interested behaviour to this extent. All this has created a gap between the standard of living and the quality of human life and between growth and development. The market, as a regulator based on performance values, inevitably weakens all the contexts of the economic sphere.

As far as economic policy is concerned, given the magnitude and importance of the theoretical externalities in modern economics, our situation appears gloomy. An all-out strategy for implementing preventive approaches for the engineering, management, and regulation of modern technology will eventually run head-on into economic theory and practice. However, preventive approaches, contrary to all conventional expectations, are proving to be economical as well as capable of creating win–win situations for all parties. Their superiority over non-preventive approaches would be much starker if economic theory and practice incorporated many of the above-mentioned theoretical externalities. In the meantime, our understanding of economic activities out of their contexts of human life, society, and the biosphere continues to affect society through economic practice based on this theory. One of the examples I have used is the extensive use of the Gross Domestic Product (GDP), which measures the value of goods and services produced from a variety of inputs. Striving for growth as measured by this indicator is to strive for importing the theoretical externalities of economic science into the structure of society. Economic growth leverages the development (or lack thereof) of all other spheres of society, and this is ultimately justified in the economic choice of performance values over context values. On the bright side, many of these problems are becoming widely recognized. Alternatives to the GDP are being proposed that begin to include a number of context issues into the indicator. The non-sustainable and sustainable components of GDP are distinguished from each other, as well as those that genuinely enhance human well-being from those that represent the costs incurred. All these developments will further demonstrate the viability of preventive approaches in the economic sphere of human life and beyond.

## 7.9 Context Matrices for Preventive Design and Decision Making

The structure of the generic map of the ecology of modern technology may now be summed up as follows: within the cultural cycle of modern societies, the reciprocal interaction of people changing technique while technique simultaneously changes people plays a central role. It leads to a way of life that is culturally constrained by our modern secular sacred and myths, and biospherically constrained by its dependency on flows of matter and energy, as described by the first and second laws of thermodynamics. With the exception of iconoclastic efforts, these constraints cannot be violated. However, the viability of any open whole can be jeopardized by the viability of the contexts on which it depends. The four open wholes of primary concern to the development of preventive approaches for the engineering, management, and regulation of modern technology are technology itself, society, local ecosystems, and the biosphere. By developing the first one primarily in terms of performance values, no recognition is given to the fact that its long-term viability as a human creation depends on the viability of the other three forming its primary contexts. The joint viability of any open whole and its contexts must be assessed in terms of context values. Thus, as every modern society turns to technique to assure its future, it restructures many activities, and thus its way of life, which in turn transforms its relations with the biosphere. At the same time, as these changes are internalized by its members, they affect their consciousness and culture, in turn reinforcing the development of technique. This self-reinforcing cultural cycle will eventually be transformed by non-cumulative changes in open wholes, in turn affecting their contexts and so on, acting either as a challenge provoking a creative response or a gradual collapse of the ability of these cultures to provide meaning, purpose, and direction to their members. In the latter case, the life-milieu cannot be sustained, and the possibility of a collapse of modern civilization cannot be ruled out. The structure described above can now be translated into a set of context matrices to support preventive approaches.

Imagine a context matrix listing all the activities that together make up the way of life of a society and tabulating all the consequences for human life, society, ecosystems, and the biosphere, including an assessment of their significance in terms of context values. Although constructing such a matrix for the entire way of life of a modern society is clearly impractical, working out its conceptual structure provides a useful framework within which specific matrices for particular activities may be developed. Such matrices can support prevention-oriented design and decision making related to those activities.

It is preferable to move from the general to the specific because of the interconnectedness and enfoldedness of reality. There are no isolated activities. Each one is connected to others through outputs and inputs, which in turn involve other activities whose consequences involve still others, and so on, potentially involving all the activities of a way of life. Depending on where we draw the boundaries of a particular analysis, different answers can be obtained. Life-cycle analysis has thus far been unable to overcome this problem, and this situation is not likely to change in the future. In other words, map-making provides us with an appropriate analogy. Prevention-oriented design and decision making must start out by mapping the most significant activities and their connections, determine and assess their consequences, and utilize this information in negative-feedback mode.

The tree structure representing the differentiation of the way of life of a society may be used as the rows of the context matrix. The columns may be derived from the tree structure obtained by differentiating the concept of sustainable development. Together, they constitute a pyramidal structure of two-dimensional matrices, of which the rows represent a certain level of differentiating human life appropriate to the design or decision under consideration. The columns represent the correspondingly appropriate level of differentiating the concept of sustainable development. In other words, the most aggregated matrix would have as rows population, consumption, technology, and eco-sensitivity, and as columns sustainability of the biosphere, life-milieu, and human life itself. It may be imagined as forming the top of a three-dimensional pyramid-like structure in which each subsequent lower matrix is characterized by larger numbers of rows and columns resulting from the next level of differentiating a way of life and sustainable development, respectively. The elements of the matrices represent the context issues that must be taken into consideration in the particular design or decision, including an assessment of their significance in terms of context values.

The above structure is incomplete, since it arranges the rows of the matrix for considering only the biosphere-related dimension of sustainable development. This arrangement is not appropriate when considering the sustainability of the life-milieu or that of human life. To consider these two dimensions, it is much more helpful to regard a particular way of life as 'producing' a variety of context issues having significant consequences for sustainable development. Since this analysis is the subject of another volume, attention in this work will be restricted to materials and production, energy, human work, and the urban habitat. The concerns to be addressed through prevention-oriented design and decision making will constitute the matrix elements. The concerns on each subsequent lower level of the matrix structure will be dominated by the excessive throughput of matter and energy for the biosphere-related dimension of sustainable development, unacceptable levels of alienation and reification in relation to the dimension related to human life, and a combination of these two for the dimension related to the modern life-milieu. In other words, the two primary categories of concerns are metabolic and sociocultural. Each higher level in the matrix structure aggregates the primary concerns from the level directly below it and anticipates those on the next-higher level. The challenge will be to effectively integrate the two matrix structures concerned with the biosphere-related dimension of sustainable development and with those related to the modern life-milieu and human life itself, while keeping in mind that the interaction between human life and technology is a reciprocal one.

The use of these context matrices are illustrated in Part Four, which deals with four areas of application of preventive approaches: materials and production, energy, work, and the built habitat. The literature corresponding to each of these areas was surveyed in terms of how current practices deal with their social and environmental implications. Scoring these practices, using the research instruments outlined in chapter 3, produced two clusters: one corresponding to what, in chapter 1, I introduced as conventional approaches, and the other as preventive approaches. The results were tabulated in the form of annotated bibliographies. A discussion of the key works that help define the emerging paradigm of preventive approaches in each area are discussed in Part Four in terms of the contribution they make to the larger conceptual framework for preventive approaches developed in the present work.

## 7.10 Conclusion

A context matrix for guiding preventive approaches for the engineering, management, and regulation of modern technology has been created by intersecting the differentiated activities that together constitute individual and collective human life with the differentiated dimensions of sustainable development. The matrix elements are constituted by the concerns that need to be taken into account during design and decision making. The context matrix can be applied to individual technologies by concentrating on those activities of a way of life that are relevant to it or the context issues it 'produces.' Metabolic concerns can thus be examined by having the rows of the matrix represent the activities involved in each of the phases of the technological cycle of that technology. Sociocultural concerns can be identified by adding the context issues these activities 'produce.' This point is further illustrated in Part Four, which deals with applications.

Four categories of context issues have been identified by examining how technique dominates modern culture. The latter traditionally sought to get at the meaning and value of anything in human experience by relating it to everything else, and thus giving it a place in human life and society. Technique is an approach making minimal use of context, relying almost entirely on performance rather than context values. This was illustrated for the case of modern economics and the practice and policy making based on economic techniques. Human life, society, and the biosphere have become theoretical externalities. The market produces context issues by delimiting to the extreme the context considerations of any economic activity. Putting this theory into practice has given rise to the same four categories of context issues identified with the spread of technique. Ultimately, human sustainability, the sustainability of our present life-milieu, and the sustainability of the biosphere are threatened by the phenomenon of technique, which dominates the ways of life of modern civilization.

We have now concluded the theoretical framework for preventive approaches. In Parts One and Two, I contrasted preventive approaches with conventional approaches for the engineering, management, and regulation of modern technology. The differences may now be summed up in another way. Conventional approaches have a very limited definition of the inputs and outputs of technology, restricting attention to the desired outputs and neglecting the undesired ones, as well as limiting consideration of inputs to those essential for the former. In this way, conventional approaches decontextualize technology, separating it from the broader implications so that performance values appear adequate. Preventive approaches, on the other hand, essentially broaden the concepts of inputs and outputs, in general, and include undesired outputs, in particular. These are then interpreted in terms of their significance for the contexts of modern technology. The conceptual structure underlying the context matrix developed in this part of the work integrates all these considerations in a way that will facilitate mapping the ecologies of individual technologies to guide preventive design and decision making. In Part Four, prevention-oriented design and decision making are considered for reducing the throughput of matter and energy in a society, redesigning workplaces to minimize alienation and reification, and improving the health and sustainability of cities.

The shift to preventive approaches for the engineering, management, and regulation of modern technology converges with current attempts to rethink one of the most fundamental institutions of modern technology – namely, the corporation. If unhealthy workers and pollutants are as much outputs of a corporation as are its desired products, it is clear that additional stakeholders must be considered over and above those legitimated through the economic world-view described above if human and biosphere sustainability are to be addressed head on. Conceptualizing the modern corporation as a community of stakeholders begins to recognize the human, social, and environmental contexts in which this institution operates and on which its ultimate survival depends. The development of preventive approaches and the formulation of a concept of the corporation as a community of stakeholders therefore go hand in hand in helping to assure our common future.[21] There is something profoundly ironic about a system that seeks to maximize the returns on investments of our pension funds when such efforts help to undermine the kind of world in which we will retire by threatening our common future.

The present development of preventive approaches also converges with the growing field of ecological economics, which seeks to contextualize modern economies with respect to human life, society, and the biosphere. Unfortunately, they converge only to a minimal extent with current trends for developing national and international standards for life-cycle assessment, environmental management, and quality management. Although such standards typically put in place good accounting systems to track the environmental and quality performance of corporations, these sytems are not well-enough embedded in a larger preventive framework to draw the full benefits from such activities. As a result, we risk burdening corporations with considerable expenditures from which many more benefits could be realized. What needs to happen is that society must hold technology-related professions more accountable for keeping the public interest as paramount. Doing so would be of great benefit to these professions, future practitioners, corporations, communities, and nations, but unfortunately that approach runs counter to the grain of contemporary cultures. This is why, in these three introductory parts, we have developed preventive approaches within their broadest context so as to overcome the limits of specialization by 'thinking globally and acting locally.'[22]

# PART FOUR

# APPLICATIONS OF PREVENTIVE APPROACHES

# General Introduction

In Part Four, I introduce four areas of application in which preventive approaches can greatly reduce the unwanted and undesirable social and environmental effects of technology. It may add clarity for the reader if I point out some general patterns shared by these four areas. A great deal about their current state can be understood as resulting from conceptually, methodologically, and practically separating in time, space, and the social what belongs together. Technological systems have been developed on their own terms by means of performance values as if they operated in a vacuum. Efforts have been almost entirely concentrated on increasing the supply of whatever these systems can produce and ensuring that this is accomplished with a minimum of inputs. Hence, the production of goods is dissociated from what to do with them once they reach the end of their useful life and become wastes. We ensure that the supply of energy can grow exponentially each year while paying almost no attention to how it is used. Work organizations are developed to maximize labour productivity, all but ignoring the fact that the ability of people to do their work depends on the institutions of a society that determine things such as their ability to acquire, apply, and adapt knowledge and skills, or the ability to physically and mentally recover from a day's work and come back the next day refreshed – in other words, on their ability to 're-create' themselves. The urban habitat separates daily-life activities such as working, eating, sleeping, and playing, and divides people among neighbourhoods according to their socio-economic roles and status.

Preventive approaches seek to reconnect things that belong together by putting technological systems and products into their human, societal, and biospherical contexts. Hence, choosing them rather than conventional approaches will have an enormous influence on these contexts, and may be said to represent a choice for the kind of society in which we want to live. For example, reconnecting production to waste-disposal activities leads to a more circular economy that could substantially reduce the pressures on human health and on the biosphere. Reconnecting energy supply with use can have such a profound effect on society that it has led to the well-known distinction between the hard path and the soft path to the future. In the former, large centralized institutions create ever more energy, much of which is needlessly wasted in distribution and use, with equally unnecessary harmful effects on human health and the biosphere. In the latter, energy systems are created and evolved to fit their contexts by using renewable resources as much as possible and ensuring energy end-use efficiency, thereby greatly reducing the need for energy production and preventing its negative effects. An industrial strategy based on maximizing the number of low-wage jobs to minimize labour costs and on reducing social expenditures does, in the short term, lower the cost of products and increase the demand for them, with the well-known benefits for producers and consumers. In the long term, it will almost certainly have the opposite effect because it creates a work force that cannot develop skills to remain productive and adapt to turbulent global markets, a situation made worse by reductions in expenditures on education and basic research. All this is exacerbated by the fact that most low-wage jobs are physically, mentally, and emotionally unhealthy, creating a greater-than-necessary demand for social expenditures, and undermining the vitality of individual and collective life. Restructuring work organizations in the recognition that competitive and viable corporations require a healthy society, and vice versa, will inevitably create a different kind of society with different kinds of enterprises. At present, we appear to be stuck in a vicious circle in which work-related problems have reduced the vitality of society, in turn leading to the kinds of work organizations that make this situation worse. Finally, creating an urban habitat that supports human activities rather than being preoccupied with designing and constructing what, in the short term, are the least expensive buildings will also help produce a much more viable society.

What may be accomplished in each of these areas by avoiding unwanted and unintended effects through the adoption of preventive approaches will set up synergies that together can have a transforming effect on human life and society. Ironically, the most difficult obstacle to preventive approaches and a humane and common future is the very influence that the current industrial-urban-information life-milieu has on human consciousness, values, and cultures, in general, and on the economic world-view, in particular. This reciprocal interdependence reproduces the status quo, and traps us in the labyrinth of the social and environmental effects of modern technology. Preventive approaches are a necessary condition for finding our way out of this labyrinth, but not a sufficient one. We must be iconoclastic with respect to our most powerful creations in order to return to a new realism based on as exact an understanding as possible of what they can and cannot do for us, and to utilize them accordingly. Exercising our technological imagination is a choice not only for a different kind of technology, but also for a society that includes technology but is not dominated by it.

The economic implications will be substantial. Reflective research and preventive approaches address the limitations of the intellectual division of labour in science and technology, thereby reducing knowledge externalities. The impact on prevention-oriented design and decision making will be the reduction of social and environmental burdens. Since these constitute a major component of market externalities, it may be expected that the operations of the market will be improved. It is important to recognize that, before the creation of market externalities, the intellectual division of labour in science and technology creates knowledge externalities that profoundly affect the engineering, management, and regulation of modern technology and, through it, economic growth. This does not eliminate the need for economic policies that recognize and compensate for the fact that markets, like any other human creation, can do certain things but not others. What I am suggesting is that the roots of some modern economic problems reach down into the intellectual division of labour in science and technology.

# 8

# Materials and Production

## 8.1 The Precautionary Principle

In this chapter, I apply the generic map of the ecology of modern technology and the accompanying context matrices to the examination of the flow of materials and production processes in order to develop preventive approaches. From the outset, it is essential to recognize that this map contains a great deal of uncharted territory, and these lacunae must be transformed into useful ignorance if responsible preventive strategies are to emerge. It is, in fact, the implicit and explicit treatment of these lacunae that explains a great deal about conventional strategies, including their limitations. The shift to their more preventive counterparts has a great deal to do with converting these lacunae from harmful into useful ignorance. This is best accomplished by means of the precautionary principle as a key to sustainable development, in general, and the development of preventive strategies for materials and production, in particular.

I have already suggested that technological development is constrained from the bottom up by its reliance on the biosphere for matter and energy, and from the top down by its cultural unity (a sacred, myths, and a hierarchy of values), which makes all alternatives to a particular way of life unthinkable and, in any case, unliveable. The cultural unities of traditional societies embedded a precautionary orientation towards anything new. Their myths eliminated the 'edge' between the reality as they knew it and the unknown, much as we have learned to 'interpret out' the edge between what falls within our field of vision and what lies beyond, because we know from experience that all we have to do is turn our eyes to see more. Hence, the edge can have no meaning, and no edge is experienced. In the same way, the unknown became an extrapolation of the known, with the belief that future discoveries and developments would simply add missing elements in a cumulative fashion. In principle, the threat of the unknown was eliminated, but, on the daily-life level, incompatibilities between new discoveries or developments and a way of life still occurred. The precautionary orientation towards the unknown was rooted in the way the myths of traditional societies related the past, the present, and the future. Put in modern terms, the way of life of a society or civilization, handed down from a distant past by one generation to the next, had proved its viability and sustainability, confirming its perfection and mythical origin. Hence, the past as embodied in a tradition became the norm for the present and the future. Although anything new was not threatening in itself, it had to be incorporated into the culture through symbolization, which provided it with its proper name, meaning, and value, and thus its place in individual and collective human life. A precautionary orientation ensured that this was done adequately; otherwise, the viability and sustainability of a tradition could be threatened. Through new discoveries a little could be gained, but much more could be lost. For these and other reasons, no traditional society needed to develop concepts of appropriate technology and sustainable development or a precautionary principle because these were enfolded into its culture.

The cultures of the industrially advanced nations continued to convert the unknown into as yet undiscovered elements of the known, but tradition was no longer a model for the present or the future. Beginning with the myth of progress in the nineteenth century, the situation was reversed: the future would be an improvement over the present and, similarly, the present would be better than the past. No precautionary orientation was present and, what is more important, none appeared to be necessary at first. The developments of the last few decades gradually necessitated the invention of the concepts of appropriate technology, sustainable development, and the precautionary principle because these could no longer be taken for granted in the cultural unities of these societies.

In essence, the precautionary principle introduces into the ways of life of modern societies the equivalent of a safety factor related to the context in which products, processes, systems, and, ultimately, the global economy must function. Given their ultimate dependence on societies and the biosphere, a safety factor must be applied to these contexts. This is no different than, as accurately as possible, calculating the load the foundations of a building impose on a particular soil. No one would challenge the use of a safety factor on the grounds that this makes technology too expensive. The same ought to be true for the precautionary principle. A safety factor must be applied to the way the global economy 'loads' the biosphere. This surely does not require any discussion, but what does require conversations between the many stakeholders is how large the safety factor ought to be. In the equation 'calculating' harm done to the biosphere as the product of population, consumption, technology, and eco-sensitivity, the last factor is a non-linear function of the load imposed and of previous damage already sustained. Would any engineer supervising the erection of a building and alerted to excessive bending or cracking of some structural members wait for a full scientific study before taking action? Surely, when faced with a reasonable probability of failure, immediate action would be taken. In the same vein, accepting the fact that ecosystems and the biosphere are severely stressed requires action ahead of a full scientific understanding.

From this perspective, the precautionary principle ought to function as a safety factor. The sciences and engineering should continue to examine the situation as best they can in the full recognition that there may be unanticipated circumstances as well as limitations to what is known. How large the safety factor should be requires an assessment of the situation, which includes the sciences and engineering but is not limited to them.

The precautionary principle recognizes that society, in general, and technology, in particular, depend on the life-support and other functions of the biosphere. Humanity should accept the formal duty of environmental care to be equitably shared between all parties. This includes taking action where there is a reasonable probability of serious or irreversible harm, even when strict cause–effect relationships have not yet been established. The burden of proof is thus shifted to those who initiate actions that may produce such harm, requiring them to show that this will not be the case and to accept strict liability for any consequences. Preventive approaches should be used in the engineering, management, and regulation of technology. Failing this, 'best available technology' (BAT) should be used, although some parties argue that costs should be proportional to their benefits, and margins of error should be allowed for, thus converting this into 'best available technology not entailing excessive costs' (BATNEEC). Restraint should not be so costly as to produce serious social problems. It should be recognized that, for a substantial domain, the precautionary principle translates into a no-regrets principle because benefits will accrue to all stakeholders. When this ceases to be the case, two attitudes appear to prevail: some nations regard the precautionary principle as a challenge that can stimulate technological and economic creativity to produce new products, processes, and systems for which there will be a growing market. Other nations, on the contrary, see the principle as an additional burden stifling creativity and entrepreneurship, which, in the long term, is bound to be a detrimental strategy.[1]

Many recent developments may be interpreted as compensating for a loss of a precautionary orientation in the cultures of the industrially advanced nations; much of this chapter is devoted to exploring ways of building it into the engineering, management, and regulation of technology, in general, and into current approaches to materials and production, in particular. The first step is the conversion of any lacunae in our knowledge of the biosphere as the ultimate source and sink of all matter and energy into useful ignorance through preventive approaches. This includes recognizing that the exponential growth of scientific knowledge has real limits and that this is equally true for other human creations such as technology. Technical and economic development cannot end the struggle for human survival, nor do they automatically lead to social and spiritual progress. The symbolization of the biosphere as a separate environment or nature converting human beings into objective observers and managers of natural capital with little risk to themselves is fundamentally flawed. Performance values cannot assure genuine development. All this implies the recognition that, as people change their technology, technology simultaneously changes people, and that this reciprocal interaction also includes the biosphere. Precaution, first and foremost, recognizes the limitations of our knowledge of the biosphere, ecosystems, society, and technology in so far as this bears on the development of preventive approaches for materials and production.

Humanity is beginning to deal with the recognition that, although the biosphere has sustained human life throughout prehistory and history, its ongoing ability to do so cannot be taken for granted under the pressure of the global economy. The complex organization of life based on processes of self-regulation and co-regulation of natural cycles, living wholes, food webs, ecosystems, and the biosphere makes up an enfolded, non-linear whole that keeps the planet far from a condition of thermodynamic equilibrium. The importance of this becomes evident when we compare the conditions on our planet with those on others. The withdrawal of the services rendered by the biosphere would substantially increase the density of the atmosphere, remove all oxygen and nitrogen, significantly raise the Earth's surface temperature, and cause many more changes, thus making the planet unrecognizably different and, in any case, unfit for life as we know it.[2] The bottom line in all of this is that non-linear systems can flip from one stable state to another, in a discontinuous manner. The extent to which anthropogenic activities increase or decrease the possibility of such an occurrence is entirely unknown. We have no idea of the eco-sensitivity of the biosphere or of how it varies with the total anthropogenic load imposed on it. There is a complete uncertainty as to how long present trends can continue before the biosphere can no longer adequately perform its regulatory functions for the planet. All this is gradually becoming acknowledged, although cultural conditions in modern societies continue to make it easy to dismiss these possibilities as extreme.

Two consequences flow from this situation. First, the significance of individual problems such as global warming, ozone depletion, species loss, deforestation, and soil erosion cannot be understood one at a time. They represent interacting alterations of the biosphere, whose individual and collective contributions to the risk of a sudden major alteration of the system are entirely unknown. Second, it is virtually impossible to scientifically establish cause–effect relationships between individual anthropogenic emissions and their effects on the health of human beings or local ecosystems. There are usually many pathways via which such emissions can reach a target population. Some of these pathways are almost certainly unknown, and many may be undergoing significant changes as a result of other anthropogenic emissions. There are bound to be positive and negative synergistic effects that are difficult to incorporate into linear causal models. The health of populations involved in these pathways and that of the target population cannot be determined one pollutant at a time, as is the case in linear systems. Finally, it is impossible to establish response functions while controlling for all other health risks. What constitutes an acceptable health risk to a particular population is a question that faces similar methodological problems. Traditional risk-assessment methodologies developed for relatively simple mechanistic systems cannot be extrapolated to more complex non-linear systems, as Perrow[3] has shown. In other words, current methodologies are at best approximations whose usefulness may be temporary, given the non-linearity of these systems. When such approximations are used to establish 'acceptable and safe levels of exposure,' there is no way of knowing how long they will be useful – assuming their initial validity. We are unable to determine the consequences of anthropogenic emissions for individual living wholes, let alone for ecosystems or the biosphere.

Gradually, various pollution problems have made it apparent that the self-regulating and organizing capacity of the biosphere is not unlimited. Anthropogenic emissions must be regulated. In the past, the policy by which this was done was based on the assumption that the ability of the biosphere to take care of anthropogenic emissions was very large but not unlimited. This was expressed in terms of ecosystems and the biosphere having a certain capacity to assimilate anthropogenic emissions. An assimilative capacity was assumed for the management of wastes in aquatic environments, a critical-load concept asserted that the atmosphere was capable of absorbing emissions up to that level without harmful effects, and safe exposure levels were established on the grounds that organisms were not affected below a certain threshold.[4] Below these levels, there would be no appreciable effect on the health of organisms, ecosystems, or the biosphere. Up to a certain point, dilution and dispersion would permit the biosphere to take care of the problem. Raising the height of smokestacks and diluting wastes in large bodies of water were expected to successfully utilize the assimilative capacities of ecosystems. Whatever could not be assimilated had to be removed from waste streams and contained or transformed. All this appeared reasonable enough at the time, but many shortcomings were gradually admitted. To scientifically establish a cause–effect relation between a certain emission and an environmental hazard requires years of study as the problem intensifies and may even become irreversible. Even then, a direct causal link was difficult to establish, so that the presumption of the environmental innocence of anthropogenic emissions until proven guilty led to a variety of serious problems. Containing hazardous materials extracted from waste streams by pollution-control devices also turned out to be problematic in many cases. Even for the best-designed landfill sites, it is impossible to extrapolate scientifically the experiences of a few decades to the length of time containment may be required. It is now difficult to believe that, except for very few cases, the biosphere can act as an effective sink for anthropogenic emissions. A more prudent assumption is beginning to impose itself – namely, that organisms, ecosystems, and the biosphere do not have a certain absorptive capacity, receiving capacity, or environmental capacity below which the input of one or more anthropogenic emissions has no appreciable effects. The scale of the networks of flows of matter and energy involved in the global economy as compared with the corresponding networks in the biosphere is now such that even naturally occurring substances have flow rates so high that it can no longer be assumed that corresponding natural cycles can act as sinks. The obvious example is carbon-dioxide emissions, which can no longer be taken care of by the carbon cycle. With some possible exceptions, environmental policies will have to be developed that no longer assume the existence of natural sinks whose capacities can be translated into permissible levels of pollution. Science will have to take on a new role – namely, to assist communities in making informed political and moral decisions as to what constitutes acceptable behaviour towards ecosystems and the biosphere. The limitations of scientific knowing will have to be recognized in the face of systems as complex, non-linear, and enfolded as ecosystems and the biosphere. Full use will have to be made of useful ignorance, that is, that there is much we do not know and that there is a great deal that we can possibly never know. An iconoclastic attitude towards science and an informed knowledge of its limitations will increasingly become essential. Out of this useful ignorance will spring the ability to constantly adjust policies as new environmental hazards are discovered.

As environmental regulations continued to be tightened in the face of the growing pollution of air, water, and land, and as global competition intensified, industry was faced with a dilemma. The contradictions of end-of-pipe strategies became apparent, and more prevention-oriented approaches were invented as a way out of the dilemma. Before seeing whether they are up to the task, it is useful to briefly review the lack of relevant knowledge we have about technology and society, in general, and the flows of matter and energy within them, in particular.

The lacunae in our knowledge of the networks of flows of matter and energy involved in technological and economic activities and of their effects on human life, society, and the biosphere are considerable. For example, it is estimated that these flows include 80,000–100,000 different substances, to which we are adding 500–1,000 new ones each year.[5] In a 1984 report, the U.S. National Academy of Science estimated that, for 65,725 (77 per cent) of these substances of possible concern to the National Toxicology Program because of their potential for human exposure, there was no toxicological data, and that for the remaining 23 per cent only minimal data existed.[6] Superimposed on these lacunae are the methodological and conceptual limitations by which toxicological data are acquired. As noted previously, tests are based on exposing laboratory animals to high dosages for short periods of time. There are no practical tests for determining the effects of long-term, low-dosage exposures, or positive or negative synergistic effects of multiple exposures, and for extrapolating animal data to human beings. There is no international consensus about the materials that belong to the category of hazardous wastes, although the so-called Basel list drawn up by the Basel Convention on the transboundary movement of hazardous wastes probably comes closest.[7] Based on the Basel list, the Organisation for Economic Cooperation and Development (OECD) estimated that the total hazardous-waste emissions in the United States were about 30 million tons, or 6.7 kilograms per $1,000 of GNP measured in 1987 dollars.[8] The U.S. Office of Technology Assessment (OTA) has published much higher estimates, based on a survey of fifty chemical companies in 1984.[9] It should be noted that the Basel list is not a comprehensive inventory of all substances capable of doing environmental harm.

Whatever the actual figures may be, they can be expected to grow steadily. Recall the equation developed in chapter 7 suggesting that the total harm done to the biosphere equals the product of four factors – namely, population, consumption, technology, and eco-sensitivity. Future scenarios can be modelled by using different projections for global population growth, rising standards of living, and various levels of technological redirection towards cleaner production. The latter produce less hazardous waste per unit of throughput of matter and energy in the economy (the hazardous-waste intensity). What ultimately matters is the accumulation of these wastes in the biosphere and not their annual production. No matter which growth scenarios are selected, the cumulative total will rise to numbing figures in the next fifty years. The implications for human life, society, and the biosphere are impossible to assess. In addition to a population explosion, there exists an explosion of harmful materials stored in the biosphere, and the overall consequences are entirely unknown.[10] The robustness of the biosphere and its resilience to disturbances as developed over millions of years is truly astounding. However, even from what little we know, humanity is putting it to increasing tests. No one knows where the limits ultimately lie.

The influence of the network of energy flows involved in the global economy on the biosphere is probably as considerable as that resulting from the material throughput. It is clear that total energy consumption cannot indefinitely continue to grow exponentially. A few details are worth highlighting. Some energy transformations, such as electricity production by means of nuclear-power plants, are alleged to consume more energy than they deliver. What this means is that a nuclear-power plant requires more energy to build, operate, and decommission than it produces.[11] Even if this is not entirely the case, such 'solutions' do not greatly help the energy situation. Modern agriculture requires much more energy than it produces in terms of the caloric energy contained in food.[12] We know too little about how current strategies for energy 'production' affect the net available energy on the planet. The problem is analogous to the one discussed earlier with respect to the GDP as an indicator for economic decision making. While we have extensive data on gross energy 'production,' we know little about net energy 'production.'

There may be thermodynamic limits to the energy throughput the biosphere can support. Its evolution was limited by the capture of solar energy through photosynthesis. Human civilizations have been able to transcend this limit by transforming materials to liberate their chemical potential energy. These energy transformations produce entropy, according to the second law of thermodynamics. The biosphere has maintained its order by exporting entropy in the form of dissipated heat through mechanisms that act as a global entropy pump.[13] There is no corresponding set of mechanisms for exporting entropy out of the global economy except through the biosphere. We simply do not know whether the global entropy is increasing or not. However, if it is, it creates a powerful force that could push some natural cycles, ecosystems, and potentially the entire biosphere towards a thermodynamic equilibrium. How well the life-sustaining ability of the biosphere can compensate for the inability of the global economy to regulate its entropy production is impossible to know. Yet it is clearly irresponsible to assume that the biosphere can indefinitely provide this service.[14]

There is another way in which the global economy is seriously interfering with the energy flow through ecosystems and the biosphere. It has been estimated that humanity now appropriates nearly 40 per cent of what is called the terrestrial net primary productivity derived from photosynthesis.[15] Jackson[16] suggests that, during the next doubling period, humanity could conceivably irreversibly damage the integrity of the biosphere. He notes that similar conclusions may be arrived at by estimating the arable land that must be found during the next doubling period of global population. In other words, the carrying capacity of the Earth may well be exceeded within the next half-century. All these are simple estimates and there is a great deal that we do not know. Yet it would appear irresponsible not to recognize the urgency of a major transformation of the global economy.

Finally, consider the role of modern societies and their cultures in transmitting, developing, and applying knowledge and ignorance. Learning to make sense of the world and to live in it by means of a modern culture is to acquire its myths and sacred, which block the transformation of harmful ignorance into useful ignorance. For example, many people have great difficulty imagining any limits to scientific knowing and technological doing, as if these human creations were omnipotent. Scientific and professional education greatly delimit the transformation of harmful into useful ignorance in their own ways. This, in turn, fundamentally affects the entire institutional framework of society. Useful ignorance of the context implications of technological and economic development, primarily guided by performance values, plays a minimal role. Political decision making and the moral discourses that accompany it have not caught up with the fact that, although we know a great deal more than ever before in human history, this does not mean that we know our world any better, particularly in relation to the enormous influence of our most powerful creations. Economic decision making is guided by inadequate measures of true wealth, genuine human well-being, and the quality of our natural surroundings. There is a moral vacuum obscured by technological values, and a technological morality masquerading as their human and social equivalents.

It may be objected that my brief overview is overly pessimistic in that it does not assess our lacunae in the context of our accomplishments. Undoubtedly this argument has merit, but I believe it brings us face to face with the roles of modern cultures of engendering a kind of secular faith in science and technology by obscuring their limits. Given what is at stake, surely the best strategy to follow is to have as clear and precise a knowledge of what science and technology can and cannot do for us, just as we know what the tools in a toolbox can and cannot do. For example, attempting to cut wood with a file is ineffective as well as messy. Undervaluing or overvaluing science and technology is equally fraught with risks. It is a question of ensuring that our accomplishments in the domain of performance values are not undermined by the problems created in the domain of context values. This requires as precise a knowledge as possible of our most powerful creations, accompanied by a useful ignorance of what we do not know and may never know.

It is essential to be as realistic as possible about the potential of preventive approaches for the engineering, management, and regulation of modern technology. Recall the equation for the total harm inflicted on the biosphere by the global economy. During the next half-century, the global population may double, and the throughput of matter and energy could easily double if the standard of living, particularly of the poor nations, improves. If the current load imposed on the biosphere were kept constant, preventive approaches would have to reduce the load imposed on the biosphere per unit of throughput of matter and energy by a factor of four. This would mean no advance whatsoever towards sustainable development over the next fifty years. In other words, if we are really serious about moving towards a more sustainable global economy, preventive approaches should do far better than reducing the environmental load per unit of throughput by a factor of four. Since there is a great deal modern societies do not know and much that they may never know, it would be more realistic and responsible to set a goal of a factor of six, which is feasible provided a comprehensive strategy for the exploitation of the potential of preventive approaches becomes the driving force behind technological and economic development. It is not a question of pessimism or optimism, but one of realism, which has the best chance at creating a humane future.[17]

## 8.2 Towards a Comprehensive Strategy

If the life-support functions of the biosphere are to be protected, a precautionary orientation must grow within the cultures of all nations. Spontaneous mutations producing such an orientation appear unlikely in the immediate future, given the continued dominance of performance over context values and conventional over preventive approaches in technological and economic growth. The alternative is to attempt to induce such mutations through deliberate policies and regulations seeking to internalize a precautionary orientation into various activities. This can best be accomplished by a strategy that simultaneously operates on two levels: the micro level, involving a bottom-up strategy, and the macro level, involving a top-down strategy.

Bottom-up strategies include the constituent wholes of technologies, societies, ecosystems, and the biosphere. Obvious examples are the incentive programs of corporations encouraging employees to suggest ways in which energy can be saved and wastes reduced; green accounting practices that attribute environmental costs to specific processes and products so that supervisors and managers can become more preventively oriented in their decision making; design for environment practices to be incorporated in the development of new products and processes; and procurement decisions to incorporate environmental performance of all inputs. Other examples include initiatives by consumers, neighbourhood groups, and non-government organizations that seek to incorporate environmental considerations in daily life. For some time to come, it is likely that such efforts can be guided by relatively straightforward criteria of success such as the material efficiency of a process (the ratio of final product weight and the weight of all inputs required to produce it) and hazardous-waste intensity (the ratio of the weight of hazardous materials contained in a product or involved in a service to the total weight of that product or total material throughput involved in the service).[18] Although such ratios are but crude indicators of context compatibility with respect to the biosphere, they have the advantage that they can be understood by everyone.

These micro-level initiatives must be further guided, coordinated, and strengthened by 'top-down,' macro-level approaches to effect coherent improvements of various kinds in the technology-society-ecosystems-biosphere whole. Examples include international environmental treaties and regulations, international agencies setting prevention-oriented environmental standards, governments developing environmental policies and regulations, industry associations examining and implementing long-term strategies for achieving environmental objectives, corporations engaged in long-term environmental planning, professional bodies setting licensing requirements to enhance environmental literacy and practice, and educational institutions reorienting their curricula in a more preventive direction. The criteria of success essential for guiding these efforts are much more complex and will have to take into account what is currently known and what may never be known. However, without imposing precautionary limits on technological and economic development, the life-support functions of the biosphere will be taken for granted and thus will be economically undervalued and overexploited.

The precautionary principle can guide both top-down and bottom-up strategies. It was first formulated in the Federal Republic of Germany, where a great deal of research had shown that preventing harm to ecosystems is almost always much more cost-effective than dealing with it in an after-the-fact manner. In 1984, it was introduced during the first conference for protecting the North Sea. It was later expanded to include any anthropogenic emission into an ecosystem that is toxic, persistent, or bio-accumulative, or has any other hazard potential (which may include naturally occurring substances). The Ministers' Declaration at the Bergen Conference put it this way: 'Environmental measures must anticipate, prevent and attack the causes of environmental degradation. Where there are threats of serious or irreversible environmental damage, lack of scientific certainty should not be used as a reason for postponing measures to prevent environmental degradation.'[19] This formulation includes all substances with a known hazard potential and does not permit the transfer of a hazard from one medium to another. Since the possibility of the biosphere acting as an ultimate sink for a particular emission without causing environmental harm is extremely limited, this formulation of the principle implies an even broader goal of reducing all anthropogenic emissions to the biosphere. A comprehensive strategy must prioritize anthropogenic emissions in terms of their hazard potential and draw up strategies for eliminating them in the most serious cases (as in the case of CFCs) or reducing their levels (as in the case with carbon dioxide). It requires a multistakeholder approach in which science can make significant contributions, but where, ultimately, political and moral decisions must be made. A no-regrets principle would impose an additional condition – namely, that there must be a positive overall benefit even if the environmental hazard does not materialize. Environmental policies based on these principles implicitly recognize that science knows things out of their usual context, that the complexity and non-linearity of ecosystems and the biosphere go well beyond our scientific grasp, and that the burden of proof should shift from society to the polluter. Their application would lead to a continuous reduction of potentially hazardous emissions, taking care that they do not increase environmental hazards elsewhere or cause unacceptable economic or social problems. The assumption of an assimilative capacity has been replaced with the assumption that virtually all emissions should be reduced as much as possible through a comprehensive strategy. The European Union adopted the precautionary principle at the Maastricht Treaty in February 1992.

There is no question that the imposition of the precautionary principle on technological and economic growth will encounter sharply divided reactions from individuals, professions, corporations, and governments. Since it goes directly against the cultural unity of industrially advanced societies and the aspirations of developing nations, many will consider such limits as yet another burden inhibiting technological creativity and economic entrepreneurship, further exacerbating current difficulties. Some may go even further and claim that implementing the precautionary principle requires that all anthropogenic emissions cease, which would bring most economic activities to a halt so that human life and society as we know it would come to an end. It has always been the case that, from the perspective of a given paradigm, the transition to another with the potential of overcoming significant difficulties makes no sense. From within the conventional paradigm for the engineering, management, and regulation of technology, the imposition of precautionary limits is indeed a burden. From within the preventive paradigm, the situation seems to be the diametric opposite: such limits will be regarded as a challenge for technological creativity and economic entrepreneurship, leading to products and processes for which there will be growing markets and a bright future. What will then happen is that technological and economic growth will be transformed into development through preventive approaches. Some corporations and nations embody both attitudes: the latter for their internal operations, quietly exploiting the preventive potential, while externally opposing any international treaties or stricter national environmental regulations on the grounds that they impose unacceptable burdens. For example, several nations continue to oppose strict limits on the emissions of greenhouse gases, even though pollution prevention is their official policy. However, if consistently applied, such a policy would significantly improve energy efficiency, satisfy these limits, and have the additional benefit of improving economic competitiveness. The preceding spectrum of reactions correlates exactly with the extent to which harmful ignorance has been converted into useful ignorance to open the road to preventive approaches. In any case, exercising a precautionary principle is in essence equivalent to building in a safety factor with respect to the ability of the biosphere to support human activity.

Suppose a society acquires the political will to impose precautionary limits on technological and economic growth (but not on development) while stimulating the diffusion of preventive approaches in the conviction that ultimately this is in everyone's interest. It must then ensure that preventive approaches do not solve some problems while creating others in different areas. This can be avoided only by stimulating such approaches across the entire spectrum of applications. It must also be recognized that a preventive orientation is a necessary but not sufficient step in creating a more sustainable way of life and a humane future because technology is but one branch of the much larger phenomenon of technique. It has led to the dominance of performance over context values in every sphere of modern society, resulting in very similar problems for human life and society over and above the ones for the biosphere.[20] For now, I will limit myself to pointing out that some nations, such as Germany, the Netherlands, and the Scan-dinavian countries, appear to be among the leaders in developing prevention-oriented strategies for dealing with current problems.[21] Devising macro-level strategies would not be so difficult were it not for the fact that the myths of modern cultures stand in the way of an accurate interpretation of our situation.

### 8.3 General Principles

The general principles for guiding both macro- and micro-level components of a strategy for improving the sustainability of modern ways of life with respect to the biosphere follow directly from the constraints expressed in terms of the first and second laws of thermodynamics. Since a way of life can neither create nor destroy matter and energy, its activities are connected by flows of matter and energy in such a way that the input flows and output flows associated with any activity connect it either to other activities or to the biosphere. Hence, a society's way of life enfolds a network of flows of matter and a network of flows of energy for which the biosphere provides the ultimate sources and sinks. The two networks sometimes overlap, as in the case of fuels and combustion products, for example. The load imposed at the society–biosphere boundary may be reduced by considering: (a) how matter and energy flow through the networks; (b) what kinds of matter and energy flow through the networks; and (c) the structures of the networks themselves. Their consideration leads to three principles for the creation and development of an overall strategy:

1 To reduce dependence on the biosphere, materials should be kept within the network of flows of matter as long as possible to render

as many services as feasible. Materials would then travel through the network in more cyclical patterns as opposed to the largely linear ones that currently dominate. Dissipative use of materials should be minimized.

2 By means of materials substitution, the hazard potential of the outputs of the network can be reduced.
3 The network itself can be restructured through preventive approaches.

From the second law of thermodynamics, it follows that additional principles govern the network of energy flows. The way energy flows through the network cannot be made more cyclical because energy transformations are irreversible. Hence, the following principles apply:

4 The maximum service must be extracted from linear chains of energy conversions within the network.
5 Such chains must be supported as much as possible from renewable sources.

The present chapter concentrates on the first three principles associated with the network of flows of materials, and the next chapter with the remaining ones related to the network of energy flows. Every initiative, whether it be on the macro or the micro level, should give concrete expression to these principles. This will create a positive synergy between them.

The emerging field of industrial ecology generally seeks to implement the first three principles. It has been defined as follows: '"Industrial ecology" is intended to mean both the interaction of global industrial civilization with the natural environment and the aggregate of opportunities for individual industries to transform their relationships with the natural environment. It is intended to embrace all industrial activity ...; both production and consumption; and national economies at all levels of industrialization.'[22] For the present work, industrial ecology must become one dimension of a comprehensive strategy to make modern ways of life more sustainable with respect to human life, society, and the biosphere without compromising one aspect for another. It can substantially reduce the pressure on scarce resources, diminish the load imposed on the biosphere by anthropogenic emissions, reduce interference in the self-regulating character of the biosphere, and rehabilitate the quality of our natural habitat.

A variety of policy instruments may be used to implement the principles noted above. One interesting proposal is to shift from taxing earnings and profits to taxing the consumption of materials and energy.[23] If this turns out to be feasible, it may help overcome the problem of growing unemployment and the overexploitation of the biosphere by reducing the material intensity with which services are rendered. By creating a better balance between net material and labour productivity within a variety of institutions, the overall performance and context compatibility of those institutions can be enhanced. Another proposal is to move towards a genuine service economy, with the understanding that it is the services technology renders and not technology itself that interests us – a subject to which I will return shortly. The load on the biosphere can be reduced further by a strategy of substituting materials that impose a lesser environmental burden for those that impose a greater one.

An effective strategy for materials substitution to improve human health and reduce the load imposed on the biosphere requires a great deal of information, much of which we do not currently possess. First, it is essential to know which substances are used by a society and how they participate in the network of material flows representing an economy, and to locate that network in the geography of a society and in its way of life. It would then be possible to examine how and where human beings, other life forms and ecosystems may be exposed to these flows to determine exposure potential. Such models of individual economies must be aggregated into a global model to examine problems occurring on a global scale. These models would reveal the precise role of the biosphere as the ultimate source and sink of all flows in the networks. This would permit detailed studies of how materials associated with an environmental concern originate in the biosphere, how they enter into a society and a global economy, the transformations they undergo, and how these materials eventually leave the economy either through dissipative use or as point-source emissions. The closest approximations we currently possess are economic input–output models in which monetary flows have been converted into flows of matter and energy supplemented by mass–energy balances of individual processes and transformations.[24] Such models would also permit the simulation of the effects of different policy instruments.

What must be understood next are the implications of these flows for human health, ecosystems, and the biosphere. It is possible to identify a subset of materials causing particular concerns for human health and biosphere protection. The Organisation for Economic Cooperation and Development (OECD) has set out procedures for systematically identifying which of the 80,000 existing commercial chemicals pose such concerns.[25] A similar study was undertaken in the United States.[26] Until such a subset of substances of concern has been identified and tested for human health and biosphere effects, societies must cope as best they can.

In the United States, manufacturing firms above a certain size and production level are required to estimate and report their discharges by weight of some 370 chemicals and chemical categories for each of their plants. The Environmental Protection Agency (EPA) publishes these data in its annual Toxic Release Inventory (TRI). Unfortunately, aggregating discharges by weight is rather meaningless since their implications for human health and biosphere protection per unit of weight vary enormously. The top producers of these waste streams are listed, but again this provides no indication of the risk these pose to human health and biosphere protection. It is possible for corporations, countries, or states to improve their performance on these terms while actually increasing the hazards they produce. A toxic-weighting factor has been applied to the TRI releases to provide a better estimate of hazard potential.[27] Despite the acknowledged limitations of weighting factors, they would make more realistic the comparisons of corporations, countries, and states. It is also possible to correct annual increases or decreases in total emissions by weight to obtain a better impression of whether the load imposed on human health, ecosystems, or the biosphere is increasing or decreasing. Similarly, the individual loads of specific products or processes could be compared, which could guide design and decision making in a preventive manner. Various chemicals and groups of chemicals have been proposed for a complete ban achieved through a phase out period or for substantial reductions. Some of these have gone into effect.[28] It is clear that we are very far from being able to develop a comprehensive strategy of substituting materials with a lower hazard potential for ones with a higher potential, or the identification of substances that should be banned outright, phased out or reduced through substitution.

A third category of knowledge that needs to be developed to support industrial ecology approaches relates to the cumulative effects on human health, natural cycles, life-support functions, or other important processes of the biosphere. The list is well known and includes issues such as global warming, ozone depletion, deforestation, and soil erosion.[29] Another category of knowledge relates to the development of environmental indicators that permit tracking the success, or lack thereof, of environmental policy instruments and initiatives.[30]

In sum, it is essential to know what we are emitting; the hazard potential of these emissions for human health and the biosphere; the possibilities of reducing such harm by means of bans, phase-outs, or reductions by substitutions as well as by preventive approaches; and the economic and social implications of any initiative. In some cases, the potential harm of a substance may be substantially reduced by ensuring that it flows in closed loops with minimal leakages through the biosphere. Those for which this approach is impractical because their use is essentially dissipative may have to be banned altogether or restricted to those applications where no satisfactory substitute can be found and where their service to the economy is indispensable. In other cases, it may be possible to extract substances from waste streams to transform them into useful resources for the network of flows of matter in an economy.

From this brief survey, it is obvious that a great deal of work needs to be done before a comprehensive strategy of substitution can be worked out. Once the preventive potential becomes better known, the political will to exploit it may arise.

A complementary component of a macro-level preventive strategy involves the industrial-ecology vision of a more circular economy. Such an economy has many potential benefits. Since a society can neither create nor destroy matter, making the flows of materials in the network as circular as possible reduces resource extraction. In many instances the environmental burden associated with resource extraction, separation, refining, and processing is considerably greater than the burden associated with recycling materials. This is particularly true when the energy requirements are lower. A more circular economy also tends to reduce the role of the biosphere as the ultimate sink of all materials. Of course, this must be checked in each case to ensure that real gains are made in reducing the role of the biosphere as the ultimate source and sink of all matter and energy. Additional criteria must also be considered. For example, it is essential to monitor and delimit further increases in the human appropriation of the net primary productivity derived from the photosynthesis of local ecosystems and the biosphere. In addition, progress towards reducing the ecological footprint of a society should be monitored.[31] All proposals should give careful consideration to the distribution of the benefits and costs of any policy initiatives: Who will pay? Who will benefit? How will the burdens be distributed? The strategy will have to be comprehensively preventive, anticipating possible conflicts and even wars that could be triggered by environmental problems. Ecosystem and biosphere health will increasingly be linked to global security.[32]

The most comprehensive way of making an economy more circular is to create a genuine service economy by recognizing that, in most instances, people are interested in the services rendered by technology and not in technology itself, as previously discussed in connection with selling illumination services as opposed to electricity. In the same vein, automobiles could become a part of transportation services, computers of information services, photocopiers of document services, and so on. An alternative proposal involves product stewardship whereby manufacturers take back their products after they can no longer perform their useful functions. A move in either of these directions could be supported by green taxes to create a better balance between material productivity and labour productivity. No policies to move towards a more circular economy or one based on product stewardship have thus far been implemented. A discussion of how such policies would influence the way materials flow through the corresponding network and how the network itself would be restructured will therefore be discussed as micro-level preventive strategies for specific products and services.

## 8.4 Micro-level Prevention

Just as input–output methodology complemented by mass energy balances can be used to examine the networks of flows of matter and energy in a society, life-cycle analysis can serve to examine those portions of these networks related to a particular process, product, or activity for the identical purpose of preventing harm. The Society for Environmental Toxicology and Chemistry (SETAC) defines the life-cycle analysis process in the following way: 'A life-cycle assessment (LCA) is a process to evaluate the resource consumption and environmental burdens associated with a product, process, package or activity. The process encompasses the identification and quantification of energy material usage, as well as environmental releases across all stages of the life cycle; the assessment of the impact of those energy and material uses and releases on the environment; and the evaluation and implementation of opportunities to effect environmental improvement.'[33] Graedel and Allenby place life-cycle analysis in the context of product stewardship, which they regard as the primary thrust of industrial ecology:

> designing, building, maintaining, and recycling products in such a way that they impose minimal impact to the wider world. Product stewardship should be broadly interpreted to include services, which should also be performed so as to have minimal impact. The way in which these tasks are addressed in a formal manner is by the process of life-cycle assessment (LCA), a family of methods for looking at materials, services, products, processes and technologies over their entire life. The essence of life-cycle analysis is the evaluation of relevant environmental, economic, and technological implications of a material, process or product across its life-span from creation to waste or, preferably, to re-creation in the same or another useful form.[34]

It is important to contrast these two interpretations. The former originally defined the process as objective, while the latter sees it as an integral part of industrial ecology, implying a more qualitative assessment. A claim of objectivity generates a great deal of harmful ignorance, as did the traditional cost–benefit methodologies that claimed to constitute a scientific and objective basis for public-policy decision making. In order to understand why the life-cycle assessment process can never become objective, it is useful to briefly outline its structure.

A life-cycle analysis involves four stages. The first sets out the goals of the study, defines the system to be examined, identifies its boundaries, decides on expected data quality, and suggests a process of internal or external review of the findings. The second stage lists and quantifies all desired and undesired inputs and outputs crossing the system boundary and includes them in a mass-energy balance of the system. The inputs specify the resource requirements of the system and all outputs are related to immediate or future anthropogenic emissions. The third stage seeks to determine the implications of all inputs and outputs for human health, local ecosystems, and the biosphere. It specifically evaluates all inputs in the light of resource scarcity and all outputs as loads on living wholes in the light of such context values as health, integrity, and sustainability. It includes a review of the reliability of the results and their sensitivity to variations in the data. The fourth stage uses the results to identify opportunities for improving the environmental performance of the system. The four stages are sometimes referred to as initiation, inventory, impact analysis, and improvement analysis.

The initiation stage is rather problematic because there is no acceptable method for determining where to draw the boundaries of the system. While in energy analysis expanding the boundaries adds an ever-smaller contribution to the analysis, this is not necessarily the case for the flows of materials. It is possible that the next branch in the network of flows associated with a process, product, or activity that was previously excluded from the system can make a contribution to the load the system imposes, which can range from negligibly small to very large. The boundary problem is, in part, related to any industry being directly and indirectly dependent on many others, and the highly enfolded character of a society. Comparing the results of environmental input/output analyses, including all direct and indirect dependencies of a system, with the results obtained by a SETAC life-cycle assessment carried out one industry at a time, shows that the contributions of indirect dependencies to environmental emissions can exceed those from direct dependencies.[35] Despite the limitations of both methods of analysis, it is nevertheless clear that, because many processes, products, and activities require outputs of many, if not all, sectors of the economy, it is difficult to define the boundaries of a system. The problem can be illustrated from another angle. Paul Hawken estimates that every American consumes approximately 136 pounds of resources a week, while 2,000 pounds of waste are discarded to support that consumption.[36] A study for the U.S. Academy of Engineering found that about 93 per cent of the materials we buy and 'consume' do not end up in saleable products.[37] It is virtually impossible to allocate these throughputs to specific daily-life activities. Allocating the environmental burdens associated with the infrastructure of a modern society to particular activities is equally difficult. Despite the problem of satisfactorily determining the boundaries of a system to be analysed and despite the fact that the results could be significantly affected by the choice of boundaries, most life-cycle manuals and standards skirt the issue.

The inventory stage essentially constitutes a mass and energy balance of the system. Although it can be rather involved for complex products, the basic methodology is fundamental to a great deal of natural science and engineering. It is the only stage that can claim objectivity for the system boundaries adopted. It is during this stage in the analysis that it may become apparent that adjustments in the system boundaries are required. In principle, the networks of flows of matter and energy that comprise the system must show its direct dependence on the biosphere as the ultimate source and sink for all flows and its indirect dependence on all flows in a society without which the direct flows would be impossible. It is often the case that some of these indirect dependencies make a negligibly small contribution to the mass and energy balance of the system, but this does not necessarily mean that they make a negligible contribution to the load a sector of the economy, a society, or the global economy imposes on ecosystems and the biosphere. This is where the previously mentioned problem arises: the total load determined by means of a life-cycle analysis of one process, product, or activity at a time or the load determined by an input–output approach may thus differ considerably.

The impact analysis examines all flows across the society–biosphere boundary (or portions thereof) that can be attributed to the system under analysis. From there, outputs and inputs affect natural cycles, food webs, ecosystems, and possibly the overall functioning of the biosphere. In the process, populations of many life forms may be exposed. These loads on a variety of wholes need to be examined in order to determine their impact, which is a function of eco-sensitivity. Particular attention is obviously focused on the effects on human health. Identifying the flows of matter and energy associated with a particular process, product, or activity within a society is complex enough; the extension of this analysis into the biosphere brings the analyst face to face with so many things we simply do not know, that the exercise of useful ignorance is absolutely essential. The claim that all impacts have a common denominator so that they can be summed into some overall index of environmental harm involves so many arbitrary assumptions and value judgments, when taken as objective, that it results in the production of a great deal of harmful ignorance. The problem is analogous to that of cost-benefit analysis, which attempts to give all implications a common monetary denominator.

The improvement analysis risks becoming hopelessly bogged down in information overload if the previous stages are given their full scope. It will come as no surprise that few life-cycle analyses have been published, and that most of these are for very simple products, such as disposable and cloth diapers, single- and multiple-serve drinking cups, and personal-care goods. The findings have been much disputed. The conclusion that imposes itself is the same as the one reached by Graedel and Allenby.[38] Life-cycle analyses are not a practical guide for industrial design and decision making because they are far too time-consuming and expensive, and their findings are uncertain. Imagine attempting to minimize the environmental burden of a complex product. Even if the design process were a linear sequence of choices between options, the number of interdependent life-cycle analyses that would have to be performed would be so large as to be impractical. It may, of course, be argued that someday all computer-aided design and computer-aided manufacturing systems will incorporate environmental data bases derived from detailed life-cycle analyses of each and every material involved in the manufacturing process, including energy sources, consumer-use patterns, and disposal or recycling. Perhaps that day will come, but it is likely to involve the creation of a great deal of harmful ignorance and a stifling of human creativity and imagination. In the meantime, design-for-environment (DfE) approaches offer most of the benefits and a fraction of the methodological problems. These are no more objective than life-cycle analyses but will furnish designers, engineers, managers, and regulators with useful approximations required for preventive practice. What I am suggesting is that life-cycle analyses and design-for- environment approaches be considered, not as an objective tool, but essentially as a mapping process of the ecology of a particular process, product, or activity so as to identify the most serious environmental hazards and thus point the way to prevention. To avoid its becoming an after-the-fact part of design and decision making, it must be quick enough that it can remain an integral part of the process. This is what a map-making approach converted into design and decision matrices can accomplish.

In principle, a life-cycle-analysis–style process, such as design for environment, can be used for four main purposes: to establish a baseline for the environmental performance of a system; to compare alternative processes, products, or activities in terms of their environmental performance; to identify particular phases in a life cycle where the greatest environmental burden is imposed and where improvements should be sought; and to preventively guide the development of new processes, products, or activities. The limitations of the process make optimization or competitive comparisons impossible. Neither can it be used for setting objective standards for green products. Its primary use is to develop preventive approaches based on map-making, operationalized in terms of context matrices. At its core, this process is identical to what is commonly known as 'design for environment,' but it is capable of integrating all other dimensions of making our modern ways of life more sustainable. This approach to life-cycle analysis is therefore an integral part of the development of preventive approaches for the engineering, management, and regulation of modern technology. It maps the ecology of a particular process, product, or activity, and uses that information to eliminate or significantly reduce environmental harm.

It is useful to distinguish processes, products, and activities according to where the greatest environmental burden occurs in the life cycle. For some, it may occur during manufacturing; for others during use; and, for still others, during disposal or recycling. The automobile is an excellent example of a product whose environmental burden is greatest during its use phase. To deal with this situation, the hypercar is being proposed as a design alternative for reducing this burden, although the use of advanced materials may increase the environmental burden during manufacturing and recycling.[39] The current approach taken by the German automobile industry, based on design for disassembly, remanufacturing, and closed-loop recycling, concentrates on reducing the environmental burden associated with the production and disposal phases. The two approaches may be combined to achieve a greater overall reduction. For washing machines, the greatest environmental burden also occurs during their use phase. Whatever the environmental profile of a product cycle may be, it is of great strategic importance in showing designers and decision makers where to focus their efforts in order to achieve the greatest possible environmental benefit with the limited resources available.

Some proposals have been made to streamline life-cycle analysis by concentrating on a few indicators of the environmental burden of a process or product. These include but are not limited to gross energy requirement (GER), global- warming potential (GWP), and solid-waste burden (SWB). This greatly simplifies the inventory stage and almost entirely eliminates impact analysis. In the case of global-warming potential, for example, all greenhouse gases are added together according to their carbon-dioxide equivalents, but their collective impact is not assessed. Although such an approach has obvious advantages in terms of reduced requirements for time and resources, it is of little use for the development of preventive approaches since it seriously truncates the mapping of the ecology of a process or product.

In sum, the environmental burden imposed by the network of flows of matter may be reduced by means of the following hierarchy of approaches. First, reduce the throughput of all materials that perform no useful function in the network and that exit as waste streams. This primarily concerns materials that are eliminated through pollution prevention, in the recognition that pollution is an incurable disease that can only be prevented. Second, wholly or partially convert all flows of materials that have ceased to or never did perform a useful function from wastes into resources that can be reincorporated into the network. Third, treat the remaining waste streams from the network so as to present minimal environmental hazards. Fourth, substitute materials with a lesser environmental burden for those with a greater one wherever possible. Together, these four approaches focus on reducing both the quantity and the hazard potential of all waste streams crossing the society–biosphere boundary. Because of the conservation of matter, this simultaneously reduces the flows into the network across the society–biosphere boundary. For the same services rendered, it reduces both the material intensity and the environmental-hazard intensity of the network, in general, and of many of its branches, in particular. In other words, the same level of useful services is maintained by transforming an essentially linear throughput into a more circular one in which flows of materials serve as many functions as possible before they exit. It is a strategy of doing more with less, which may be expected to have the additional benefit of substantially reducing the throughput of energy in the associated network of energy flows in a society.

After all products and processes have been designed to be as compatible with their contexts as possible, the remaining wastes should be converted into resource flows by adding as few additional links to the network as possible. This may be undertaken by the following hierarchy of approaches arranged in a decreasing order of expected environmental benefits. First, all scrap produced in a process should be recycled by converting it to useful inputs. Second, processes can be symbiotically linked together into industrial eco-parks where some or all of the wastes of one process become useful inputs into another, preferably by fixed links or by means of transportation. The most discussed example is that of the Kalundbörg Industrial Eco-park in Denmark,[40] where a number of businesses in a small community jointly transform wastes into useful resources. In its development, there were few, if any, technological or economic barriers. The others were gradually overcome by the fact that the managers and decision makers as members of the same community gradually recognized that they had common concerns about remaining competitive without degrading the quality of their community and ecosystem, on which they vitally depended. They learned to trust one another and to cooperate, with considerable benefits for each business and for the viability of the region. Kalundbörg represents a web of multidimensional recycling and waste exchanges between different companies as well as the town itself, bringing enormous economic benefits to all participants. The heart of this artificial ecosystem is a coal-fired power station that pipes steam heat (which would normally be wasted) to the entire town, as well as to a biotechnology firm and an oil refinery. Recovered heat is also diverted to a fish farm. The power plant's scrubbers create gypsum that is sold to a plasterboard manufacturer, obviating the need to import gypsum from open-pit mines in Spain, 3,000 kilometres away. The plasterboard manufacturer uses the light gases from the refinery (which are usually burned off as waste) to fire its drying ovens. The power station sells its fly ash and clinker for road building, and to a cement manufacturer, instead of disposing of it in a landfill. The refinery operates a desulphurization plant for its gas, and this plant yields a liquid sulphur residue that is sold to yet another plant manufacturing sulphuric acid. The refinery also recycles its cooling water to the power plant, where it is used for cleaning and as feed water for the boilers. Organic sludge from the fish ponds and the biotechnology company (used for culturing microbes) is sold as fertilizer to a thousand neighbouring farms.

A third approach is to close as much as possible the technological cycle of a product by means of product stewardship mandated by take-back legislation, under which the manufacturer accepts responsibility for the product at the end of its useful life. This encourages the manufacturer to engage in design for disassembly and in remanufacturing, with the former having the additional benefit of facilitating repairs and thus extending the useful life of the product. This closed-loop recycling first seeks to reuse as many subassemblies and components as possible by returning them to their original state through remanufacturing. The remainder is separated into material flows that are converted into resources, replacing inputs of virgin materials. Additional benefits can be realized by recognizing that what consumers ultimately are interested in is not the products themselves, but the services rendered by these products. By having the manufacturers sell services by leasing products, the throughput of materials in a closed technological cycle of a product can be reduced, with considerable additional environmental benefits. It is now in the manufacturers' interest to prolong the useful life of a product as much as possible by improving its quality, durability, and ease of repair. This can be accomplished through a modular design, permitting the upgrading of subassemblies and components to incorporate the latest advances. When remanufacture is no longer possible, design for disassembly will ensure that services have been rendered with the lowest possible material and energy intensity.

The next approach turns to open-loop recycling for cases where the use of materials degrades their engineering properties in a way that cannot be cost-effectively remedied. In such cases, other uses for the materials must be found that are less demanding from an engineering perspective. In this way, such materials can cascade in the network of flows of materials to perform a series of increasingly less demanding services, with their final use being a long-term, low-demand application such as a building material that can remain in the network for a very long time.

A fifth approach, to be practised in parallel to those outlined above, is to reduce the requirements for materials that dissipate during the use phase of the technological cycle of a product. The hypercar is an interesting example, since it will eliminate many fluids whose use is partially or wholly dissipative. Many agricultural practices are also in urgent need of restructuring so as to reduce their dissipative use of fertilizers, pesticides, and herbicides, and to lessen dependence on genetic engineering and biotechnology, whose risks are far from clearly understood. Of course, organic farming remains the sole comprehensively preventive approach.

A sixth approach reduces waste streams by mining them for useful materials, thus partially converting them into resource flows. Whatever remains must be treated to reduce or largely eliminate the environmental burden imposed by their discharge from the network. For example, this may involve processing wastes to ensure that they will degrade within the time span in which the integrity of a landfill can be expected to be maintained.

Interventions through environmental policies and regulations should gradually eliminate all materials from the network that cannot be successfully dealt with in one of the ways outlined above and that, according to the precautionary principle, represent an unacceptable environmental hazard. In addition, the substitution of certain materials for others should further reduce the overall environmental burden imposed by the network. All the approaches described here should be supported by a policy framework that seeks to facilitate, enable, and accelerate the removal of any technological, economic, social, or cultural barriers that needlessly impede these approaches, without compromising a community's ability to evolve its way of life according to its fundamental values and beliefs. Instruments for doing so may include removing technological barriers by providing research grants; removing economic barriers by eliminating certain subsidies and imposing others; making changes in taxation and imposing penalties; and lifting institutional and cultural barriers through education and informed public discussion.

The overall results will be to transform the throughput of matter in a society away from the present linear pattern, whereby virgin resources are extracted, processed, manufactured into products, distributed, used, and disposed of in a linear sequence. Of this, 93 per cent is directly converted into waste, according to the previously quoted study, and the remainder into products, many of which have an extremely limited life span before they also become waste. Graedel and Allenby suggest that the development away from a linear throughput may resemble what happens in natural ecosystems.[41] As an ecosystem matures, it converts more of its wastes into useful resources by expanding its food webs, by improving its use of natural cycles, and by its symbiosis with other ecosystems and the biosphere. It should be noted, however, that no ecosystem is closed (i.e., having no flow of matter across its boundaries), and Jackson[42] suggests that the implications of the second law of thermodynamics are that no individual economy or the global economy can ever be closed. There are simply too many instances of inherently dissipative use of materials. Theoretically, it is always possible to collect materials dispersed through various activities, but the energy and resources involved rise exponentially with the degree of dispersion. The situation is analogous to mining: the lower the concentration of the ore, the higher the costs of extraction. Nevertheless, the notion of a network of flows of matter evolving in ways that resemble those of ecosystems is a fruitful analogy.

It should be noted that the gradual conversion of linear throughput patterns towards more circular ones will transform the role of performance values. Consider the influence industrial eco-parks and waste exchanges may have. The state-of-the-art processes for manufacturing particular products are likely the most efficient and cost-effective in a stand-alone mode, but the predecessors they displaced could turn out to be more efficient and cost-effective in an industrial eco-park network because more of their wastes can be turned into useful inputs. In other words, the stand-alone efficiency and cost-effectiveness of a process must be re-evaluated within a network constituted by an industrial eco-park or waste exchanges. The same pattern may arise for materials. The trade-off between desirable engineering properties and cost is likely to broaden to include compatibility with closed or open-loop recycling within the network of flows of matter in a society. Corporations may expand their product lines to increase the efficiency and cost-effectiveness of their existing operations by creating internal industrial eco-parks. In other words, context compatibility will increasingly be assessed with respect to the networks of flows of matter and energy within: a corporation, an industrial eco-park, a region that links processes through waste exchanges, a nation, and a global economy. This is one way of increasing compatibility with ecosystems and the biosphere. It should be noted that a comprehensive preventive strategy includes but is not limited to the vision of industrial ecology. All products and processes must be designed using preventive approaches before they become the 'parts' in a 'system' based on material flows that are as circular as possible. In other words, industrial ecology must not function as a kind of end-of-pipe solution to the limitations of current methods and approaches for the engineering, management, and regulation of modern technology.

## 8.5 Applying Context Matrices

The context matrices developed in the previous part of this work can be applied to guide both macro-level and micro-level components of preventive strategies. Here the analysis converges with the design for environment strategies so well developed by Graedel and Allenby,[43] which are an essential component of making modern ways of life more sustainable with respect to the biosphere. The precautionary principle demands that care be taken to ensure that environmental problems not be traded for economic, social, or cultural ones. Hence, this component of making modern ways of life more sustainable with respect to the biosphere must ensure compatibility with other components related to human life and the life-milieu.

Consider a context matrix to guide the preventive design of a product, process, or facility with respect to the biosphere-related dimension of sustainable development. The rows of the matrix represent the differentiation of the way of life of a society into a chain of activities directly and indirectly involved in the technological cycle of a product, process, or facility, beginning at the society–biosphere boundary, where all requisite virgin materials are extracted, and ending at the same boundary, across which all materials are eventually returned to the biosphere. For a product, the rows may include resource extraction; refining and processing of virgin materials; recovery, separation, and processing of recycled materials; manufacture of components and subassemblies, and assembly of complete systems; packaging, shipping, and distribution; product use, including routine maintenance and repair; any extension of product life through remanufacturing of components or subassemblies for the entire product; reuse; disassembling the product at the end of its useful life, recycling the materials, and disposing of any waste not recycled. Possibly, with the exception of product use, each activity in the chain is carried out on the basis of a process requiring indirect inputs such as tools, manufacturing equipment, lubricants, and other products. In turn, each process is generally housed in a facility requiring the next tier of indirect inputs, such as factories, office buildings, and service facilities, in turn requiring other inputs, some of which are supplied by means of infrastructures. For example, the system for production may be an assembly plant requiring direct inputs flowing in and out on a regular basis, indirect inputs that remain in the system for a much longer time such as the equipment required for the manufacturing process, and a second tier of indirect inputs remaining in the system for a still longer time related to the facility itself. The latter may house several generations of production processes and equipment. As long as the chain of material flows through the network representing an economy remains linear, the rows of the matrix reduce to resource extraction of materials; processing, manufacture, packaging, and distribution; use and discarding of product. Each indirect input is itself part of a technological cycle.

The columns of a product matrix result from differentiating the biosphere-related dimension of sustainable development. These include effects on the biosphere in its capacity as ultimate source of all materials; effects on the biosphere in its capacity as the ultimate source of all energy flows, including solar energy 'prepared' by the biosphere; effects on the biosphere as the ultimate sink of all materials; effects on the biosphere as ultimate sink of all energy flows; effects on biosphere life-support functions; and effects on the biosphere as habitat. The matrix elements can be thought of as checklists, alerting product designers to possible harmful effects, and including suggestions for how they have been and may be prevented. The column structure could be simplified to: resource and materials issues; energy issues; effects of solid, liquid, and gaseous emissions to air, water, or land.

In addition to their use in drawing up checklists of possible harmful effects and how they may be prevented, context matrices may also be used to undertake a qualitative comparison of the environmental-hazard potential of alternative designs of products, processes, or facilities. Each matrix element then contains a score reflecting the seriousness or absence of such hazards; and a cumulative score may be computed for the matrix of each alternative product, process, or facility in which the contribution of some columns may be weighted relative to others. Such a scheme has been developed by Graedel and Allenby.[44]

It should be recalled that any context matrix is derived from a three-dimensional structure generated by differentiating a way of life and sustainable development. Micro- to macro-level issues can be focused on, depending on the level of differentiation. For example, the activity of manufacture in a product matrix may be differentiated into product, process, and facility-related dimensions that each, in turn, can be differentiated into a chain of activities reflecting their technological cycle. This would generate two additional matrices, located underneath the product matrix, whose results can be incorporated into the product matrix. Similarly, the context matrices of several products may be integrated into a 'higher' context matrix representing an activity in which these products are involved. Furthermore, the column structure need not be restricted to the biosphere-related dimension of sustainable development. Additional columns related to the physical, social, and mental health of the human beings involved in the particular activity can be added. For example, as I will analyse in a later chapter for a manufacturing process, it is essential to consider its effect on the physical, social, and mental health of factory workers.

Macro-level preventive strategies can make important contributions to streamlining the use of context matrices. For example, materials selection may be guided by an index of biosphere related hazard potential. Taken as essentially quantitative, such indices would be open to serious criticism, but interpreted as a scale ranging from very severe to no hazard potential, they could make materials selection more preventive, although this must never be taken for granted. Graedel and Allenby report that Volvo uses such a system.[45] Such indices could also be used to assess the hazard potential of waste-streams associated with a particular product, process, or facility and could even be used to compute an overall score by summing up tonnages weighted by such indices. A more simplified form would divide materials into several categories according to their hazard potential. For example, one category may be substances that have already been banned in many jurisdictions. A second category may be substances that are potential candidates for future bans, phase-outs, or restrictions. Still other categories may contain substances that are potentially hazardous to human health under certain conditions. Grouping substances in this way would be a good beginning to improve the preventive orientation of materials selection and the minimization of the hazard potential of waste-streams. There are many ways in which larger frameworks could be created within which the preventive orientation of design and decision making can be improved. This also extends to drawing up check-lists to be considered in conjunction with different matrix elements. Examples are: Have recycled materials been used to the greatest extent possible? Has the use of toxic or otherwise harmful substances been minimized through substitution? Have all waste-streams been examined for possible materials recovery to create useful inputs for other activities? Have more abundant materials been substituted for less abundant ones? Has suitability for recycling been considered as a criterion for materials selection? Have the energy implications of materials choices been examined? Have human-health implications been checked? Have materials that contribute to particular concerns such as ozone depletion or global warming been substituted with others as much as possible? A similar range of questions can be devised for indirect inputs such as solvents, cleaners, lubricants, or catalysts. From subsequent chapters, additional lists can be drawn up for implications related to energy use, work, and the built habitat.

It goes beyond the scope of this work to develop all this in detail. However, this research into preventive alternatives has produced an intellectual by-product in the form of annotated bibliographies.[46] Once the decision is made to fully exploit the potential of preventive approaches by a corporation, municipality, state, or international agency, many possibilities open up. It is the very nature of making technology more context-compatible that precludes prescriptions that are universal. They themselves must be appropriate for the institution, the cultural-value context, and the ecosystems involved. It goes without saying, for example, that the hazard potential of a given waste-stream varies with the kind of ecosystem into which it is discharged. Similarly, certain social implications may be more acceptable in some cultures than others. Nevertheless, context matrices may be used to develop design-for-environment procedures that can become integral to DfX, where 'X' represents the first letter of a particular aspect, such as manufacturability (DfM), materials logistics in a plant and complementary processes in the field (DfMC), ordering from the customers' perspective (DfO), reliability (DfR), serviceability (DfS), and safety and liability considerations (DfSL). In other words, DfE further extends the family of preventive approaches by looking ahead to the environmental implications of particular design and decision choices in order to adjust them so as to eliminate or greatly reduce potential harm.

### 8.6 Societal and Strategic Implications

The basic map of the ecology of the technology of a society represented earlier as four concentric circles reminds us that the network of flows of matter associated with a way of life, and particularly the products that result from it, must be regarded from at least four perspectives because of the way technology, the economy, society, and the biosphere are enfolded into one another. The network and its products simultaneously participate in all four of these wholes.[47] As a consequence of the economic view of society as structured by the invisible hand of the market, the flows of matter in the network and its products are regarded, first and foremost, as *commodities*. It plays down the fact that the economy was made for people rather than vice versa, so that these products should be primarily regarded in terms of the services they render to maintain a way of life, which does not exclude the possibility that, through the workings of the invisible elbow, they may simultaneously render a disservice. Products should, therefore, be regarded, first and foremost, as *social resources*. This perspective must be complemented by products as *ecological resources* since they embody flows of matter and energy temporarily borrowed from the biosphere. In addition, products represent *strategic resources* in so far as they can promote or hinder a transition towards a more sustainable way of life. The application of the precautionary principle demands that improving the environmental performance of the network of flows of matter and its associated products must not be achieved at the expense of creating economic or social difficulties. Combining this with the principle that the economy is made for people suggests that products are to be judged in terms of the social services they render with minimal disruption to society and the biosphere, and that this should be reflected in the markets for these products.

There appear to be three ways of accomplishing this. The most radical is a restructuring of the economy by shifting its focus from products to services (similar to Edison's desire to sell illumination rather than electricity). The second is to implement across-the-board product stewardship by which manufacturers accept full responsibility for their products after these are no longer capable of rendering a service. The third is based on implementing the best available technologies for reducing the material and energy intensity of economies. The first two distinguish themselves from the third primarily in the way they would accelerate a current trend of corporations extending their control over the technological cycles of their products. In the past, resource extraction and materials processing were mostly under the control of suppliers. The same was true for manufacturing to the extent that components and subassemblies were obtained from suppliers. Packaging and shipping have always been under the control of manufacturers, but the use phase (with the exception of leasing arrangements) and the disposal phase have not. Lean production is leading to much closer symbiotic relationships between manufacturers and suppliers. Product take-back legislation is making manufacturers responsible for their products after their useful life. Shifting the focus of an economy from products to services and fully implementing product stewardship (except where use is inherently dissipative) will further extend this control. I will briefly discuss these three options.

Tim Jackson[48] advocates a shift towards a service economy. Generalizing Edison's idea that people are primarily interested in the services products can render, as opposed to the products themselves, would create a market incentive towards delivering services with the lowest possible material and energy intensity. It would not resolve the problem that the most competitive services may be those whose private costs are lower because of greater social costs since they are delivered by products manufactured in places with lower labour and environmental standards. Thus far the profitability of corporations has been based on the sale of goods, with the result that increased profitability means an increase in the throughput of materials and energy in the economy. By selling the services provided by goods, the incentive of increasing throughput frequently turns into its opposite. For example, the energy sector is structured as if people were interested in energy as opposed to the services it supplies. Utilities have to become service providers, rewarded for helping their customers obtain electricity-based services as cheaply as possible. This concept can be extended to the operational leasing of consumer durables, computers, photocopiers, and cars. Offering competitively priced computing services would require producers to minimize the throughput of materials and energy by designing equipment so that it can be easily upgraded, repaired, remanufactured, and recycled. Providing services as opposed to producing products could permit corporations to get involved in demand-side management by offering alternative ways of providing the service, or by providing another service that might reduce the need for the first. A transportation service company might offer a client assistance in relocating so as to reduce transportation needs. Centralized laundry services could be made more competitive by reclaiming detergents and warm water through membrane-based technologies and using waste heat for preheating water. A corporation could lease degreasing agents to clients on a use-and-return basis, made economical by recovering them in a degreasing plant. Farmers could be offered a crop-protection service as opposed to being sold pesticides and herbicides. These kinds of ventures must be sharply distinguished from others that merely have the appearance of dematerializing the economy. For example, the spectacular growth of financial services increases speculation in commodities and the trading of investment capital, both of which directly and indirectly increase the throughput of matter and energy in the economy.

The shift to a service-based economy only partially addresses one of the fundamental weaknesses of the present economic order – namely, that it overutilizes the biosphere and underutilizes people. This is the case because materials are cheap and labour is expensive, so that manufacturers continue to focus on labour productivity. The balance could be further shifted by the introduction of ecological taxes, making materials relatively more expensive. The revenues from such taxes should then be used to finance a reduction in labour-related taxes. This would encourage a better balance between investments in materials and energy productivity and labour productivity. Some obvious problems will have to be dealt with – namely, the impact on some industrial sectors such as mining, energy, and chemicals manufacturing, and the impact on poor and low-income people as a result of the regressive nature of ecological taxes. A gradual shift towards a service economy could help in overcoming fundamental structural problems in the present system. Reducing the material intensity of the economy will almost certainly expand employment opportunities. Transforming manufacturers into full-service providers in order to make the economy more circular in terms of the flows of materials will affect the technological cycles of products in a way that will increase the workforce: preventive approaches will expand the efforts that go into developing, manufacturing, and managing new products; extending product usage will require highly skilled maintenance, repair, and remanufacturing; and closing the loops on materials flow will create entirely new categories of work. As a result, mining and energy companies, along with chemical manufacturers, would be transformed into full-service providers with a profit basis decoupled from increasing the throughput of materials. It is also not difficult to imagine schemes by which the regressive nature of ecological taxes could be offset. Hawken[49] shows that business can thrive without jeopardizing societies and the biosphere. At present, many corporations are so stuck in the conventional paradigm for the engineering, management, and regulation of technology that they simply cannot see that they are acting against their own long-term interests.

Jackson's wonderfully forthright assessment gets to the root of our present situation. We simply have no idea where technological and economic growth are taking us in terms of the quality of human life. Quoting the late Robert Kennedy, he argues that the measures of our economic accomplishments are so flawed that they measure everything except what makes our lives worthwhile. The material and economic view of society and human nature has become so deeply rooted that it is extremely difficult for us to take seriously the fact that all earlier cultures and civilizations were almost unanimously convinced of the essential but limited contribution material things could make to human well-being. The assumption of *homo oeconomicus,* so central to this world-view, is now contradicted by the latest findings of the social sciences and humanities. Greed and avarice have always been vices that earlier societies attempted to keep under control, and it is only our modern civilization that has institutionalized them in a wholescale fashion. Jackson points out that, just before his death, John Stuart Mill made a crucial distinction that, had he incorporated it into his theories, might have moderated the modern economic view of society and human nature. Mill said, 'Those are only happy who have their minds fixed on some object other than their own happiness: on the happiness of others, on the improvement of mankind, even on some act or pursuit, followed not as a means, but as the ideal end.'[50] In other words, the pursuit of profit makes one wealthy, but it will almost certainly not deliver happiness. In the same vein, the invisible hand cannot create happy societies.[51]

I doubt, however, that the wholescale acceptance of the utilitarian view of human nature that underlies our economic world-view can be credited to those who invented the basic theories and concepts. These might have had little or no impact were it not for the fact that the relationship between people and their technology is a reciprocal one: as people change their technology, they are simultaneously changed by it. As I argue in a separate analysis, with the emergence of industrial societies began the building of a new life-milieu that profoundly influenced human consciousness and cultures. It is this influence that was 'in the air' and made explicit in utilitarian theories and concepts. Once formulated, they resonated in the minds of many who felt that this described what was happening to themselves and others. The current transformation of industrial societies under the influence of technique is transforming *homo oeconomicus* into *homo informaticus* – a change no more receptive to the kinds of values required to transform the bases of economies from products to services. To put it briefly: industrialization, rationalization, and now technicization have created mass societies in which the traditional cultural unities have been eroded by a process usually referred to as 'de-symbolization.' There are no longer any clear values or moral reference points for orienting human life enfolded in modern cultures. As a result, these cultures have to be fed technical information from the outside to meet the existential and social needs of their members. It takes the form of the mass media immersing people in a bath of images and messages informing us of who we should be, what we should wear, what we must eat and drink, what we must use to ensure normal bodily functions, what we must look like, what we must own, how we must live, who our friends should be, and a great deal more. These are the equivalents of what in a traditional society was accomplished through customs, traditions, values, morality, and religion. De-symbolization has all but depleted our cultural and moral capital, as is most obviously manifested by performance values all but replacing human and social values. It is the problem of alienation and reification by the 'system' that has enormously weakened the 'set point' by which we regulate technological and economic growth on human terms. This is the ultimate threat to our common future. It is therefore the cultural dimension of our present situation that must be added to Jackson's analysis.

I am far from saying that the present situation is hopeless. Preventive approaches are one of several potential catalysts that could turn things around. The second approach is that of industrial ecology, seeking to make modern economies more circular in terms of the throughput of materials. Product stewardship, mandated through appropriate legislation, is a cornerstone of this strategy. It undoubtedly has the potential of substantially reducing the pressure on the biosphere in its role as ultimate source and sink of all matter and energy. From a strategic point of view, it is also likely to have an important catalytic effect on transforming modern economies, but the question remains as to whether it will be enough. A World Bank report suggests that, when developing nations achieve the per-capita income levels of the industrially advanced world, a forty-six-fold increase in efficiency would be required just to hold resource consumption and environmental emissions constant.[52] If such a spectacular feat could be accomplished, it would bring the global economy no closer to becoming more sustainable. Others have argued that modern economies will have to dematerialize by a factor of ten to achieve a more sustainable course.[53] Jackson argues that neither of these estimates allows for continued economic growth of the industrially advanced nations, which would eventually begin to increase throughput once again unless we could continue to dematerialize almost indefinitely. The second law of thermodynamics suggests that this is impossible. The point for now is simple enough: an industrial-ecology strategy is a necessary condition for making our modern ways of life more sustainable, but it is almost certainly not a sufficient one.

These first two approaches call for two additional observations. First, both imply an assumption that making modern economies more circular will reduce the throughput not only of materials, but also of energy. This is almost certainly the case, but it needs to be carefully examined. Second, both will require significant transformations of some of our basic institutions. Expanding the control corporations have over the technological cycles of their products further extends the threat to modern democracies. At present, corporations represent growing islands of central planning in a shrinking sea of market-based economies.[54] This political challenge can be overcome through legal reforms transforming corporations into communities of stakeholders, as is currently being advocated. As the effects of the invisible elbow on human life, society, and the biosphere continue to increase, the rights of owners and shareholders will have to be balanced with obligations to other stakeholders. Neither of the above approaches is able to deal with the problem Jacques Ellul has called the 'political illusion,' which stems from the influence of technique on society.[55] This may well become economically, socially, and politically feasible once a strategy of restructuring modern economies gets under way. As we rebuild the 'system,' we will change its influence on human consciousness and cultures, which offers the possibility of creating a civilization that includes technique but is not dominated by it.[56] Once a transformation gets under way, things could change quite rapidly and unexpectedly. Jackson[57] reminds us that we have become trapped in a narrow view of our self-interest, essentially defined in economic terms. It has undermined our communities, societies, and ecosystems, and brought a whole range of new poverties: the poverty of identity, of community, and of spirit. He quotes Lewis Herber, who remarked that we have reached 'a degree of anonymity, social atomization and spiritual isolation that is virtually unprecedented in human history.'[58] The spread of such a recognition could empower democratic societies to sustain strategies of necessary changes once the influence of technique on human consciousness and cultures becomes weakened.

The third strategy is proposed in the book *Factor Four: Doubling Wealth – Halving Resource Use: The New Report to the Club of Rome.*[59] Based on fifty examples of improvements in resource productivity by a factor of four or more, an efficiency revolution is proposed by rapidly adopting the best available technologies for resource productivity. This would permit humanity to do everything it does now with half the resources while doubling the standard of living, which is particularly important for poor people. Ways of unleashing this revolution are set out, and its possible contribution to resolving some of our most important problems is also discussed. The analysis suggests that growth has more to do with turning over resources than with human well-being, that current free-trade policies make it more difficult for nations to implement their own social and ecological policies, and that non-economic and non-material values must be clearly understood since they point the way to human satisfaction. In other words, this book argues that we should get on with what we can currently do in terms of the best available technologies for improving resource efficiency and utilize the results to leverage genuine development. This strategy may well be the best entry point into the more comprehensive strategy advocated in this work. The implementation of the precautionary principle by the European Union, along with the concept of strict liability, could greatly accelerate these kinds of developments because compensation may be sought for unnecessary harm done to people or ecosystems in terms of what could have been prevented by using the best available technologies.

## 8.7 Getting Started

In conclusion, I wish to emphasize that the above macro-level strategies can, in part, be implemented immediately by taking the following steps. Corporations that have not already done so can set out on a more sustainable course by carrying out an environmental audit on their operations. It begins by completing a mass–energy balance of all operations to account for all inputs, as well as desired and undesired outputs. On this basis, a flow diagram is drawn up connecting all inputs and outputs. All costs associated with these flows are identified. Waste-reduction opportunities are sought by better housekeeping, materials-input substitutions, process modifications, and product redesign. Technical, economic, and environmental assessments of these options must be undertaken in order to prioritize them. Finally, on the basis of this information, a waste-reduction strategy is developed, including target dates for each component or phase. New products, processes, and facilities should be developed using preventive methods based on context matrices. Corporations can investigate the strategic advantages of selling services as opposed to products, including the implications this has for the throughput of materials and energy.

Engineering and management schools can begin to reorient their curricula to make them more preventive. Governments can tighten environmental standards in the full confidence that the industries that will strengthen a nation's economy will creatively respond, while industries that know only how to respond through lobbying have no long-term future in any case. Democratic societies should devise ways in which governments can be protected from excessive lobbying to enhance the possibility of developing and carrying out policies that are genuinely in the public interest. Implementing the precautionary principle and strict liability will accelerate the diffusion of cleaner technology. Together these measures will begin to create a climate that makes their benefits evident, and thus encourages taking further steps.

# 9

# Energy

## 9.1 Basic Principles

In chapter 8, I showed that, since a society can neither create nor destroy matter and energy, the activities of its way of life are interconnected through networks of flows of materials and energy. All inputs and outputs into these networks come from and return to the biosphere. The metabolic aspect of the way of life of a society can be made more sustainable by reducing the reliance on the biosphere, which is accomplished by transforming linear throughput patterns into circular ones for the network of flows of materials. Such a restructuring will also affect the network of flows of energy because both result from the fabric of activities that make up the way of life of a society. A reduced throughput of materials translates into lower quantities of virgin materials that must be extracted from the biosphere, then refined and processed into useful materials. If the reduction in the energy input required to sustain this chain of activities is greater than the input into the activities needed to close loops, including remanufacturing and recycling, the total energy necessary to sustain a way of life will decrease and its biosphere-related sustainability will improve. This is generally expected to be the case unless the transportation requirements for the latter chain of activities are excessive. Unfortunately, the industrial ecology literature does not pay as much attention to the energy implications as it might.[1] Nevertheless, one of the major consequences is likely to be the reduction in our dependence on the biosphere for energy.

The network of flows of energy embodied in the way of life of a society resembles a network of flows of materials. For example, fossil fuels enter the network as flows of materials distinguishable from others in terms of their ultimate purpose of being converted into energy and mostly useless materials. This energy supports most activities in a modern society and is therefore closely linked to the network of flows of materials. For example, it may be temporarily absorbed into flows of materials (such as steam leaving a boiler or molten iron leaving a blast-furnace), stored in the chemical bonds of materials (such as in plastics), stored in energy carriers (such as electricity or hydrogen), or degraded and dissipated in the performance of useful work associated with the technological cycles of materials and products. Every material and product has a gross energy requirement equal to the total of all energy inputs into the chain of activities producing that material or product. It includes several additional components: the energy required to sustain the chain of activities delivering each unit of fuel or energy, commonly referred to as its 'fuel-cost value'; indirect inputs required to produce the process equipment and facility for each activity; and another tier of indirect inputs to produce this equipment and these facilities, and so on. It is evident that the structure of the network of flows of energy is as complex as the fabric of activities that makes up the way of life of a society, and hence almost impossible to model in its entirety. Nevertheless, such a structure exists, and it must be reckoned with as such, and not merely on the level of its constituents.

There are two lessons we are gradually learning from the experience of building the present energy networks. First, energy policy should aim to involve the entire network of energy flows and not just the supply for the network. Second, energy policy should be preventively oriented and recognize that the energy network is simultaneously enfolded into an economy, a society, local ecosystems, and the biosphere. It will therefore have major economic, as well as social and environmental, implications, which must be dealt with preventively in accordance with the precautionary principle. Had this been done much earlier in the development of energy networks, a great deal of investment capital could have been saved, and many social and environmental problems related to energy use could have been prevented or at least substantially diminished. These two lessons could have been learned directly from the general principles derived from the first and second laws of thermodynamics, set out in the previous chapter. Energy flows through the network cannot be made more cyclical because energy transformations are irreversible. Hence, each linear chain of energy transformations and the associated chains on which it depends can be made more sustainable by supplying them as much as possible from renewable sources and by extracting as many services as possible before the chain crosses back into the biosphere. The corresponding restructuring of the network is the focus of this chapter.

## 9.2 The Development of Energy Networks

In most countries, electricity and natural gas have been distributed through government-regulated monopolies. The reasons are obvious. It makes no technological or economic sense to have a variety of competing suppliers each run their own electric grid or network of pipelines to their customers. Imagine six sets of hydro poles and wires lining our city streets or having to dig up those streets each time the occupant or owner of a house changes gas companies. In addition to being impractical, such arrangements would also be very expensive since distribution costs, already a significant portion of energy costs when delivered through a single infrastructure, would have to be significantly higher. The creation of monopolies regulated in the public interest therefore amounts to the creation of a partial energy commons. For it to work well, utilities and their customers would have to arrive at a mutually satisfactory arrangement. Since customers have no alternative supplier to turn to if dissatisfied, they have to be assured of a totally reliable supply, receive electricity or gas at a price that provides utilities a reasonable return on their investment, thus ensuring their long-term viability, and have access to an accountable regulatory body with the expertise to adjudicate disputes in the public interest.

For electric utilities, this arrangement presents a particularly difficult challenge. There is no cost-effective way of storing electricity, so that it has to be produced on demand, and predicting that demand has always been a major preoccupation of utilities. Generating capacity must be adequate to meet daily peak demands (which usually occur in the morning when people are getting ready to go to work and in the evening when they return home), as well as anticipate future requirements. Failure to do so would not permit utilities to supply customers with whatever quantity they require at any time. Keeping up with an exponentially growing demand during rapid industrialization and economic growth has led to a single-minded focus on supply, which dominates energy planning and policy making.

Government or utility forecasters begin by carefully examining past trends. How much primary fuel of various kinds have they used in the recent past, and how much of this requirement is met through importation? Who are their industrial, commercial, and private customers, and how much electricity have they purchased? How has this varied over the last few years? Can the same trends be expected to continue in the next few years? What factors could change such trends and by how much? In the 1950s, demand was growing exponentially and appeared to correlate rather well with economic growth measured in terms of GNP. Energy planning increasingly reflected the growth-oriented cultures of the industrially advanced nations, deeply rooted in their cultural unities (sacred, myths, and hierarchy of values). Performance measures masquerading as human and social values dominated the energy business, as it did most other sectors of society.

Technological trends reinforced these developments. Economies of scale led to ever-larger power stations, requiring ever-larger amounts of investment capital; ever-longer lead times for their planning, construction, and being made operational; with ever-larger environmental effects. All this further complicated energy planning. One thing, however, appeared certain to almost everyone: more economic growth would require more energy, and to be caught short would jeopardize practically everything modern societies stood for. Needless to say, anyone who challenged anything to do with energy policy was essentially treated as a heretic might be in a traditional society. Many governments entered the race to keep up with energy demand by developing nuclear programs expected to make electricity 'too cheap to meter.'[2] Conventional wisdom in engineering was all but ignored as the size of nuclear power plants was scaled up in ways unheard of in the chemical industry.[3] The OPEC oil embargo further reinforced the nuclear programs as a way of reducing dependency on foreign oil supplies. Soon after, the nuclear enterprise went into a decline and today stands as one of the most monumental failures of planning and forecasting. Enormous cost overruns, design problems, environmental and safety considerations, and – last but not least – the dangers of nuclear proliferation all served to bring the bandwagon to a halt. Utilities generally operated nuclear-power plants at considerable losses that had to be covered by governments. All this despite the fact that in every country the nuclear industry was legally protected through legislation governing liability, and the need to plan and pay for long-term disposal of hazardous nuclear wastes was deferred.

Until the mid-1970s, energy planning by utilities, corporations, and governments was based on the premise that energy use would continue to grow and that no limits were in sight. More economic growth, and thus energy use, would simply go on for ever. Common sense was in scarce supply, with a few notable exceptions. As far back as the 1950s, E.F. Schumacher, then adviser to the British National Coal Board, showed that the arithmetic of exponential growth which energy planning took for granted led to completely absurd results in a surprisingly short time.[4] In the late 1960s, M. King Hubbert of the U.S. Geological Survey correlated the rate at which we were depleting oil stocks with the rate at which we were adding to these stocks through new discoveries, and concluded that oil could continue to play a role in human history for about two centuries.[5] Three influential projections of future energy needs were the 1977 report of the Workshop on Alternative Energy Strategies (WAES),[6] the 1978 and 1983 World Energy Conference reports on World Energy Demand (WEC),[7] and the 1981 report of the Energy Systems Program Group of the International Institute for Applied Systems Analysis (IIASA).[8] All pointed to slower growth of energy demand in the future and anticipated a shift away from non-renewable resources to renewable or more abundant resources, but this still led to implausible futures. For example, to achieve the IIASA projected energy demand for 2030 would require expanding energy-production capacity at an average rate between 1975 and 2030 of one new central station (coal or nuclear power plant) of one gigawatt electricity-generating capacity every one to two days, one new nuclear plant of one gigawatt's capacity every four to six days, and new fossil-fuel capacity amounting to bringing on line the equivalent of an Alaska pipeline with its capacity of 2 millions barrels of oil per day every one to two months.[9] This is clearly economically, socially, politically, and environmentally absurd. The capital costs would be so enormous that energy policy would have to take priority over everything else, while the environmental impact would be beyond comprehension.

The fundamental flaw in energy planning since the end of the Second World War has become increasingly apparent. The industrially advanced world had walked into a trap, which became obvious in hindsight, but not so from the perspective of the electric and gas utilities. They had concentrated entirely on supply, based on the assumption that the demand side of the energy networks was beyond their control, being influenced only by the prices utilities were allowed to charge. Future energy use was predictable only by extrapolating trends, and these, in turn, were essentially determined by economic growth. As far as the suppliers were concerned, energy demand was autonomous – a non-negotiable demand from society at large. The utilities simply had to supply it. The role of society in all of this was to cooperate with the regulators and utilities, pay the necessary costs, eliminate barriers, and offer incentives. This was more or less what happened until the middle 1970s, despite the fact that predictions were frequently inaccurate. Gradually, some adjustments were made, recognizing that past trends would gradually change in the direction of a less rapid rate in the expansion of energy demand, but this did not alter the fact that the only solution to the energy problem being contemplated was to produce even more energy. Any reduction in energy demand could be seen only as a reduction in the quality of life enjoyed by most citizens of the industrially advanced nations. Little consideration was given to the implications for the developing world or the biosphere.

The energy-policy project set up by the Ford Foundation in 1971 was one of the first to move away from the supply-side focus. The final report, published in 1974,[10] declares its intent in its title: *A Time to Choose*. At this point, the OPEC oil embargo had driven industrial economies into a severe recession. The report was premised on the recognition that a nation has many possible energy futures, and it presented three plausible scenarios that might be achieved through different policy choices. It recognized that the real energy future may not resemble any of the three scenarios, but this was not the point of the exercise. The report's intention was to suggest that societies should begin to consider the implications of different rates of energy growth for the economy, society, and the biosphere. For example, what are the implications for a society's foreign policy, social equity, lifestyles, and human health? What policies and resources are required to bring about a desirable scenario? What is remarkable about this report, apart from looking at the broader implications of energy use, is the fact that it reverses the hitherto prevailing relationship between forecasting and policy making. A society should not first make a forecast and then generate the necessary policies. The relationship should be reversed: policies should affect forecasts by steering in a direction that conforms to a society's values and aspirations.

The three scenarios selected were historical growth, technical fix, and zero energy growth. The first extrapolated energy use from 1950 to 1970, increasing at about 3.5 per cent per year, and thus almost doubling by 2000. The second scenario assumed the same mix of goods and services supplied by the economy but involved a conscious national effort to use energy more efficiently through best available technology and its expected evolution. This cut the rate of increase of energy use to about 1.9 per cent, so that, in 2000, energy use would be two-thirds of that projected in the first scenario. No detrimental consequences for people's lives were expected. The third scenario shifted the balance between goods and services provided by the economy somewhat towards services: better public transportation, health services, parks, and so on. It also included an emphasis on energy efficiency through best available technology, as in the second scenario. The result would be zero energy growth by the year 2000, while at the same time people would experience an improvement in the quality of their lives. Given the cultural conditions of the time, it is not surprising that the report was widely greeted with disbelief despite its extensive documentation. Nevertheless, it was a sign of things to come – namely, the recognition that economic growth is possible along with a dramatic decline in energy use.[11]

Contemporary thinking about energy owes much to the pioneering works of Amory Lovins, beginning with his 1973 publication, which was soon followed by his classic work entitled *Soft Energy Paths*.[12] His distinction between the hard and soft energy path became a lightning rod for controversy. He suggested that society had two fundamental, and mutually exclusive, energy options. The hard path was essentially U.S. policy in the 1970s, based on a supply perspective to be ensured by ever-larger facilities converting ever-larger quantities of primary fuels (mainly coal and uranium) into electricity. He argued that this required more capital than society could reasonably supply, provided few jobs despite this enormous investment and thus led to unemployment, and threatened the biosphere through the extraction of resources and the disposal of wastes. The development of these energy systems required the suppression of the values of freedom and democracy, and increased the risk of nuclear war. He was particularly critical of a U.S. plan to shift from uranium to plutonium using fast breeder reactors, of which 400 were planned to be in operation by the year 2000 according to the U.S. Atomic Energy Commission's report on nuclear-power growth.[13] A single large-scale breeder reactor would require at least 10 tons of plutonium for its operation. Not only is plutonium exceedingly toxic, but only 10 kilograms (a volume approximately the size of an orange) can make a bomb similar in destructive power to the one used on Nagasaki. Then imagine processing, shipping, and storing 4,000 tons of plutonium by the year 2000, while ensuring that not even 10 kilograms falls into the wrong hands, and the reason for his voice of protest becomes self-evident – it is amazing that the American people did not take to the streets. Fortunately, many others began to recognize that something was going fundamentally wrong.

The soft path was rooted in the observation that there was a considerable discrepancy between energy use in the United States and in European countries with similar standards of living. As Lovins noted, the United States was not more civilized or more comfortable than countries like West Germany, Sweden, or Denmark. However, the average German, Swede, or Dane used only about half as much fuel and electricity as the average American. Energy use in Canada was higher still, despite a climate comparable to that of Sweden. Also, various studies suggested that the Europeans had far from exhausted the opportunities for energy efficiency. Thus, the soft energy path could be imagined, based on doing more with less through energy efficiency and supplying the energy by means of very different hardware designed to be flexible, resilient, sustainable, and benign. These soft-path technologies rely on renewable resources provided by energy income as opposed to depleting energy capital. Such an infrastructure would help to build and be built by a democratic and more egalitarian society than would be possible with the hard path. In sum, a soft-path energy policy could eliminate or vastly reduce many problems associated with the hard path, more easily realize Western aspirations and values, and greatly reduce the negative impact on the biosphere.

Lovins's work was complemented by a new field called 'energy analysis,' developed by Peter Chapman in his book *Fuel's Paradise.*[14] Its object is to assemble and examine data on the energy conversions associated with every activity of a way of life, including farming, running a household, extracting or processing resources, constructing and operating buildings, manufacturing goods of any kind, transporting goods, and so on. This data can then be assembled into a comprehensive picture of energy use in a society. This led to another technique called 'energy accounting,' which is analogous to financial accounting but, instead of tracking cash flows and financial investments, tracks energy flows and investments. For example, the energy investment in a power plant represents the total energy required to build the equipment and the facility. Its operation needs a continuous flow of energy to convert into electricity. The distribution of electricity requires additional equipment and facilities. Tracking all these flows makes it apparent what is involved, in energy terms, in an activity such as reading a book by lamplight. In other words, energy accounting seeks to understand where the energy in a society is coming from and where it is going. Its point of departure is how people use energy and for what purpose. The value of this approach was further demonstrated by Gerald Leach in his book *Energy and Food Production.*[15] Its complexity becomes evident when we take into account that the use a litre of diesel fuel has is not limited to the gross heat content but includes all the fuels consumed in exploring for oil, extraction, shipping, refining, and delivery to a final consumer. This also includes fuels used to provide all the materials, equipment, and facilities involved in this chain of activities. On this basis, the book examined the energy budgets of food-producing activities in the United Kingdom, including entire farms of different kinds, and compared them with the budgets of earlier food-gathering and agriculture. His conclusion is worth quoting:

> No technical fix will alter the inequity that allows, for example, fifty-five million Britons to consume enough primary foodstuffs to feed half the Indian subcontinent, five hundred and fifty million people approximately, or to use almost a tonne of oil per head in doing so ... The food supply systems of the rich, represented here by the UK, have become heavily dependent on large injections of fuel subsidy. While this energy prop has improved the productivity of the land and of labour, the overall improvement when one looks at the entire food system has not been all that dramatic ... the apparent 'success' of these micro-improvements, and the dominance of Western advice and technology ... has persuaded much of the world that the Western way is also the best development path to follow.
>
> Fortunately, a new vision that other development paths are both more promising and more practical has now begun to seize the minds of scientists, planners and politicians throughout the developing world. Carbon copies of Western methods are increasingly seen as irrelevant due to their high cost, high resource use and labour-saving characteristics, and the persistent tendency to widen the development gap between the cities and the country-side, the haves and the have-nots. Instead, the new emphasis is on decentralized rural development, using appropriate small-scale technologies in the context of self-help and self-reliance.[16]

The International Institute for Environment and Development later published a report on further research that had a profound impact on

British energy policy and influenced energy studies around the world. Its title, *A Low Energy Strategy for the United Kingdom*,[17] shows that the exponential growth of energy use can be modified. The book demonstrates how the United Kingdom could have fifty years of prosperous material growth while using less energy than today. It challenges the warnings of traditional energy forecasting that the only alternative to ever-rising energy use is an economic decline and a reduction in the standard of living. It also shows that there is no correlation between energy use and GDP. For example, in the United Kingdom the non-manufacturing sector generates over 50 per cent of the total GDP but uses only 12 per cent of all energy because it essentially depends on human activities and skills. Energy use is primarily related to the heating and lighting of buildings, which could be significantly reduced as GDP could continue to rise. The energy–GDP correlation also breaks down in domestic activities. A detailed analysis showed that for some twenty-five years domestic fuel and electricity use grew in proportion to income, but that does not mean it will continue, because houses will then become so hot that they are uninhabitable. The correlation here would imply that the more you earn, the hotter your house. Leach referred to this as the 'roast the rich' theory, which is one instance of what has become known as 'saturation effects.' Energy analysis can detect saturation effects and other limitations, and thus comprehend the development of energy usage from the inside out. The analysis must always start with the ultimate purpose for which energy is used and work back to the primary energy supply, permitting policies based on energy end-use planning and demand-side management. In other words, it is possible to examine in detail the energy implications of a way of life and its activities.

The basic approach, pioneered by Leach, works as follows. It begins with the ultimate purpose for which energy is used and works back to the primary energy supply, fuel by fuel and subsector by subsector, showing in detail how different fuels are used in different sectors to provide useful energy that sustains the activities of a society. Using these categories, Leach and his colleagues broke down energy supply in 1976 into some 400 categories, determined by end-users, fuels, and appliances. The likely evolution of energy demand was then analysed by selecting an array of activities, each associated with particular fuels and electricity. For example, for housing they considered average interior temperatures, dwelling size and volume, quantities of hot water used per person, the amount of cooking, and the use of seven categories of electrical equipment. They then tracked these activities and their associated use to the year 2025, assuming business as usual but more of it. Houses were assumed to become warmer so that more people enjoy the comfort levels of today; most families were expected to acquire freezers, dishwashers, clothes dryers, colour televisions, and other heavy users of electricity. Car ownership and air traffic were also assumed to grow rapidly. These assumptions were made to better understand the shortcomings of how official energy forecasts are made. This led to a startling conclusion. Using these assumptions, and building up energy use sector by sector, they found that, in their high scenario, the total primary energy use of fuel and electricity in 2025 in the United Kingdom would be 8 per cent lower than in 1975. In their low scenario, it would be 22 per cent lower.

The analytic approaches pioneered by Lovins and Leach are gradually gaining acceptance around the world. In 1988, a report of the end-use oriented Global Energy Project concluded that, with the right focus, it was possible within forty years for everyone on earth to have the material well-being now enjoyed in the industrial countries while global use of fuels and electricity would increase by only 10 per cent and then start a downward trend as higher efficiency gains were implemented.[18] This involves bringing close to four-fifths of the world population up to the level of material well-being of one-fifth of the population, and of these four-fifths a significant percentage have yet to progress beyond the use of firewood as the primary energy source. Of course, this will not happen by chance. It will require a sustained effort and clear policies and incentives, but the rewards are enormous, and the conventional alternative could not even come close to accomplishing the same thing with such positive effects for so many people, while at the same time making our modern way of life steadily more sustainable with respect to its dependence on the biosphere. Probably nothing can do more for many of our current problems than an all-out effort to improve energy efficiency.

Many events in the last few decades have heightened our awareness of how deeply energy use affects society and its ecosystems. One of the most obvious examples was the OPEC oil embargo and the subsequent economic recession. It painfully heightened our awareness of the impact of oil imports on the balance of payments of many nations, the differential effects of energy costs on rich and poor countries and their citizens, the impact on lifestyles, the improvement of air quality in cities where car use was restricted, the influence on agriculture, and a great deal else. Goldemberg and colleagues[19] show how energy usage is intimately related to a range of global problems, including the wretched living conditions of half of humanity, nutritional deficiencies and food shortages, widespread unemployment and underemployment, staggering national debts, scarcity of development capital, environmental degradation, the threat of global warming and associated climate changes, and global security, including the risk of nuclear war. I will limit myself to a few of these linkages. The current global economic order was essentially established during the colonial period. The role of the south continues to be heavily based on the extraction, refining, and processing of resources, while the manufacturing activities, accounting for the greater part of the value of goods, is carried out or controlled by the north. The former activities tend to be much more energy-intensive and more of a burden on the biosphere than the latter. Commodity prices tend to fluctuate much more than those of manufactured goods, and the former tend to decline relative to the latter. This creates obvious problems for those nations whose foreign earnings largely depend on the export of one or more commodities. Transnationals continue to control the flow of capital, manufactured goods, technology, and many commodities. The attendant preoccupation with maximizing performance values creates frequent conflicts with the interests of their host countries and the communities in which they operate. This is clearly not a satisfactory arrangement for the majority of humanity.

When perturbations occur in this global economic order, they tend to have a much greater effect on the south than on the north. For example, when OPEC countries sharply increased oil prices, many nations in the south had difficulty financing their oil imports. When interest rates rise sharply, many poorer nations have trouble servicing their debts. Much of this debt load is in U.S. dollars, making these countries sensitive to the performance of the U.S. economy. Energy policies in the north are likely to have increasingly global effects. Already the north consumes some two-thirds of global energy, and this proportion continues to grow. This will undoubtedly lead to increased competition for dwindling non-renewable energy resources, which is bound to further disadvantage the south. Energy shortages will continue to be a major bottleneck in the south. International agencies encourage, and sometimes impose, what are essentially conventional energy strategies, even though many nations in the south are much less prepared to deal with their negative implications. Such strategies tend to benefit the urban sectors much more than the rural ones, further enlarging the development gap. The widespread adoption of northern agricultural methods has made the production of food much more dependent on energy, frequently in the form of imported oil. The dependence of the north on oil is so crucial to its survival that it will almost certainly go to war if its umbilical oil cord is cut. The rapidly growing population of the south will increase pressure on dwindling oil reserves, creating a growing threat to global security.

What we are beginning to learn may be summed up as a growing recognition of the extent to which the network of energy flows is simultaneously enfolded in the economy, social fabric, and local ecosystems of nations, as well as the global order that binds them together. This means that energy, like materials, must be simultaneously regarded as a commodity, a social necessity and resource, an ecological resource, and a strategic resource. Thus far, we have behaved as if energy were no more than an economic commodity so that we could entrust it to the invisible hand of the market to put our energy house in order. As a result, the invisible-elbow effect has been enormous because energy is more than a commodity, even though the market externalizes much of this effect. Conventional economic wisdom would have us believe that, if alternative energy strategies were as advantageous as claimed, they would have been implemented a long time ago. This would be true, provided that conventional economic wisdom did not treat much of human society and the biosphere as externalities, and if energy conversion did not involve many such externalities. It is time to put the assumptions underlying conventional economics to the test. It is increasingly obvious to all but conventional economists and the governments that rely on them that people do not behave as *homo economicus* when they purchase appliances, for example, or make other energy-related choices. Nor is it the case that if governments seriously engaged in reducing greenhouse gases, the economy would suffer a serious blow. This would be true only if conventional engineering approaches were used. Most economists have not yet appreciated the fact that preventive approaches take into account many things that conventional economic wisdom considers mere externalities. The same wisdom argues that governments should keep their hands off markets while ignoring that massive subsidies now flow to energy options that are least economic. In the dominant economic world-view, markets can do no wrong, and they are treated with a certain secular religious awe, obscuring the fact that they are mere human creations that are very useful for certain things, useless for others and irrelevant to still others. Markets function best when the limitations are clearly understood and compensated for by means of appropriate policies. Conventional economics would greatly benefit from a reality check that would open it up to the recognition that an economy is enfolded into human life, society and the biosphere. In the meantime, it is important to get on with the choice of an energy strategy. This will be made much easier if the short-term and long-term implications for human life, society, and the biosphere of each option are clearly understood and explained. Because of their vast implications, energy choices are of great strategic importance to make modern ways of life more sustainable.[20]

## 9.3 Steps in the Right Direction

Events in the 1970s brought U.S. utilities to a crossroads. Their technological infrastructure was quantitatively and qualitatively changing: the new power plants were larger, costlier, and less reliable, and the nuclear technology was unproven and frequently much more expensive than anticipated because of cost overruns. The economic and regulatory framework within which the utilities operated also underwent significant changes. The oil embargo in 1973 and the Iranian Revolution in 1979 contributed to higher oil prices, high interest rates, and inflation. In 1978, the U.S. Congress obliged utilities to purchase at a fair price privately generated electricity (produced by co-generation plants in industry, for example) and distribute it on their grids, thus effectively opening electricity production to some competition. These kinds of challenges stimulated some utilities to rethink their business. Edison's concept helped some of them return to basics by recognizing that their customers were not interested in electricity per se but in the services electricity provided. This opened up entirely new horizons as the focus of their business now shifted from supply to end-use.

Several utilities entered into what Amory Lovins had called the 'negawatt revolution.'[21] Negawatts, so they discovered, were almost always cheaper than megawatts, meaning that it was advantageous to negatively generate additional capacity by reducing demand through improving end-use efficiency rather than by continuing to increase supply by means of building new power plants. By selling energy efficiency to consumers, the saved electricity could be used to offset increasing demand.

The negawatt revolution could not succeed, however, without some significant changes to the regulatory framework within which utilities operated. From the beginning, public utility commissions have set electricity prices based on a paradigm that saw utilities much like any other producer of a commodity. It determined what return on investment utilities would need to attract investment capital, given current markets. The necessary income had to be generated from sales to households and businesses in accordance with their fair share of operating costs. The fundamental problem with this arrangement was that utility profits depended on how much electricity they sold. If they sold more than what was projected, their income would increase, or vice versa. This kept all parties focused on demand forecasts because, at public utility commission hearings, it was in the interest of utilities to understate expected sales and for customers to overstate them. Once the hearings were over and the electricity rates set, it best served the utility to sell as much electricity as possible. It therefore had no interest whatsoever in advising customers on end-use efficiency, since this would undercut sales and the financial viability of the utility. Some of the limitations of market forces had to be overcome by a subtle but all-important shift – namely, from a market for electricity to a market for energy services. Once this was accomplished, market forces could drive the negawatt revolution.

In the late 1970s and early 1980s, some U.S. states recognized this problem and changed the regulatory framework within which utilities operated so that their profits no longer depended on how much electricity they sold. The temptation for different parties to 'play around' with forecasts was eliminated by setting up a balancing account into which a utility paid if it earned more than was anticipated and from which it was compensated if it earned less. In addition, utilities were now permitted to derive income from advising customers to lower their electricity bills by improving energy end-use efficiency. Since saving electricity was almost always cheaper than producing it, utilities no longer had an incentive to maximize electricity sales and build new power plants. The shift from markets for electricity to markets for energy services could now begin to grow. The potential impact on utilities was enormous. For example, the Pacific Gas and Electric Company is the largest investor-owned utility in the United States, serving most of northern California.[22] In the early 1980s, this utility was planning to build some ten to twenty power stations to keep up with demand. By 1992 it changed all this, and by 1993 it permanently disbanded its engineering and construction division. It decided that, for the 1990s, it could cover three-quarters of new demand through energy end-use efficiency increases and the remaining quarter would be generated by privately bid renewable resources. As a fall-back position, it would consider purchasing the newer combined-cycle plants with gas-fired, steam-injected turbines. Conventional power plants were considered too costly. In 1992, the utility invested in excess of $170 million with a lump-sum benefit in excess of $300 million, which was split 85/15 per cent between its customers and the utility. Obtaining energy end-use efficiency had become the most profitable activity for the utility. In 1993, the California Public Utilities Commission (PUC) reported that, from 1990 to 1993, utilities under its jurisdiction saved their customers almost $2 billion after costs, and in 1994 the PUC reported that field studies confirmed that efficiency programs had made the predicted energy savings at a far lower cost than generating the electricity.[23] Energy-efficiency approaches can also help reduce the appetite for vast hydro-electric projects with their enormous economic, social, and environmental consequences. 'Virtual dams' can, in fact, be built through energy-efficiency projects. This was proposed to Hydro-Québec when it was planning to construct the 450-megawatt Great Whale Hydroelectric Project in the early 1980s. The virtual dam could be built in Montreal by saving this capacity through energy-efficiency projects, costing only a fraction per kilowatt hour. This was not accepted, but the project was eventually shelved.[24] The success of electricity- conservation programs in the United States has been examined in some detail by Howard Geller.[25] His research indicates that between 1973 and 1987 the energy efficiency of the U.S. economy steadily improved. Energy end-use efficiency programs were cutting load growth up to 1.4 per cent per year at a much lower cost than energy-supply alternatives, even when social and environmental costs were not taken into account. The study points to considerable improvements in energy end-use efficiency. For example, at the end of the period, refrigerators consumed 42 per cent less electricity than 1972 models, despite an average 14 per cent increase in size. The efficiency of air conditioners improved by 25–35 per cent. Electricity requirements per square foot of commercial buildings fell by an average of 18 per cent, and high-efficiency motors improved about 5 per cent. The study also suggests that many other devices improved hardly at all, that the potential of energy-efficiency improvements were nowhere near their cost-effective potential, and their market penetration required improvement. The report suggested that underinvestment in energy-efficient devices by consumers can, in some cases, be explained by the fact that landlords decide on capital-cost acquisitions while tenants pay the operating costs. Utilities can finance supply-side options at interest rates that are not available either to landlords or to consumers. In 1987, 80 per cent of utilities had conservation programs to help remove barriers of this kind. Such programs involved cash rebates, mostly on heat pumps and air conditioners, but some programs also included efficient refrigerators, water heaters, and lighting. Commercial and industrial programs also existed, which included heating, ventilation and air conditioning, lighting, and (less commonly) efficient electric motors and industrial process modification. The scale of these programs was rather modest. There were only five residential and two commercial programs with budgets over $5 million per year. Energy efficiency enabled utilities to avoid increases in peak demand for one-tenth to one-half the cost of alternative supply options. Much publicity was given to the so-called free-rider problem of people who would have bought the more efficient equipment in any case. Nevertheless, careful studies with a control group appeared to indicate that financial incentives increase investments in energy efficiency.

The same study also reports that some utilities are using demand-side bidding by landlords, industries, or independent energy-service companies to undertake conservation projects. Since bids include details of all financial arrangements, utilities are able to evaluate technical feasibility, reliability, and cost, and then enter into contracts with the most competitive bidders. This demand-side bidding permits utilities to acquire negawatts in large chunks in a cost-effective manner. Some utilities, however, use direct installation, where a utility or energy-service company hired by a utility installs energy efficient equipment in the residential, commercial, or industrial sectors. The study suggested, however, that a number of issues remain unresolved.

There is one obvious factor in demand-side management programs (to which I referred in the previous chapter) – namely, that selling services rather than products or commodities gets corporations more directly involved in the daily lives of people and institutions. Hence, how these kinds of programs are marketed and delivered is of great importance. It is not surprising, therefore, that the study noted above reported that demand-side management programs worked much better where there was community involvement. The progressive weakening of communities and the erosion of moral capital that parallels industrialization and the excessive reliance on the market as an ordering principle of society stand in the way of fully realizing the potential of preventive approaches in these areas despite obvious economic, social, and environmental advantages for all parties. In some cases, however, intrusions can be avoided, again to the advantage of all parties. For example, von Weizsäcker and colleagues[26] suggest that better results can be obtained by paying rebates for prevention-oriented design. Utilities could pay for better design or building standards. The Swedes pioneered this approach. In the United States, a design competition was held for refrigerators, aimed at doubling the efficiency without CFCs, with no compromise in convenience. The manufacturer who won the competition was rewarded in accordance with the number of units of the prize-winning design actually being sold. This has the advantage that utilities pay only for the savings they will incur, while at the same time leveraging other manufacturers to follow suit. Such competitions could have an even greater pay-off if government procurement for public housing and other facilities guaranteed purchases of the award-winning design. Another variation suggested by von Weizsäcker and colleagues is for utilities to lease highly efficient equipment and finance it in exactly the same way as they finance new power stations, which could make it very competitive indeed. If financial rebates are to be awarded, they should be paid to the manufacturer so that there will be a multiplier effect on wholesale and retail prices. On a larger scale, in addition to marketing megawatts, markets for negawatts could be created by making saved electricity a tradeable commodity, just as this is done for other commodities. Specific markets could be changed by following the example of British Columbia Hydro, which transformed the market for electric motors. At first, high-efficiency electric motors were a special-order item. They were not being purchased since, when there is a motor failure, long delivery times are impractical. By offering generous rebates, the market was reversed, so that low-efficiency motors became the special-order item. It is worth noting that rebates were paid only where 'death certificates' accompanied proof of purchase so that these motors would not end up somewhere else – the purpose was replacement, not displacement.

In many cases, however, the way of delivering a particular service may have to be rethought. For example, realizing the potential of preventive approaches for refrigerators would require that heavily insulated refrigerator cabinets be built into kitchens, just like the other cupboards. The compressor and required coils would have to be a separate unit, so that it could be located far enough away from the refrigerator to avoid heating it. Further advances could be made for colder climates, in so far as thermostatically controlled cold outside air could be used in winter, backed up by the compressor whenever required.

Once market reversals making highly efficient equipment the standard are achieved, rebates can be withdrawn provided governments make these trends permanent by mandating higher efficiency standards. However, to bring about a comprehensive and systematic restructuring of the electricity portions of energy networks will require a level playing field between supply- and demand-side options.

**9.4 Integrated Resource Planning**

To further drive a preventive restructuring of the electricity portion of energy networks, integrated resource planning or least-cost planning can force utilities to invest in energy efficiency or demand-side management where this is more cost-effective and environmentally benign than supply-side alternatives. It also is an essential step in incorporating the precautionary principle into decisions regarding the evolution of energy networks. Integrated resource planning seeks to provide societies with energy at the lowest possible cost, and can be extended to include both social and environmental externalities. In 1992, the United States adopted the National Energy Policy Act in order to improve the overall efficiency of the energy network. Utilities must use this approach, which the act defines as 'a planning and selection process for new energy resources that evaluates the full range of alternatives, including new generation capacity, power purchases, energy conservation and efficiency, co-generation and district heating and cooling applications, and renewable energy resources to provide adequate and reliable service to its [the utility's] customers at the lowest cost ... [IRP] shall treat demand and supply resources on a consistent and integrated basis.'[27] The act also requires the Western Area Power Administration to include environmental externalities in the life-cycle system costs of the options considered. The act is an attempt to improve the energy efficiency of the American economy because western Europe and Japan use 57 per cent and 44 per cent, respectively, of the energy the United States uses to produce one unit of GNP. Everything must be considered to find the best tool for the job. For example, to provide one unit of energy from a fossil-fuel–fired power plant at the point of end-use requires three to four times the primary energy that would have been necessary if the fossil fuel had been burned directly to do the job. It is therefore advantageous to, wherever possible, burn the primary fuel and apply the heat directly. It is these kinds of considerations that the act encourages so as to reduce the overall primary fuel necessary to sustain the energy network to lower costs and minimize externalities.

Schweitzer and colleagues[28] suggest that the following approach is generally adopted. A strategy is developed to fill the gap between the forecasted future demand and existing resources. Both supply- and demand-side options are investigated, and the ones that are technically and economically feasible are retained. Next, the social (non-market-priced) costs of environmental impacts are added. Options are then combined into desirable strategies that are subjected to sensitivity analyses to determine what would happen if key parameters such as demand growth, fuel prices, interest rates, or consumer response to demand-side programs turn out differently than expected. The final outcome is a strategy for meeting future energy demands giving equal consideration to supply- and demand-side options to do the job with the least cost and minimal externalities.

Demand-side management options include peak-clipping by means of direct control of the utility over customers' use of appliances, valley-filling by increasing demand when it is low, load-shifting by moving equipment use from peak to off-peak times, and improving energy end-use efficiency. For example, the load-shaping options can be facilitated by installing meters that record consumption in differently priced categories according to time of use to distribute demand and thereby lower the overall capacity required. Water heaters and space heaters equipped with means to store heat can be charged at night. Some customers may be prepared to shift certain activities around so as to take advantage of more favourable rates. Some measures, such as boosting end-use efficiency, can reduce peak load demand as well as conserve electricity. Renewable resource generators installed in homes, offices, or factories can do the same. The above survey of how integrated resource planning is being implemented at U.S. utilities notes that, when social costs are added to market prices of electric power, many renewable technologies become cost-effective because they have lower externality costs than conventional fossil-fuel–fired power plants.

It should be noted that integrated resource planning was inserted into an economic system that essentially rewards producers of any commodity who lower their costs by increasing the associated externalities or social costs. This means that gains in performance values are partly or wholly offset by losses in the domain of context values. The thrust of integrated resource planning runs counter to that of the system within which it operates. It should not be surprising, therefore, that in the name of 'free' competition and the right to choose one's own supplier, some large corporations lobbied for a revolutionary change called 'retail wheeling.' It would allow utility customers to choose their own suppliers and arrange for a mutually satisfactory price. The consequences are obvious. First, it would once again permit social costs to be entirely passed on to the community and future generations. The invisible-elbow effect would be maximized within the regulatory framework. Second, large customers would avoid paying their fair share of what was beginning to resemble an energy commons, stranding small and thus weaker customers with the burden of paying for the existing infrastructure. This includes all the uneconomic decisions that were made in the past and that led to unnecessary expansions of supply and distribution capacities resulting from rewarding utilities for maximizing sales as well as the adoption of nuclear-power technology. Third, the level playing field between demand- and supply-side options would practically be eliminated as the focus shifts back to supply. All this is heralded as deregulation, an expression of economic fundamentalism which has faith in the market as a supernatural institution that has no limitations. Retail wheeling negates many advances that have been made in permitting the market to do what it does best, on the condition that as a community we recognize its limitations and behave accordingly by setting an appropriate framework.

As could have been expected, in 1994 a small group of California PUC commissioners advocated retail wheeling rather than the existing scheme whereby all customers paid their fair share of the costs of the system, which integrated resource planning sought to keep to a minimum. Wholesale competition requiring utilities to obtain the cheapest source of electricity was already part of the system, as mandated by U.S. federal law since 1992. So all we can expect from retail wheeling are possibly some small short-term gains and much larger long-term pains because investments in energy efficiency will be penalized despite the fact that this is generally the cheapest source of electricity, with enormous social and environmental benefits. Fortunately, there are enormous obstacles to be overcome in implementing retail wheeling, but in the meantime uncertainty and confusion reign, leaving the potential for preventive approaches largely untapped once again. The deregulation bandwagon is simply absurd. The electricity network and its regulation were beginning to encompass both the benefits of wholesale competition and the much larger benefits of demand-side options. There would be competition to ensure the most competitively priced electricity inputs into the grid while rewarding utilities for lowering their customers' bills, as opposed to rewarding them for selling more electricity. This would be in the public interest, since all parties, including customers, shareholders, society, and future generations, would win. It may still happen if common sense prevails as all the details are considered and worked out. Unfortunately, voices for the public interest are weak, so that the outcome is not at all certain.

Developments in the United States are being closely watched in many other parts of the world. In Canada, the idea of privatization has taken hold, which, as in the United States, has more to do with faith in the omnipotence of the market than in economic realism for more sustainable development. Von Weizsäcker and colleagues[29] report that the Europeans are moving towards a 'free' market for electricity that will eventually lead to the dismantling of the regional power monopolies following 1987 legislation. Fortunately, the Danes have inserted provisions to allow for the financing of energy efficiency and renewables from a modest portion of sales revenues. The European version of integrated resource planning is being stalled under pressure of the German energy-intensive industries.

Thus far I have primarily concentrated on the electricity-related part of the network of energy flows in a society, in part because it accounts for a substantial portion of primary fuel input and thus has massive social and environmental implications. The additional reason for focusing on this portion of the network is that it may well become the energy commons of the future. Many current developments point in the direction of an energy future that resembles in many respects what Amory Lovins described years ago as the 'soft' path.[30] The energy requirements of many end-use components in the network will drop dramatically if economic realism continues to produce large gains in end-use efficiency. For example, preventively designing new buildings – as well as their heating, ventilation, and air-conditioning systems; lighting systems; office equipment; appliances; and entertainment devices – for end-use efficiency will lead to a substantial drop in energy requirements. At the same time, these buildings can be designed to generate a certain amount of energy by using passive solar shells wherever possible, with super-windows, photovoltaics, devices to store excess heat, and active solar systems. This avoids distribution losses on the grid, thus making renewable resources more cost-effective, with obvious social and environmental benefits. Large electricity consumers on the grid can frequently improve their energy efficiency by turning themselves into producers as well as consumers through some of the new innovations in co-generation and in renewable resources. These kinds of developments will turn the electric grid into a system of components that simultaneously produce and consume electricity. This, in turn, will have substantial implications for the role conventional power plants play, making some of the newer and smaller innovative alternatives extremely attractive. All these developments are synergistic and have the potential of making modern ways of life significantly more sustainable.

This brings us to the role gas utilities could play in the immediate future. This fossil fuel is the best suited for the transition towards a more sustainable energy future. Integrated resource planning could be expanded to identify energy end-uses for which gas continues to be more effective than electricity from the perspectives of thermodynamics, economics, and sustainable development. If gas utilities seize present opportunities for transforming themselves into energy-service companies, they could play a crucial and profitable role in making the economy more energy-efficient, competitive, and sustainable by assisting customers in lowering their energy bills. This could be done by maximizing energy end-use efficiency, using gas and electricity where most appropriate, and by beginning to tap the potential of renewable energy resources. The considerable advances in photovoltaics, energy storage devices other than batteries, and passive solar building designs could in a short time radically transform the network of energy flows in modern societies. If electric utilities turn away from energy services, gas utilities could fill the void to help bring this about.

### 9.5 Context Matrices for Prevention

Fortunately, as the battles over deregulation and privatization rage, many benefits of preventive approaches can continue to be realized by those daring to be economic realists. Preventive design and decision making in the energy sector can be guided by context matrices. For particular materials, products, processes, and facilities, the rows of the context matrices represent the activities in the chain delivering the energy required for a particular phase in their technological cycle. Such chains begin and end at the society–biosphere boundary. The matrix columns represent the consequences of these activities for sustainable development differentiated to an appropriate level to reflect the influence on the role of the biosphere as ultimate source and sink, life support, and habitat, as well as the consequences for human life and society, and the present technical life-milieu. Once such matrices for each of the phases of the technological cycle of the material, product, process, or facility are drawn up, they can be integrated to show the overall implications on a life-cycle basis.

Energy accounting is an essential tool in identifying the activities that form the rows of these matrices as well as their relative significance in the overall energy account of a particular material, product, process, or facility. To illustrate this, I will briefly describe the energy accounting of the use of passenger vehicles as undertaken at the Rocky Mountain Institute and leading to the concept of the hypercar.[31] It is paradigmatic of how preventive design and decision making must be guided by energy accounting, and can be applied to advancing the design of almost any entity. The modern car is an excellent example of a design process by specialists. This has greatly contributed to end-of-pipe solutions as safety standards and pollution regulations were tightened and consumers' appetite for a wide range of options increased. It has made the modern car needlessly complex and, from an energy point of view, rather inefficient, despite substantial improvements in fuel economy in the 1970s and early 1980s. The automotive industry claimed that future improvements would be more modest and slower in the making, which is true if business as usual continues. However, when the car is analysed and designed as a system, it becomes obvious that many components are based on an end-of-pipe approach to design as opposed to dealing with the roots of a particular problem.

An energy account for the operation of a passenger vehicle illustrates the point. Only 15 to 20 per cent of primary energy reaches the wheels, and even less moves the driver. The Rocky Mountain Institute claims that, of the energy reaching the wheels, one-third overcomes aerodynamic resistance (doubling at highway speeds), one-third heats the tires and the road, and one-third heats the brakes. Rethinking the engineering of modern vehicles should begin with the recognition that they provide a transportation service. The service should be delivered with minimal material and energy intensity. Hence, the car should be designed using modern high-strength, lightweight materials that can be cost-effectively recycled at the end of the useful life of the vehicle, thus reducing the weight of the vehicle by 50 per cent or better. This would have a multiplier effect in that a smaller engine and power train could be used, further reducing the weight of the vehicle and the energy lost by the drive-train, wheels, tires, and brakes. Safety can be improved by designing the vehicle to absorb energy on impact while protecting the passenger compartment, the effectiveness of which has been demonstrated by spectacular accidents with racing cars. To improve the handling and stability, the aerodynamics of the vehicle should be designed so as to reduce air resistance as well as sensitivity to crosswinds.

Having thus re-examined the car as a system, delivering a service with minimal material and energy intensity opens the road to the redesign of many components. A small engine can drive an electric generator that powers electric motors in each of the four wheels. This would eliminate the need for the transmission and drive-train, further lowering the weight of the vehicle. It would also facilitate the recovery of energy lost during braking. All these gains do not merely add together, but multiply, because the overall efficiency of the system is the product of the efficiencies of each component in the chain. The hybrid drive-train not only reduces the weight it is required to move, but eliminates the friction losses of the now redundant transmission and drive-train, reduces tire friction because of reduced weight, and recovers a good deal of the energy previously lost to heating the brakes. The design is now attracting a lot of attention and has been nominated for several awards. As is always the case with a paradigm shift, many experts are sceptical, but some companies are quietly exploring the possibilities of this new design approach. The concept of product stewardship should be integrated into the approach to facilitate repair and maintenance, remanufacturing of critical components to extend the life of the vehicle, ease of disassembly and remanufacture of other components, and the closed- or open-loop recycling of materials. All this would be greatly facilitated by the fact that preventive design tends to reduce system complexity as problems are dealt with in a preventive rather than an end-of-pipe fashion that adds additional components. These vehicles require fewer fluids, with additional implications for maintenance costs and environmental impacts. From the perspective of making our modern way of life more sustainable, these hybrid vehicles, first with an internal-combustion engine and later with a fuel cell, appear far more promising than the so-called zero-emission electric vehicles that merely displace the pollution from the car to the power stations that convert primary fuel into electricity with a very low efficiency. Contrary to common belief, transportation services based on these vehicles rather than on conventional cars will greatly increase the overall burden imposed on the biosphere. The same is true for fuel-cell–powered vehicles if the energy carriers used by these cells are produced by means of electricity. If, on the other hand, these energy carriers are produced by solar energy, transportation services are likely to become much more sustainable.

In briefly reviewing how energy accounting can help us rethink the modern car, I am not suggesting that this is an adequate solution for our transportation problems. The exponential growth in the number of vehicles and the ever-greater proportion of them being used in metropolitan areas are primary reasons why cities are increasingly unsustainable in all three dimensions. The hypercar can help us buy time as we rethink transportation and the urban habitat, beginning with the services they render and then finding the best means to perform them with the lowest material and energy intensity and the greatest context compatibility.

The same preventive approach can be used to rethink the design of buildings. Their mechanical systems have become much less efficient than they might be, given the fact that, to a large extent, they are end-of-pipe solutions to problems created by inadequately designing the building shell and layout of space to fit human activities and the physical surroundings. In many areas, properly situating a building, designing a passive solar shell, and modifying the shape of the building can greatly reduce, if not eliminate altogether, the need for heating, ventilating, air conditioning, and illumination. Money saved on these systems can be spent on a more expensive building shell and, until this approach becomes commonplace, higher design costs. Once the building is completed, a great deal of money will be saved every year through lower operating costs. It so happens that reducing the material and energy intensity with which buildings support a variety of activities generally goes hand in hand with greatly enhancing the quality of the space. Although it is impossible to assess the financial value of physical surroundings that better support human activities, it is certainly going to be considerable. Improved air quality with better humidity and more natural light can reduce health problems and, coupled with a general improvement in the quality of the space, can improve hu-man well-being and productivity. When all this includes designing the building for disassembly and takes into account human health and environmental factors when selecting building materials, mod- ern ways of life are made more sustainable in all three primary dimensions.

Passive solar building shells make use of super-windows equipped with special filters that allow light and solar energy to enter the building while permitting little heat to escape. Making the shell virtually air-tight, heavily insulated, and ventilated by means of heat exchangers reduces heat loss to a level where it can largely be replaced by the occupants, the lights, and the various devices being used. It is often possible to equip the shell with a thermal mass to make inside conditions less sensitive to fluctuations in the weather. What can be accomplished with such building shells is remarkable. For example, the Rocky Mountain Institute headquarters, at an altitude of 7,100 feet above sea level in the Rocky Mountains near Aspen, is situated in a climate where winter temperatures can dip to –40°C, and where the sun can be covered by clouds for up to thirty-nine days in a row. The shell and its highly efficient equipment save 99 per cent of the space and water-heating energy, 90 per cent of the household electricity, and 50 per cent of the water for an area of 4,000 square feet. Compared to local housing of the same size, the energy savings are some $7,000 a year, so that, even if these kinds of building shells and equipment are a little more expensive, the difference is recovered in a very short time. After this, the savings can help to pay off a significant part of the mortgage. All this was accomplished with 1982 technology, and today it is possible to do better.[32] Such building shells may also be the answer to the problem of indoor air pollution that is making so many people sick. The virtual absence of electromagnetic fields, mechanical noises, outside noises, the abundance of daylight, and the quality of the air help keep people alert and cheerful. There are other buildings around the world that have proved the viability of this design approach in a variety of climates: the passive house in Darmstächt (50 kilometres south of Frankfurt), a home in Davis, near Sacramento in California (designed for 40°C), and the head office of the International Netherlands Group are among some of the examples described by von Weizsäcker and colleagues.[33] Toronto now also has a home that is not hooked up to the electric grid, city water, and sewers.[34] Once again, the pattern holds: preventive design is often more cost-effective, as well as context-compatible, and contributes significantly to making modern ways of life more sustainable in all primary dimensions.

Context matrices can also be used at the macro level. In this case, the activities in the way of life of a society that require energy services are grouped into meaningful categories to constitute the rows of a matrix. These categories may include residential, commercial, industrial, transportation, agriculture, resource extraction and refining, and construction. The columns could contain an aggregated estimate of their contribution to problems that must be eliminated or reduced to make a way of life more sustainable in all its three primary dimensions. This mapping could direct policy making, the setting of standards, and other such activities to those areas of a way of life where attention is urgently needed. Such matrices can further benefit from the kinds of energy end-use analyses of the way of life of a society as those carried out by Goldemberg and his colleagues.[35]

## 9.6 Conclusion

Once again, it appears that preventive approaches for the engineering, management, and regulation of the network of energy flows associated with the way of life of a society can simultaneously benefit the four wholes into which it is enfolded: the economy, society, local ecosystems, and the biosphere. Energy policy so profoundly marks the way of life of a society that reorienting it in a preventive direction can reduce many significant problems humanity currently faces, as well as making modern ways of life more sustainable. The synergies with the preventive approaches discussed in the previous chapter and the two chapters still to come will further enhance its benefits.

A strategy for improving the biosphere-related dimension of sustainability involves several components. The first is to represent the networks of flows of matter and energy involved in the way of life of a society. All or parts of these networks may be represented by means of methodologies such as input–output analysis, life-cycle inventories, ecological economics, industrial ecology, industrial metabolism, environmental-impact assessments, energy analysis, and energy accounting. The second component involves implementing the general principles set out in the previous chapter by means of initiatives affecting the entire networks or substantial portions thereof; the entire technological cycles of materials, products, processes, and systems; or specific phases of these cycles. Initiatives on the first level include international and national social and environmental standards; initiatives to make the economy more circular in terms of the flows of materials by means of green taxes, product-stewardship legislation, industrial eco-park development, and open-loop recycling strategies; pollution-prevention strategies; integrated resource planning and energy-efficiency approaches; standards based on best available technologies; research initiatives and knowledge-transfer strategies; alternative indicators for net economic performance; and fundamental changes in professional education and licensing for the technology-related professions. Initiatives on the second and third levels include pollution prevention through the redesign of products and processes; methods such as design for environment, design for repairability, and design for disassembly; product stewardship (shifting the emphasis from products to the services they provide); remanufacturing; industrial eco-park development; mining waste streams; reducing use of dissipative material; green audits and green accounting; and green labelling. Many of these approaches can be combined into comprehensive strategies for improving the biosphere-related dimension of the sustainability of corporations, communities, regions, and nations.

Care must be taken that improvements in the biosphere-related dimension of sustainability do not cause problems in the other dimensions. For example, will a move towards a more circular economy expand corporate decision making and further weaken the ability of communities to have a democratic say in their future? Will a further linking together of the activities of a society involving flows of matter and energy expand the technology-based connectedness of society at the expense of its connectedness based on individual and collective human life? How will a further reinforcement of the global 'system' for materials, production, and consumption affect the relationships between north and south? It is essential to ask these and other questions to ensure that making modern societies more sustainable in one dimension does not negatively affect the other two. Care should also be taken that our expectations of the potential of preventive approaches related to industrial ecology are realistic. For example, comparisons between an industrial ecology and a natural ecosystem have been severely criticized.[36] If industrial ecology becomes primarily an ideology that makes us feel better about ourselves, much is lost and little is gained. We know all too little what a move towards an industrial ecology would mean for the energy component of the biosphere-related dimension of sustainability. The results will most likely be positive, but the energy-related implications of industrial ecology are very poorly developed. However, there is no shortage of possible initiatives and ideas whose potential could be tapped if specialists also became generalists capable of thinking about the broader implications of their work.

# 10

# Work

## 10.1 Work, Life, and Society

In the previous two chapters, sustainable development was linked to the reciprocal relationship that exists among human life, society, and the biosphere in terms of flows of matter and energy. The flows are required to constantly renew and replace the cells in our bodies, to support the many activities that constitute individual and collective life, and to maintain and evolve the built habitat. They come about as a result of networks of activities integral to a way of life and involve a great deal of human work. Via these activities, these flows contribute to individual and collective experience, and thus to human consciousness and culture by means of the cultural cycle. When individual activities or their networks are guided predominantly by performance values, this reciprocity is treated as a set of externalities. The economic system rewards individual and institutional behaviour that reduces private costs even when these are partially or wholly offset by social costs. In the absence of social and environmental standards to ensure that human lives, communities, and ecosystems are not 'mined,' global competition has intensified this development, with the result that the effects and benefits of the invisible hand are gradually undermined by the effects and problems resulting from an ever larger invisible elbow. This undermines the life-support functions provided by the biosphere and by society. Additional recent evidence of this includes the continuing drop in male fertility that threatens the biological sustainability of humanity[1] and the possible collapse of insect populations, including critical pollinators, with the implications this has for the sustainability of all life.[2] The invisible elbow has created an economy that is successful in terms of performance values but that undermines the integrality of a society through unemployment and underemployment. Working families and individuals have to work longer hours, and even double up jobs, to maintain living standards. Work intensity is increasing, which correlates with growing work-related illness and disability, and the frequency of accidents. Modern people are much more likely to succumb to depression than were their grandparents, and require a host of mental-health-care practitioners and other social and health services. If the United States and Japan are any indications of what lies ahead, we face a serious situation. In Japan, surveys show that over 70 per cent of workers are mentally exhausted when they come home, and job insecurity is also on the rise.[3] This occurs at a time when the traditional social support for these workers, and hence their capacity to recover and re-create their ability to do another day's work, is diminished. The consequences of the emergence of mass societies with their weak social relations, families, neighbourhoods, and communities are exacerbated by the spill-over effects of work on people's lives.

When examining the patterns of technological and economic growth, particularly since the end of the Second World War, in terms of their implications for the human and societal dimension of sustainability, the same patterns become evident, as observed in the previous two chapters. As economies boomed after the Second World War, societies responded to negative implications by creating a complex patchwork of end-of-pipe social and health services. Since these often did not address the root problems, the patchwork had to be expanded until public expenditures overtook wealth creation, thus contributing to the debt crisis. Governments from all across the political spectrum felt compelled to cut back on these health and social services, again without going to the root of the problem. The industrially advanced nations became caught in the labyrinth of technology, from which they have been unable to emerge. For example, without belittling the accomplishments of modern medicine, it has all but neglected prevention, which is possible only when people are actively involved in their own health. The benefits of being so involved are becoming evident to a great many people who do not cease to use conventional medicine where appropriate, but who, when conventional medicine fails them or does not allow people to be proactively involved in their own health, turn to alternative medicine. Work-related illness is an excellent case in point. As we will see, socio-epidemiology shows that modern work is at the root of a great deal of disease, to which conventional medicine mostly responds by end-of-pipe health services.

Present ways of life and the global economy are undermining their own foundations. This is gradually being recognized for the life-support functions rendered by the biosphere, but not so for the ability of societies to support individuals. On the one hand, there is a great deal of talk about how increasing global competition requires research, innovation, and development as well as entrepreneurship to ensure ongoing technological and economic growth, and a continual rise in the productivity of intellectual and physical labour. On the other hand, governments are backing away from conceptions of the welfare state by cutting educational and social expenditures, which is bound to have a significant effect on the 'social capital' accumulated through a historical process of cultural development, and without which no economy can be viable and no community can provide a healthy, creative, and active workforce. Hence, the present system is undermining the life support it derives from society and the biosphere. It overuses nature and underutilizes people. The struggle for democracy against authoritarian, arbitrary, and hierarchical regimes is a struggle to empower citizens to shape their individual and collective futures. For two centuries these developments have been undermined by disempowering people through work organizations with ever larger numbers of meaningless, repetitive, and unhealthy jobs. Adam Smith warned humanity against the trap it was walking into when he said:

> In the progress of the division of labour ... the great body of people come to be confined to a very few simple operations; frequently to one or two. But the understandings of the greater part of men are necessarily formed by their ordinary employments. The man's whole life is spent in performing a few simple operations ... naturally loses the habit of [solving problems] and generally becomes *as stupid and ignorant as it is possible for a human creature to become*. The torpor of his mind renders him not only incapable of rational conversation ... generous, noble or tender sentiments ... [judgment about] the great and extensive interests of his country ... he is equally incapable of defending his country in war ... But in every improved and civilized society this is the state into which the laboring poor, that is the great body of the people, must necessarily fall *unless government takes some pains to prevent it* [emphasis added] .[4]

Apparently, the loss of human qualities was a small price to pay for the wealth of nations. Since this wealth was to be realized from a technical division of labour, the advice that governments ought to do something amounts to advocating end-of-pipe solutions since the root of the problem is, as Adam Smith realized, in this very division of labour. He foresaw the road that lay ahead: an invisible hand attached to a growing invisible elbow. The latter has 'knocked over' more lives than almost anything else in this century.

In other words, the reciprocal interdependence that must be explored with respect to human work is that individuals growing up in a society learn to make sense of, and live in the world by acquiring its culture and way of life, which embodies the 'social capital' from which all economic activities draw. A healthy and vital society empowers people, who will then create a dynamic economy; conversely, an economy that disempowers people will degrade the very social foundations on which it depends. The former society lives off the interest of its 'social capital' while the latter 'mines' that capital. Similarly, a healthy and vital society will generate a healthy and vital workforce in which the latter helps to maintain the former. An overemphasis on performance values will inevitably lead to a degradation in the reciprocal interdependence among human life, society, local ecosystems, and the biosphere (as measured in terms of context values). A healthy and vital society will balance performance values with context values and evolve preventive approaches, which further reinforce the vitality of the reciprocal relations on which it depends. The opposite will occur with an overreliance on performance values.

The choice of preventive approaches is a choice for a vital and healthy society with a sustainable relation with the biosphere. Integrating performance and context values into economic behaviour ensures gains that will not degrade the health and vitality of individual and collective life and its dependence on the biosphere. Consequently, the preventive approaches developed in the previous two chapters must be complemented by preventive approaches related to human work. As will become increasingly evident, the one is impossible without the other.

Before briefly examining how modern work came to take on its present forms, it is useful to remind ourselves of one of the fundamental facts regarding its origins. Industrialization separated work from the family context and all the activities that came with it. It sought to create a separate part of human life that would be 'pure work' as free as possible from interference from all other activities that made up the other parts of life. The work-related part would be regulated by labour markets. It was all but forgotten that work can never become a separate part of someone's life in a mechanistic sense, despite separate workplaces and training based on technical rationality (i.e., not based on experience and culture). Work is enfolded into a human life, so that the effects work has on a person's so-called working life are really effects on the whole of his or her life. Each one does not have its own distinct health and well-being. Similarly, such effects spill over into all social relations that together weave the fabric of a human life. Western civilization has lost the keen sense it inherited from the Judaeo-Christian tradition, based on the recognition that hiring someone for a wage is more than simply buying some services from that person and is, in effect, taking possession of something fundamental in that life, with profound ethical implications. This is why the sociology of work recognizes the possibility of alienation, to be possessed by something or someone to the point that you can no longer be yourself. For example, compelling someone to work long hours may not leave him or her any resources for having healthy relations with spouse, children, friends, and neighbours. It can make a person into someone he or she does not wish to be. Hiring someone for a wage opens the possibility of enslavement, which in the Judaeo-Christian tradition was the model for sin. Hence, those who were without power and compelled to work to survive had to be protected from alienation.

It should not be surprising, therefore, that the ongoing attempts to improve labour productivity through the conventional two-stage approach exhibit the same patterns we have seen earlier: spectacular successes in the domain of performance values and equally spectacular problems in the domain of context values. Few developments in the last hundred years have had a greater influence on human health and well-being than the restructuring of work. It is difficult to exaggerate the spillover effects on almost everything else, including the fabric of social relations in families, groups, communities, and nations, in general, and the workforce and economic institutions, in particular. Corporations spend a great deal of time and money on end-of-pipe fixes that attempt to restore their 'social capital.' Work is, therefore, a particularly good test of the thesis of this book since it is integral to so much of human life and society. This means that preventive approaches in the domain of human work should be able to do more for the sustainability of human life and society than almost anything else I can think of.

## 10.2 The Emergence of the Technical Division of Labour

The Industrial Revolution shifted the focus from improving the yield of 'natural capital' to that of 'social capital' (particularly labour) and monetary capital. Adam Smith, in his *Wealth of Nations*,[5] published in 1776, recognized that a nation's wealth would greatly increase if everyone could produce more, and that this could be accomplished through a technical division of labour. Such a division of labour is entirely different from a social division of labour through which a way of life is maintained by the members of a society performing different but entire tasks or services. A technical division of labour transforms the domain of work the way science had previously transformed the domain of knowledge and intellectual work – namely, by dividing the world into its fundamental constituent elements. Similarly, human work was disassembled into its fundamental constituent skills, and reassembled into a logical sequence to perform the task or service.

The advantages appear obvious. First, workers were assigned tasks that were as small as possible so that they become very good at what they were doing and no time was wasted in preparing for the next task. These constituent elements of work could now be optimized to eliminate wasted motions in order to ensure optimal performance. Second, it became possible to prepare, optimize, implement, and evaluate a production plan in the form of the optimal sequence of the optimal constituent elements. None of this was possible in traditional craft production. Third, in this de-composition and rational re-composition of work, it was possible to assign constituent tasks either to human beings or to machines to obtain the greatest efficiency, productivity, and profitability. It opened the road, first, to mechanization and, later, to automation. Fourth, the tasks assigned to human beings could be performed by those people having just the required skill level, thus greatly reducing labour costs. This was impossible in craft production, where the skill level of a person had to be adequate to meet the most difficult facet of a particular craft. Fifth, workers performing distinct tasks could be isolated and group structures eliminated, including time spent with a group, which was considered a waste since it was not available for production. In sum, a technical division of labour seeks to increase labour productivity and reduce labour costs by dividing work into its constituent elements, optimizing these elements, and combining them in an optimal process.

What are the costs and inefficiencies associated with the technical division of labour? First, Adam Smith himself, as previously noted, acknowledged that it would make human beings as stupid as they could possibly become. The monotony of carrying out the same trivial operation over and over again would atrophy human creativity, skill, problem-solving ability, and a great deal else related to their humanity. Second, in so far as people still used skills, these were of an entirely different nature from craft skills because they were not optimized in the person's mind through the accumulation of experience, but by someone else, external to the work process, who was to determine the one best way for doing the job. This required the suppression of the self in human work, leading to nervous fatigue and a range of psychosocial consequences. Normally, in the past, even the most routine human activity, such as handwriting, was an expression of a person's life and experience, as shown by the possibility of correlating characteristics of handwriting with personality traits. Third, the more the technical division of labour developed in conjunction with mechanization and industrialization, the more workers lost control over their work: the methods to be used, the time to be allocated, how work was to be paced, how problems should be looked at and solutions attempted, what deadlines were to be set and met, how to report, and when and how to assess progress. In other words, a separation of knowing, doing, and managing occurred, making the regulation of work through negative feedback more difficult and barring the worker from effectively dealing with problems. Those who knew and planned the production process were not the same people who did the work, thus also requiring supervisors to ensure that the work was done according to the plan. Fourth, as the technical division of labour and mechanization expanded, greater demands were placed on workers by the work setting and organization, including the passive monitoring and tending of machines. Problem-solving demands increased because of the gap that inevitably occurred between the ideal work plan and how all this worked out in reality, including the problems that originated upstream in the work process. Even if workers could solve some problems, they almost never had the authority to do so. There were also the demands associated with machine pacing, quotas, and reporting responsibilities. All these demands, coupled with little control, negatively affected a worker's health and mental faculties. Fifth, workers were increasingly isolated, losing the social support of the group while demands rose and control diminished.

There were two landmarks in the development of this technical division of labour that are worth highlighting. The first was the invention of scientific management, as formulated by Frederick Taylor in 1911, and the second was the introduction of the assembly line by Henry Ford in 1913. Both these developments removed significant barriers to the further development of the technical division of labour. This has two dimensions. Separation of the work process into a logical sequence of production steps is called a 'horizontal' division of labour, while the separation of hand from brain is referred to as the vertical division of labour. One of the limitations to both was the difficulty of rationally de-composing and re-composing highly skilled work such as tool-and-die making and machining. Another limitation was the problem of producing identical parts that made the assembly process dependent on a great deal of skill, since each part had to be custom fitted. Once a machine was developed to the point that the operations were repeatable, predictable, controllable, and measurable, the manufacture of interchangeable parts became possible, thus permitting the de-skilling of assembly work. This led to Fordism, as a way of assembling complex products from interchangeable parts by means of an assembly line and a low-skilled workforce.

The limitations to objectifying the work process were pushed back by Frederick Taylor. He founded scientific management as a way of doing this, which was intended to result in a full-blown vertical division of labour. It would determine the 'one best way of doing things' and convey this to the workers in the form of precise instructions on how to carry out various tasks within an allotted time. Scientific management sought to extract the knowledge from skilled workers in order to objectify it and bring it under the control of management. In the past, management could base its authority only on ownership and paternalistic leadership. Authority was now to be based on supposedly indisputable scientific procedures that objectively established the one best way to do work. It should be noted that, later on, these principles were also applied to mental work, breaking down complex operations into their constituent elements and recombining them into logical rules, algorithms, and programs. Thus the mental operations required to plan, execute, control, and coordinate the horizontal division of labour were complemented by a vertical division of labour striving for a kind of vertical intellectual assembly line. This would provide top management with all the required information for the execution of its orders.

The two developments reinforced each other. The more precise vertical division of labour allowed for an extension of the horizontal division of labour, while the latter reinforced the former. This interaction led to the Fordist/Taylorist system, which became the pre-eminent manufacturing system in the world, making the United States the leading industrial nation until the invention of the lean production system. It overcame one of the greatest obstacles the United States faced, since, unlike Europe, it had an abundance of natural resources but a great shortage of skilled labour. The system was so successful that Taylor promised higher wages to workers, and greater profits and control to management. Ford was so confident of the benefits of the system that he offered to double the wages of workers. As the most advanced form of the technical division of labour at the time, it had several advantages. First, work would now be more rigorously paced and controlled by means of the assembly line. Second, the average skill of workers was so reduced that they could be easily trained and replaced. This latter advantage created significant problems in a society where the average level of education was steadily increasing, and the human cost was great. There was fierce resistance from highly skilled workers. The negative impact on workers was so great that wildcat strikes, sabotage, high turnover rates, and poor workmanship were common, but higher wages and benefits, made possible by greater labour productivity, led most U.S. unions to accept the system by the early 1950s.

The assault on the humanity of the workers may be appreciated by examining how the Gilbreths, a husband-and-wife team, analysed and improved the way a worker operated a machine.[6] They mounted three cameras that viewed the work along three perpendicular axes, and attached a small lightbulb to the wrists of the worker operating the machine. With a lowered illumination level, photographs were taken that resemble the ones we have all seen of the traces of head- and tail-lights of vehicles moving around the cloverleaf intersection of a super-highway at night. From the three orthogonal projections of the path traced out by the wrist of the worker, three-dimensional wire models were made that could be analysed to identify redundant movements, as well as better ways of performing the task. The analysis could be improved further by including time markers produced when the lightbulb was momentarily switched off at regular intervals. This permitted the measurement of the average accelerations and decelerations of the worker's hands. After the one best way for doing the task had thus been determined, the worker could be taught to execute it. It should be noted that this is essentially no different from programming a robot.

Everything that is normally enfolded into human behaviour (as a result of its spontaneously originating in the brain–mind system that enfolds something of a person's personality, experience, feelings, emotions, culture, and way of life) was stripped off in this 'one best way' behaviour and was to be suppressed by the worker seeking to execute it. As noted above, this led to the fundamental characteristic of modern work – namely, nervous fatigue having largely displaced physical fatigue. Hence, the improvement of human work guided by performance values leads to the same kinds of patterns encountered previously: spectacular gains in labour productivity, profits, and GDP at the expense of the compatibility between the new work methods, on the one hand, and people in work organizations, on the other. This incompatibility spills over into the whole of the lives of workers, affecting their families and communities.

Such context incompatibilities were exacerbated by the fact that the horizontal and vertical divisions of labour necessitated ever-larger bureaucracies that analysed, planned, implemented, monitored, and assessed work. However, these developments were not limited to corporations and physical work. Max Weber examined the rise of bureaucracies at the beginning of the twentieth century, showing that they were the most rational and efficient way of performing collective work, but warned that, by striving for rationality in all spheres of life, humanity was locking itself into an iron cage.[7]

Maximizing the efficiency of work involves many inefficiencies. However, this changed little in the evolution of the horizontal and vertical divisions of labour because most of these inefficiencies are market externalities. The 'human relations' school discovered that meaningful social interactions in the workplace were a more important determinant of work efficiency than the technical division of labour and the accompanying specialization. The organization theorists noticed that bureaucracies were exceedingly rigid and resisted creative adaptation to an ever-changing and evolving context. Bureaucracies tended to get bogged down in power struggles that obstructed organizational and economic efficiency. If this were not enough, they underutilized the creative resources and skills of people. In addition, the hand–brain separation in individual workers has a collective parallel in Taylorism's separating the collective brain from the collective hand, thus dividing an enterprise into two relatively distinct segments, operating, as I show shortly, on the basis of very different kinds of knowledge. The resulting dividing line marks the locus of many organizational failures and antagonisms, and in a sense divides the organization against itself. Enormous costs have been incurred to implement end-of-pipe fixes devised to deal with the problems of production, including industrial relations, job enlargement, ergonomics, human factors, most quality-of-working-life movements (not including sociotechnical design), health care, stress-reducing therapies, and counselling. These all deal with significant symptoms, but none deals with the root of the problem – namely, the suppression of the human self in work. In sum, the conventional two-stage approach to the engineering, management, and regulation of production optimized gross labour productivity rather than net labour productivity, gross profitability rather than net profitability, because of hidden inefficiencies. What the real or net efficiency gains may have been remains an open question since these are impossible to calculate. However, it is obvious that the social costs associated with the development of the horizontal and vertical divisions of labour are very large indeed. Had the labour market been appropriately constrained, the outcome would have been far different. Now we have divided individuals, organizations, and societies against themselves.

## 10.3 The Demand-Control Model

Modern conceptions of health suggest that the health of human beings depends on the demands placed on them by their contexts and the resources available to meet those demands. From a biological perspective, this means that since all body cells (except those of the brain) are replaced at least every seven years, there is a fundamental dependence on a throughput of matter and energy and, to the extent that this is contaminated, a higher load is placed on the immune system. If the load is greater than the immune system can handle, people become ill. From a sociocultural perspective, the brain–mind system depends on an ongoing and varied symbolic exchange. If monotony, repetition, or sensory deprivation severely curtails this exchange, or this exchange overloads people's resources acquired from experience, education, and culture, significant negative health effects occur as well. Problems in one dimension of human health affect all the other dimensions. For human work, this means that jobs are uninteresting when their demands are so weak that people cannot creatively use their resources. Work is interesting and stimulating when the demands do allow this, while work is crushing when the demands overwhelm available resources. It is assumed here that people have the possibility and the authority to use their resources as they see fit. Inhibiting their ability to do so, in effect, diminishes available resources.

This simple topology of the relationship between human health and work is confirmed by the findings of social epidemiology. Karasek and Theorell[8] have synthesized their findings by creating a demand-control model that divides jobs into four basic categories according to the demands made in a work setting and the control people have over their resources to meet these demands. They identified five independent variables (not of equal importance), by which work settings may be characterized: decision latitude, psychological demand, physical demand, social support, and job insecurity. Not all of the demands imposed by work organizations produce stress and lead to disease. For example, they cite the findings of a study showing that engineers experienced qualitative challenges as satisfying, but quantitative challenges (such as time pressures) as stressful. The difference is not difficult to comprehend. With sufficient control, the former permit engineers to creatively apply their resources, expand them by learning from these situations, derive satisfaction from success, and reduce stress levels. Quantitative challenges, on the other hand, permit learning, but there is little satisfaction and stress is not reduced. The two must therefore be distinguished in a model attempting to predict unhealthy work situations that produce disease. Thus, decision latitude, which includes skill discretion and decision authority, permits learning, the expansion of one's skill set, and the reduction of stress. Decision latitude is labelled as 'control' and plotted on the *y* axis while psychological demand is labelled as 'demand' and plotted on the *x* axis. Their orthogonality is rooted in the fact that demand may represent a risk to human health, while control does not. The remaining independent variables – namely, physical demand, social support, and job insecurity – were found to be less important for creating healthier workplaces.

This left Karasek and Theorell with a two-dimensional demand-control model. The four kinds of work experiences they identified were labelled as 'active' jobs (high demand, high control), 'low strain' jobs (low demand, high control), passive jobs (low demand, low control), and 'high strain' jobs (high demand, low control). The active and passive jobs have medium strain. The model predicts that these four categories of work experiences have different effects on human health, non-work activities, and community participation. I explain this system of classifiction for each of the four job experiences, moving from the least healthy to the most healthy ones (highest to lowest strain-producing jobs). The discussion and examples are based on Karasek and Theorell's landmark work.[9]

Jobs that produce high psychological strain occur in work settings that so constrain human behaviour that no effective response to increases in demand are possible. Such situations occur when an assembly line is speeded up; in restaurants at lunchtime, when everyone needs to eat quickly, so that waitresses can barely cope and are being run off their feet; in sweatshops where people are afraid to use their usual ways of blowing off steam to cope with severe time pressures out of fear that they may be singled out and laid off; or when telephone operators are severely constrained in how they may respond to a wide range of questions. Most machine-paced jobs are so characterized as well, with the rigid rhythms of the traditional automobile assembly line providing a typical example. Depending on the severity of the situation and how long a person has been exposed to it, the results include fatigue, anxiety, depression, physical illness, aggressive behaviour, and social withdrawal. Strain can be reduced by periodically returning to some kind of psychological equilibrium during coffee and smoke breaks, or by blowing off steam with others (some of the wasted motions that Frederick Taylor sought to eliminate). It is ironic that, in present attempts to reduce smoking and alcohol abuse, few people are talking about the need to create healthier workplaces. Nor is the medical profession raising the alarm over the staggering increase in the consumption of a whole range of antidepressants such as Prozac. The relationship between stress at work and smoking or drinking alcohol was established as far back as the late 1960s by the ground-breaking work of members of the Tavistock Institute in England.[10]

Passive jobs result in average levels of strain. They occur in situations where both demand and control are low. Such situations are faced by people who tend highly automated machines, or night watchmen who must punch a clock at different points on their hourly rounds. There are no challenges, and no opportunities to initiate anything or try out better ways of doing things, yet one's behaviour is frequently rigidly controlled. People tend to lose previously acquired skills, their motivation, and their initiative. Many people in these situations become withdrawn and have a tendency to daydream.

Active jobs also produce medium levels of strain because few, if any, constraints are placed on meeting challenging demands. These situations occur in operating theatres where surgeons perform difficult operations; in stadiums where professional athletes accomplish tremendous feats; or in offices where managers, engineers, and accountants may be engaged in solving some difficult problems. There is almost complete freedom in how to meet challenging demands, and near-total control over coping mechanisms. People in these situations can usually accomplish their tasks and return to some kind of psychological equilibrium with the satisfaction of having done a difficult job and having learned some interesting new things. Hence, jobs can have high levels of demand but, when this is accompanied with high levels of control, the negative implications for human health and well-being are less severe than for high-strain jobs. Indeed, such jobs may have positive psychosocial outcomes because they are accompanied by learning and growth. Research indicates that people in such jobs tend to engage in more leisure activities outside of work, despite heavy work demands.

Low-strain jobs are found in situations where there are few psychological demands and high levels of control. In these so-called leisurely jobs, people are actually made healthier and happier at work. The high level of control allows such workers to optimally respond to their duties with minimal psychological strain. The work of most natural scientists, senior professors, and architects falls in this quadrant. Some lower-status jobs, such as repairman, lineman, and foreman, also fit this category.

The distribution of work in modern societies among the four categories identified above is undergoing constant change, but a few trends are apparent. Mechanization, automation, and computerization tend to replace physical fatigue with nervous fatigue and severely reduce decision latitude. The latter is particularly serious in light of the increasing levels of education required for many jobs. These and other trends, including pollution and the contamination of food and water, are transforming the spectrum of health problems, including the frequency of their occurrence. Modern medicine is almost exclusively responding by means of end-of-pipe approaches, thus contributing to rapidly rising health costs.

The demand-control model also predicts the spillover effects from people's work into their lives, where they are free from the constraints of work, and in principle free to be themselves. Unfortunately, few appear to be able to escape the influences of their work. The highest incidence of passive leisure-time activities occurs in the population of people with passive jobs, and the lowest incidence is found among people with active jobs. The same trend holds for participation in political activities. It is worth noting that people with high-strain jobs have the highest levels of participation in mass political activities, usually interpreted as a protest of their plight. As people move into more active jobs, leisure-time activities tend to become more active, and social and political participation tends to increase. The reverse also appears to be the case. What this means is that the trend away from active jobs as a result of mechanization, automation, and computerization bodes ill for democratic societies, where workers in lower socio-economic strata withdraw, so that people in higher strata tend to have a disproportionate influence. This weakens the integrality of a society and intensifies 'us–them' sentiments. In other words, if we move away from the end-of-pipe treatment of work-related problems and engage in a more preventive approach to work redesign, a wide range of problems can be reduced in scope and magnitude. Benefits include improvements in human health and well-being; a reduction in aggressive or withdrawal behaviours; a reduced dependence on a variety of drugs to keep bodily and mental functions going; greater psychological, emotional, and mental resources for participating in social relations, with significant positive effects on communities, marriages, families, and the nurturing and education of children. Healthy workplaces can empower people, with positive effects on every aspect of the way of life of a society.

It may be objected, of course, that all this is of little interest to modern corporations. This is true only if we can see no further than the next quarterly statement. For many large corporations, health-care agencies are among the largest 'suppliers.' Absenteeism from work because of health or social problems, disempowered workers, and the additional tax required to defray the national costs of work-related problems directly affect the long-term viability of a corporation. Its customer-base is also negatively affected by the present trend of underemployment and unemployment, and a growing proportion of jobs that pay a minimum wage and no benefits. The design of healthy workplaces is essential for the economic, social, and political vitality of societies. This will radically change the balance among the following three work situations. If a worker's resources are underutilized, the job is not stimulating and no learning occurs, with negative implications for the worker, the corporation, and the community. If these resources are fully utilized, the job may well be demanding but also satisfying, with many positive spillover effects. If these resources are completely swamped, the worker cannot cope and strain occurs, resulting in negative implications for the worker, the corporation, and the community.

It is essential to design work organizations that do not constrain workers in the way they use their resources, so that they can build them up through learning and effectively cope with demands in a variety of ways, including reorganizing their own work. The creation of healthy workplaces should be a top priority in economic and social policy to improve the economic viability, social vitality, and political effectiveness – in short, the true wealth of nations. Adam Smith and his successors have it wrong: the true wealth of nations sustainably utilizes and does not degrade their 'social and natural capital.'

## 10.4 Support for the Demand-Control Model

The demand-control model is an important exemplar of the preventive paradigm developed in this work. It was designed to predict the effects of different kinds of work on human health and their spillover effects on individual lives, work organizations, and communities for the purpose of using this understanding to design healthier workplaces. Its elegance and simplicity derive from having empirically identified the principal independent variables that determine the health or lack thereof of jobs and work organizations. It represents a deliberate trade-off with alternative models having a far greater complexity. The overwhelming majority of subsequent research supports the model.[11] Exceptions to the rule appear to confirm the fact that a greater complexity, taking into account additional parameters, could not significantly affect the usefulness of the model for prevention-oriented work design.

Within the context of preventive practice, the strengths of the demand-control model become further apparent when it is compared with alternative ones. For example, the Michigan Occupational Stress (MOS) model[12] essentially supports end-of-pipe approaches because it emphasizes how workers perceive their environment, including problems they face, and how this may produce stress. The conceptual framework makes it essentially subjective, with the result that solutions are sought not in the prevention-oriented redesign of work, but in end-of-pipe fixes, such as psychologically screening job applicants or providing workers with coping mechanisms such as counselling and therapies of one kind or another.[13] Such fixes do not address the objective reality of work, making them compatible with traditional occupational health and safety approaches. Siegrist's Effort–Reward Imbalance model[14] focuses on the reciprocity of exchange in work in general, and how high cost and low gain occupations are particularly stressful. These are poorly paid jobs with little prestige or occupational status that nevertheless demand a great deal from workers. Hence, there is an imbalance between the rewards and the effort demanded, leading to negative emotional states that induce stress. Important as economic injustices related to work are in a modern society, the model does not aim to support prevention-oriented job redesign and does not nearly have the empirical support the demand-control model does.

Considerable advances have been made in operationalizing the demand and control constructs in the demand-control model by a research team at the Institute for Work Psychology at the University of Sheffield.[15] The research program seeks to more rigorously define and operationalize these constructs, which were largely based on survey responses. Job control has been analysed in terms of timing and method controls. Timing control includes the ability of workers to decide on the order in which things are done, when to start or finish a piece of work, and to set the pace. Method control includes the ability of workers to plan their own work, to vary their work, to control the quantity and quality of what they produce, and to decide on how to get the work done and the methods to use.

Demand has been analysed in terms of monitoring and problem-solving. Monitoring demand includes the extent to which undivided attention is required by the work, the necessity to keep track of more than one process, the need to constantly watch for things to go wrong and to rapidly intervene to prevent problems from arising. Problem-solving demand includes the need for workers to deal with new problems that are difficult to solve, solve problems that have no obvious correct answer, use the knowledge of the production process to solve job-related problems, have a knowledge of the production process in their area, and solve problems they have not encountered before.[16]

The research program is continuing the testing of these constructs for a variety of job settings. The findings thus far include the following: Higher levels of job control generally correlate with higher levels of job satisfaction and lower levels of strain; they negatively correlate with job-related anxiety and depression. Low timing and method control were associated with higher levels of job-related anxiety.[17]

Next, a single demand scale was constructed by combining the monitoring and problem-solving demand constructs and a control scale from timing and method-control constructs. The findings of the demand-control model were confirmed for work settings characterized by high demand and low control. It was also observed that timing control did not significantly interact with monitoring demand or problem-solving demand. However, method control did. This led the researchers to the important finding that method control, much more than timing control, could fend off the adverse affects of high demand with respect to producing job satisfaction and job-related depression.[18] In sum, healthy work requires that monitoring demand and problem-solving demand should not be excessive, and may be partially offset by increasing method control. Increasing timing control has a smaller benefit. Extending this kind of research to the other three quadrants of the demand-control model can lead to similar useful refinements for prevention-oriented work organization.

There is another category of evidence that indirectly supports the demand-control model. The end-of-pipe solutions applied to the problems resulting from the technical division of labour have had only small positive effects. These include improving the physical surroundings by better temperature and humidity control, optimal illumination levels, improved colour schemes, Muzak, and improving associated facilities such as cafeterias and washrooms. Ergonomics and human factors have improved the functionality of items such as tables, chairs, controls, and instrument panels. Job-enlargement and job- enrichment schemes have also had some positive effects. Health care, therapy, and counselling have helped workers cope with the consequences of their work. Human relations and a quality of working life movement (distinguished from social-technical work design) have also had their effects. However, none of these addresses the root of the problem: work alienation resulting from the technical division of labour. This is exactly what the demand-control model would predict.

The implications of the research described above for prevention-oriented work design may be summarized as follows: A work organization places certain demands on workers, who seek to meet them by means of the resources they have acquired. If these resources are underutilized, the job is not stimulating, and negative learning and deskilling may occur. A job is demanding but satisfying if the resources are fully utilized. If the resources are completely overloaded to the point that the worker cannot cope, job strain occurs, and this situation is exacerbated if the work organization constrains the ability of workers to use their resources, making these essentially unavailable. If, on the other hand, it permits their full utilization, workers can learn to deal with increased demand by ongoing skill development, reorganizing their work, or other means that allow them to cope better. Increasing the control workers have is therefore paramount in creating healthier work. Over the years, there have been a number of work-redesign efforts that include aspects of the preventive paradigm, such as the QWL approach used in the Shell Sarnia plant;[19] human-centred design approaches, particularly the ESPRIT 1217 Project;[20] sociotechnical systems design;[21] Total Quality Management; industrial democracy; business-process re-engineering; lean production' and Swedish production. These and others have significantly modified the technical division of labour, with many beneficial effects.

Unfortunately, the message is not getting through. The report of the Committee on Foundations of Manufacturing of the National Academy of Engineering puts it as follows:

> The American manufacturing environment is now in a rapid state of change. Yet our business schools and engineering schools have not yet begun to provide the leadership that this restructuring ... demands ... It appears to me that Taylorism is alive and well in the minds of engineering faculty throughout the nation. Furthermore, it appears that the unresponsive change-resisting attitude exhibited by many engineers in North American manufacturing practice is, in large measure, due to this primitive and ineffective education paradigm.[22]

Karasek and Theorell share the same concerns: 'Taylor's principles are still at the core of industrial engineering curricula ... In spite of the entirely different and glitteringly successful examples of alternative work design ... Taylor's principles still probably represent the predominant direction of job design in the modern world.'[23] Benda[24] confirmed these two quotes by carrying out a survey of manufacturing textbooks and handbooks. His findings further corroborate those of the study of undergraduate engineering education discussed in chapter 3. The social and environmental implications of work design are either not mentioned at all or receive only brief and superficial attention. Performance values reign supreme, and engineers and managers are kept ignorant of the vast implications of work design, creating a major threat to the wealth of nations.

### 10.5 Collective Work and the Technical Division of Labour

To obtain an overall picture of the impact of the technical division of labour on work organizations, it is necessary to understand how it ignores, underutilizes, and atrophies certain human skills while allowing others to develop through learning. This, in turn, requires an understanding of human skill acquisition before an overall picture can emerge. In a previous work, [25] I suggested that two processes appear to dominate the functions of the brain–mind system – namely, the process of differentiation (classification without any rules or taxonomies) and the process of integration (the mapping of all experiences within a life). These processes can explain the acquisition of skills by babies and children before and after the emergence of language. During formal education, essentially two kinds of knowledge are encountered, which I call 'knowledge embedded in experience' and 'knowledge separated from experience.' I have suggested that, in human memory, lived experiences form a symbolic structure that enfolds metaconscious knowledge. To explain this further, I will give several examples of metaconscious knowledge related to the acquisition and use of human skills in work.

While erecting the steel structure for a high-rise office building, an experienced worker observes the crane bringing up a beam that, to him, does not look adequate for the span it must bridge. Before riveting it into place, he calls his supervisor, who will do one of two things. She may remind him that he gets paid to erect steel and to leave the thinking to the engineers. Alternatively, knowing that this worker has a great deal of experience, she may quickly run to the field office to check the drawing and possibly call the design engineer.

The worker has acquired a metaconscious knowledge of the strength of materials that permits him to look at a particular situation and declare that it just does not look right, or that it appears to be fine in the context of countless work-related experiences. His knowledge is embedded in experience, and no matter how well it develops it will never turn into the knowledge engineers have of the strength of materials. This knowledge is acquired in a classroom and through books that begin, not in the real world, but in a highly abstract world of finite elements. It is similar to learning physics in high school. Students do not start out in the real world but in an abstract world of point masses, frictionless planes, weightless pulleys, free fall without air resistance, and so on. As students advance, they gradually learn to take into account more effects that make the world of physics approach but never coincide with the real world.

This knowledge, acquired by engineers and physicists (or any other specialist, for that matter), is 'knowledge separated from experience.'

The experiences of classroom learning, reading, and problem-solving also form symbolic structures that generate metaconscious knowledge, but these are of a different kind from the ones associated with knowledge embedded in experience. Metaconscious knowledge embedded in knowledge separated from experience has, at best, an arm's-length relationship with experience as lived through a culture. This metaconscious knowledge is not fully contextualized into the life of a knower and, via it, into the way of life and culture of his or her society. Metaconscious knowledge embedded in knowledge separated from experience is contextualized only within the 'world' of a discipline or specialty. For example, in an extreme case, someone could learn stress analysis, design a bridge, and never set foot on a construction site. Such a person's only experiences of bridges would be the culturally mediated ones related to the daily-life use of bridges. These are clearly very different from the experiences of bridges obtained via drawings and technical specifications.

On the shop floor we find the same coexistence of knowledge embedded in experience of machinery and processes and the corresponding knowledge separated from experience. Engineers who design machinery and processes do so in a highly idealized fashion. For example, they assume that the properties of materials are uniformly distributed throughout space. No such materials exist in the real world. For example, a batch of identical parts heat-treated in a furnace will be affected differently, depending on their exposure to the heat. Sometimes these differences have no practical importance, but at other times machining such parts could break tool bits, requiring adjustments that experienced tool-and-die makers would, in many cases, make without thinking because they are constantly bridging the gap between the idealized world of engineering drawings and specifications, and the reality of actual materials, parts, machines, and processes. In other words, tool-and-die makers, machine-shop operators, and shop-floor workers know the technology in the shop differently from how engineers know it. Each kind of knowledge has limitations, and thus has a unique role to play in conjunction with the other. The following example illustrates this.

A manufacturer of photocopying machines found that equipment assembled during a particular shift had a well-below-average incidence of customer complaints and service requirements. An investigation of the situation led to the finding that the foreperson on that shift would not release any machine until it sounded right. Unfortunately, we train engineers to write this performance indicator off as nonsense, but it is quite the contrary. Once again, the engineering drawings and specifications of the photocopier depict an ideal machine with ideal dimensions, with the only concession to reality being the tolerances within which real dimensions should lie. This means that when a large sample of machines is examined in terms of the distribution of their tolerances, a few machines will have most of their tolerances under the specified dimensions, and a few over those dimensions. Most of them will have a fairly random distribution. It is entirely possible, therefore, that machines whose tolerances were predominantly over or under did not sound right because they were either a little too tight or too loose. Apparently, the shift supervisor had noticed this and learned to make some adjustments, with beneficial results for the company and its customers. The shift supervisor had acquired a knowledge of these photocopiers that the engineers did not have, and vice versa.[26]

Zuboff[27] interviewed workers to gain a better understanding of how they perceive the effects of automation. Pulp-and-paper mill workers reported that after computerization they had little physical contact with the process, whereas in the past they could handle the pulp to see if everything was working properly. They were now obliged to watch numbers on a computer screen. The new control technology thus significantly curtailed the development of their knowledge embedded in experience of the process, which could negatively affect their performance as operators. The contribution knowledge embedded in experience makes to industrial and other processes is unknown. I am not aware that any studies have been done of it and, with few exceptions, managers and engineers are unwilling to acknowledge its existence, let alone allow it to play an essential role.

The development of metaconscious knowledge plays a fundamental role in human skill acquisition. I will combine the model developed by Hubert and Stuart Dreyfus[28] with my model of the acquisition of culture and daily-life skills[29] to deepen our understanding. In the former model, five stages of skill acquisition are distinguished: novice, advanced beginner, competent, proficient, and expert. During the first stage, people learn to recognize and detach specific features of situations that are relevant to a particular skill in order to place them in the context of some simple rules for determining action. At this point, attending to these features and rules fully engages their attention, so that anything else that may be happening is largely distracting. Rules must be context-free, applicable to a whole range of situations. Initially, the experiences of situations are based on a foreground–background distinction in which the features relevant to the rules constitute the former. However, this will not last long. The processes of differentiation and integration begin the construction of a symbolic structure of these experiences, making people more aware of both their similarities and their differences without being able to articulate what these are.

During the second stage, that of advanced beginner, the processes of differentiation and integration apply to a considerable number of real situations, gradually permitting the consideration of more features in each case. A sense of the need for more complex rules emerges. All other features not relevant to these rules remain a distraction with respect to carrying out the skill. However, the process of differentiation, now based on a more extensive range of prior experiences, permits advanced beginners to recognize some situations as essentially similar to previous ones, thereby facilitating the application of the rules, and others as dissimilar, to which the rules are still applied but with some sense of caution, apprehension, or reluctance. For example, in stop-and-go traffic, advanced beginners may still shift into second gear at the speed they were told but begin to recognize that this may not make any sense if traffic appears to be heading for another stop. In such an obvious case the person may ask the instructor if it would make more sense not to bother shifting, but, in many other cases, it will be much more subtle because the contribution of the processes of integration and differentiation is still rather modest. Little if any metaconscious knowledge has been accumulated in the symbolic structure of experience. This will soon change.

When people become competent in carrying out a particular skill, they no longer follow the rules automatically. They begin to recognize a range of situations where this makes no sense, leaving them with no other option but to analyse such situations in order to determine an appropriate response. They begin to develop plans for dealing with situations where the routine application of rules is not satisfactory. In other words, the processes of differentiation and integration have expanded the symbolic structure of experience to the point where there is a growing set of situations to which routine, rule-based responses are not appropriate. However, the experiences of such situations have not yet been sufficiently structured by the processes of differentiation and integration for prior experiences to function as paradigms. There are now two kinds of behaviour patterns: more or less routine responses to some situations, and others that require analysis and a plan for dealing with them. People become much more involved in what they are doing and begin to have a greater sense of responsibility if something goes wrong. Routine responses also become affected. It is now much less a question of mechanically applying rules that you did not invent to situations over which you have no control. There is the beginning of a greater emotional and intellectual involvement in applying the skill. This is essential for learning because a successful response to a somewhat unusual situation brings satisfaction, and failed responses are less easily forgotten. In other words, the symbolic structure of experience is beginning to imply metaconscious values capable of guiding the development of the skill. The processes of differentiation and integration enfold these values into similar past and present developments in the entire symbolic structure of experience of a person's life. As noted by Hubert and Stuart Dreyfus, during this stage people tend to behave like stereotypical puzzle-solvers, but this does not mean that human intelligence is, first and foremost, problem-solving that can be captured by heuristics, algorithms, and rules. This becomes evident during the next two stages.

The transition from being competent to being proficient is marked by a qualitative change. Carefully thought-out behaviour based on considering various alternatives for the application of the skill to accomplish a goal still occurs, but it is much rarer. Instead there is a growing incidence of responses to situations that reflect a fluidity, rapidity, and level of involvement not seen before. It is as if people have an almost immediate intuition of how to tackle the situation, and only occasionally do they have to think it through. This can be explained by my theory of the acquisition of culture and daily-life skills. The number of experiences a person has accumulated in a particular skill domain has grown to the point that the processes of differentiation and integration have been able to create an extensive symbolic structure of these experiences, which is now beginning to act as a repertoire of paradigms that are 'intuitively' fitted to new situations. It is only when a situation is sensed to be different for some reason or another that conscious analysis comes into play. In other words, the metaconscious knowledge built up in a symbolic structure of experience related to the skill domain is now so extensive that fluid and intuitive responses to many situations can be made with an apparent ease that characterizes proficient behaviour. There is a kind of holistic recognition of a variety of situations on the basis of the symbolic structure of differentiated and integrated prior experiences. Hubert and Stuart

Dreyfus call it 'knowing-how' or intuition, as distinguished from 'knowing-that,' which is based on rules and sets of isolated features deemed relevant for the application of a skill. Only when a situation is intuitively grasped to be different from what has been encountered before does detached decision making occur.

The behaviour of the expert is not qualitatively different from that of a proficient performer. The skill has now become such an integral part of a person's structure of experience that only very rarely do conscious choices and decisions need to be made. Under normal conditions, experts do not have to solve problems or make decisions, much as I have shown in the acquisition of culture and daily-life skills. The development of the symbolic structure of experience is now so extensive as to be able to cover almost any situation that is encountered. In so far as analysis occurs, it is no longer addressed to the situation directly, but to the metaconscious sense a person has of it (intuition). The processes of differentiation and integration have, as it were, built up an enormous repertoire of distinguishable situations on the metaconscious level. In other words, human intelligence is ultimately based on making sense of and living in reality by giving everything a place in our life in relation to everything else. It is this relational character of our being in reality, and not conscious problem-solving analysis, that is the basis for human intelligence.

Decades of heavily funded research in artificial intelligence attracting some of the best minds ultimately failed to capture daily-life knowledge and skills by means of rules and algorithms.[30] There is a significant implication for expert systems. Based on rules and algorithms, they are unlikely to advance beyond the third stage, but this does not mean they are useless. On the contrary, Hubert and Stuart Dreyfus[31] specify the conditions under which expert systems can be useful. At present, there is no reason to believe that neural net–based expert systems will go any further. In a later paper,[32] Hubert and Stuart Dreyfus applied the five-stage model to the skill of morally coping with the world. My theory of the acquisition of culture and daily-life skills suggests that ultimately the learning of a skill is giving it a place in our life and being in a way that is individually unique and yet culturally typical. Hence, neural nets, while constituting a fascinating development, are not likely to capture human knowledge and skills in a machine.

Hubert and Stuart Dreyfus[33] estimate that a chess master can recognize roughly 50,000 different positions and that expert automobile drivers can probably intuitively respond to a similar number of situations. When this is multiplied over all the daily-life skills that human beings acquire, it is obvious that our language does not have a sufficiently large vocabulary to make these distinctions explicit. Hence, much of human intelligence cannot be brought into words – a problem knowledge engineering continues to struggle with.

The skill-acquisition model outlined above implies a fundamental distinction between expertise acquired through knowledge embedded in experience and that acquired through knowledge separated from experience. The skills of the welder, shift supervisor, and pulp-and-paper mill operators referred to above are acquired through culturally mediated daily-life experiences. That means that their meaning and value are relative to everything else in their lives in a way that is individually unique yet culturally typical. This contextualization involves all of reality, including the unknown, through myths and the sacred, which is why traditional technologies tended to be appropriate.

In contrast, the skills acquired by scientists, engineers, and all other scientific and technical specialists, do not begin with daily-life experience in an apprenticeship-like relationship with others, but with the experiences of a highly abstract world to which they can only be introduced in classrooms or through the reading of books. These experiences of a different world constitute a subset of the symbolic structure of experience of a person. This subset is not shared with the members of their culture but with the members of their invisible colleges or fellow practitioners in a particular domain of expertise. In other words, these correspond to what Thomas Kuhn[34] first called a 'paradigm' and later a 'disciplinary matrix.' Metaconscious knowledge is built up through this structure by means of the processes of differentiation and integration, and plays an important role. In science this has been recognized by Polanyi as tacit knowledge.[35] The role of this metaconscious knowledge is exemplified by the many experiences of the mathematician Poincaré, frequently cited in the literature on creativity.[36] Similarly, an expert in circuit design can look at a particular diagram and immediately recognize some possible problems or stand amazed at its ingenuity.

The metaconscious knowledge built up by physicists and engineers (or by any scientific or technical expert) results from the processes of differentiation and integration operating on the subset of experiences derived from participating in a specialty, and is only indirectly connected to the larger symbolic structure of the culturally mediated experiences of a person's life. The experiences of this subset can be contextualized only within the domain of expertise. Hence, it can have only a scientific or technical meaning and value. It stands in sharp contrast with the metaconscious knowledge built up through knowledge embedded in experience, which results from the role of culture giving everything a name, meaning, and value in relation to everything else in human life, including the unknown. I have already referred to a great deal of evidence to this effect. The diagnoses of the experts as to what to do about the hunger problem in a Colombian valley could neither be mutually integrated nor be a part of the daily-life world of the non-expert. The engineer trying to diagnose the problems of one of her company's plants arrived at an interpretation of the situation that tended to put those features corresponding to her domain of expertise in the foreground. As technological knowledge was separated from experience at the beginning of the twentieth century, it began to develop its own values, which I have called 'performance values.' Because of the unique cultural unity that emerged, these values were adopted as cultural values despite the fact that they were merely technological measures and had nothing in common with what until that point in human history had been values. Finally, the term 'appropriate technology' had to be invented because technology was no longer spontaneously adapted to its context.

In sum, expertise based on knowledge embedded in experience is fully contextualized in the individual and collective life of a culture, while scientific and technical expertise are contextualized only within their own domain or specialization. The latter, therefore, need to 'borrow' elements from a culture, which greatly limits their ability to be objective. The whole question of the implications for human life and civilization resulting from the coexistence of knowledge embedded in and knowledge separated from experience is the subject of a separate work.[37] But for now it is clear that expertise based on knowledge separated from experience presents an obstacle to preventive approaches that must place technology in the human, societal, and biospherical context. This the case is because scientific and technical expertise are decontextualized from the world of experience and culture. No expert is aware of the 'edges' of this world, which indicates the presence of myths and makes it very difficult for the expert to identify the limitations of his or her expertise. As pointed out earlier in this work, scientific and technical knowledge therefore include intrinsic ignorance, which is built into this knowledge, leaving the door open at any time to a kind of Kuhnian revolution. Expertise based on knowledge embedded in experience is attached to experience and culture, and therefore to the world, while expertise based on knowledge separated from experience is detached from the world from the very beginning and can only approximate a very small portion of it one domain of specialization at a time. The fundamental discontinuity between the two kinds of expertise, since the evolution of one cannot produce the other, has important implications for work organizations.

It is now apparent that a work organization based on a technical division of labour has two segments: one mostly based on knowledge separated from experience, and the other mostly based on knowledge embedded in experience. A dividing line runs through the work organization where the two kinds of knowledge and expertise confront each other, producing power struggles, antagonisms, mistrust, and dislocations. However, this need not be the case if the limitations of each kind of knowledge and expertise are clearly understood. It would then be recognized that, although it is possible to know either with knowledge separated from experience or with knowledge embedded in experience, they are different and to some extent complementary since each one can do things the other cannot. By modifying education to recognize this situation and by redesigning work organizations, a complementarity and a symbiotic relationship could be established. For now, the dividing line between the two segments of an organization, according to the kind of knowledge and expertise that is dominant, divides the organization against itself and causes a great deal of trouble.

It is essential to conceptualize a work organization as a network of people having a diversity of knowledge and expertise that constitute the inputs into that network. The work organization can ignore these inputs, underutilize them, or allow them to atrophy. Alternatively, it can nurture and expand these inputs through learning, or overutilize and crush them. It can either set the two dominant forms of knowledge and expertise against each other, or establish a symbiosis so that each one can to some extent cover the limitations of the other. Thus far, the technical division of labour (with its hand–brain separation) has sought to maximize knowledge and expertise based on knowledge separated from experience and as much as possible eliminate knowledge embedded in experience. Had this been successful, the dividing line that undermines present work organizations would have been eliminated, but this is impossible. No one can cross this line by doing more of whatever is done on one side of it, and attempts to do so lead to problems related to the break-up of the integrality of the organization.

In other words, preventive approaches will have to address the fact that Taylor, and his successors, seek to eliminate the dependence of a work organization on knowledge embedded in experience, thus opening the road to complete mechanization, automation, and computerization. Had artificial intelligence succeeded, it would have provided the tools to complete the task. The rational de-composition and re-composition of human skills sought to bring everything within the domain of knowledge separated from experience, but this failed. People on the shop floor continued to bridge the gap between the ideal and the real, between knowledge separated from experience and knowledge embedded in experience. There are two responses to this situation, well analysed by Brödner,[38] who suggests that there are essentially two options – namely, the anthropocentric and technocentric alternatives. Of these, the former is inherently more preventive in orientation.

## 10.6 The Technocentric and Preventive Alternatives

The technocentric option involves continuing the strategy of displacing human workers by machines, based on the conviction that people are the problem, and machines the answer. It has driven management–labour relations into a vicious cycle in which one end-of-pipe solution after the other benefits neither party. It has produced work organizations with a highly technical infrastructure, operated by a network of human beings having the pyramidal structure familiar from organization charts. This particular structure is the result of the horizontal and vertical division of labour created, developed, and maintained by a variety of techniques. The productivity of manual labour has in part been achieved at the expense of human skill development and learning, the health of the workforce, management–labour relations, and the community. The productivity of intellectual labour has not increased commensurate with the large investments in the technical infrastructure, negatively affecting the creativity and health of the people involved, and thus the viability of work organizations. Gains have been realized in the domain of performance values, and losses in the domain of context values. The technicization of work enlarges the technical infrastructure and improves its performance, but degrades the performance of the work organization necessary to operate it. This pattern creates a variety of problems to which a technical response is sought. A further technical intervention leads to further problems, and so on. Management interprets this cycle as proof of non-cooperation, irresponsibility, and even sabotage. On the receiving end, the actions of management are seen as a failure to understand what is happening on the shop floor and in the office, a lack of trust in their employees and arbitrary authority. For many managers, this vicious circle can be broken only by further automation and computerization, which creates a positive-feedback system producing more problems, taken as evidence that further advances in the same direction are required.

This situation is no different from the way corporations respond to government actions limiting their ability to pursue performance values at the expense of the biosphere. Corporations have created environment departments within the corporate structure that lock them into end-of-pipe solutions. A preventive approach seeks to break this vicious circle by going back to the root of the problem by means of negative feedback. While a number of corporations are beginning to break out of this cycle of restructuring their relations with the biosphere via positive feedback, the parallel problem of structuring relations with the economy and society by means of positive feedback is less well recognized. Nevertheless, the problem is equally acute.

Thus far, I have examined how the technical division of labour produces both efficiency and inefficiency. The horizontal division of labour established through the rational decomposition and re-composition of human skills is a source of as many efficiencies as inefficiencies. The vertical division of labour increased labour productivity while at the same time negatively affecting the capacity of people to work effectively by degrading their skills and suppressing the self. The separation of knowledge from experience, first in engineering and later in management and accounting, further weakened the integrality of work organizations.[39] It also weakened the integrality of corporations, creating an 'us–them' attitude manifested in varying levels of mistrust and non-cooperation, and used by each side to pursue their course. The fundamental question this raises is obvious: How has the net efficiency of work organizations evolved with the technical division of labour? Alternatively: to what extent have gains in the domain of performance values been offset or possibly been undermined by losses in the domain of context values? These questions are impossible to answer because nearly all the inefficiencies are economic externalities not tracked in accounting practices. This does not make them less real. It is clear that a corporation depends on societies as much as it depends on the biosphere. It requires a continuous throughput of people whose capacity to work is affected by the work organization and by the communities in which they live. In addition, these people are the customers of corporations and, in many cases, shareholders through their pension plans. In sum, the consequences of 'the people are the problem and machines the answer' for human life, communities, and societies are pushing us towards a human and societal crisis that is as profound as the environmental crisis. A great deal of our life is spent working for organizations of one kind or another. The quality of our working life affects our whole life: our families and social relations, our communities, the fabric of our society, and our culture. All this, in turn, affects the organizations for which we work because we depend on viable communities from which to recruit a healthy workforce and on which their vitality and ability to respond to change depend. Techniques for organizing work so as to limit dependence on a highly skilled workforce as well as free-trade agreements have weakened the reciprocity in the relationship between organizations and their host communities and nations, leaving the latter increasingly vulnerable. However, shifting plants and offices from one part of the world to another only masks the fundamental problem and avoids dealing with it, at least in the short term. Quarterly statements tell us very little about the long-term viability of corporations, given that these externalize almost everything about the viability of their relations with their contexts.

Brödner[40] suggests that, if we continue with the technocentric strategy of which the asymptote is a nearly fully automated office with fully automated factories, several contradictions in this approach will be intensified without any possibility of a long-term solution. First, pursuing the technocentric alternative of replacing human beings with machines increases the complexity of the production plan, which must be objectified in a hierarchy of machines and computers, but this does not enhance the capacity to deal with its own imperfections, and with the gap between the plan and reality. Even under the best circumstances the plan requires regular modifications, and this becomes particularly problematic where there is rapid technological innovation in turbulent markets requiring regular product modifications. It does not take long before the plan is so full of patches as to become unmanageable, and shutdowns become increasingly frequent and longer in duration. This failure in planning leads to further planning, thus setting in motion a development guided by positive feedback. The only way out of this dilemma is to make the plan more flexible, which can be done only through people, thus negating the technocentric strategy.

The second contradiction is closely related to the first. The more management seeks to increase its control over the work process by replacing human beings by machines, the greater the resistance from the remaining operators. Yet it is on these workers that management must rely to bridge the gap between the plan and reality. The validity of this becomes all too apparent when workers decide to go by the book, carrying out the letter of the plan and nothing more. Attempting to decrease the dependence of the production plan on the remaining workers paradoxically leads to a greater dependence on their ability to mediate between that plan and reality. There is a loss in productivity, to which management responds by attempting to create a 'commitment organization' on the assumption that the problem is that workers do not identify with the goals of the organization, while at the same time management is sending a clear message that it is not committed to its workers. If management decides that the only solution is to push ahead with a technocentric strategy, it risks exacerbating these problems and intensifying the positive-feedback mode for evolving its production plans.

The third contradiction stems from the fact that the technocentric alternative requires ever more rigid and central planning, resulting in limited flexibility, longer throughput time, and delayed delivery dates, while markets increasingly demand the opposite. Adding more computers to the system tends to reinforce rather than alleviate the problem. Greater flexibility can be obtained only by decreasing the technical division of labour by means of a strategy other than the technocentric one. Brödner sums up the situation as follows:

> The market demands a flexible production process with short throughput times, but the structure of the process organized on the basis of the division of labor militates against this. Conversely, the demand for transparency and control is satisfied by precisely that form of division of labor which obstructs production economies and market demand. Development has thus reached a point where prerequisites for maintaining control are in conflict with those for deploying capital. Resolving this conflict, or at least taking the edge off it will have to be the result of a quite different production concept.[41]

Within the Taylorist paradigm, it is impossible to recognize that human beings and machines make almost diametrically opposite contributions to production. Machines and computers can deal with situations only through rules and algorithms, which is highly effective for routine or semi-routine tasks. This pays off on long production runs. The human contribution to the production plan, on the other hand, is based on skills that are fundamentally non-algorithmic. All the psychological evidence shows that human health and well-being are severely undermined by situations rendered meaningless because of sensory deprivation, monotony, and excessive repetitiveness. The failure of artificial intelligence to capture daily-life experience and human skills in rules and algorithms and the inability of expert systems to move beyond the third stage of human skill development call into question the Taylorist vision. What is required is an alternative production strategy that is preventive in orientation (which includes what Brödner calls the 'anthropocentric alternative' but is not limited to it). Preventive approaches can help establish a healthier, more reciprocal relationship between management and labour, between knowledge separated from experience and knowledge embedded in experience, and between the work organization and the community and ecosystems on which they depend. Recognizing the reciprocal interdependence between work organizations and their contexts requires that their development be guided by both performance and context values. Only in this way can an optimal symbiotic relationship be created between the human and the organizational segment of a work organization (capable of providing maximum flexibility) and the machine- and computer-based segment (capable of providing the benefits of algorithm-based continuity or incremental change). In a global economy based on a highly non-linear technological development, the contribution of human intelligence and skills to the work process needs to be rethought.

## 10.7 Building Blocks for a Preventive Strategy

Prevention-oriented production strategies must be based on the following principles. First, the strengths and limitations of both knowledge separated from experience and knowledge embedded in experience must be recognized, so that the two kinds of knowledge can be made as complementary as possible. This requires a different management–labour relationship. Second, the recognition that metaconscious knowledge associated with knowledge separated from experience and knowledge embedded in experience can be captured neither in expert systems nor in production or organizational techniques of any kind implies different relationships between human beings and machines (including computers), so that human beings can continue to develop their skills without obstructing or diminishing the contribution made by machines or computers. Third, the limitations of the technical division of labour must be pushed back. The horizontal division of labour must be structured around machines extending and enhancing human skills through design islands, production islands, and human-centred information systems. The vertical division of labour must be a spectrum that blends knowledge separated from experience with knowledge embedded in experience, where the latter serves the former near the top of the organizational hierarchy and the reverse near the bottom. Fourth, the engineering, management, and regulation of work organizations must also be preventive with respect to their interdependence on their contexts (human life, society, and the biosphere). This means that the methods of industrial ecology, energy efficiency, and those described in the next chapter dealing with a sustainable habitat, should be an integral part of the overall strategy. Integral to a healthy work strategy is the evolution of the knowledge and skill base of a work organization through learning, keeping open the possibility that some parts of human skills can best be delegated to machines, and others not.

Implementing these principles will, first and foremost, depend on a greater trust between all parties. Is it possible for corporations to evolve towards a community of stakeholders? Would such an evolution be in the best interest of all parties? Is there a place for the values of modern societies (such as democracy, human freedom, and the dignity of the individual) in modern corporations that have evolved into authoritarian islands of central planning? Is the growing gap between the highest and lowest salaries in modern corporations another sign of the breakdown of their integrality because of a widening gap between the values of the institution and the values of most of the people working for it? There is a lot of historical baggage to be overcome. From the beginning, industrialization was primarily driven by a desire to establish a greater control over the work process. A first general observation, however, appears reassuring. When with a broad brush an attempt is made to correlate viable management–labour relations with the opportunity to learn from countries that industrialized earlier, it would appear that the most reciprocal relations are more common in the nations that industrialized later. Much more can be accomplished if we can learn from our mistakes by, as realistically as possible, analysing the present situation while avoiding the usual ideological stereotypes and narrow self-interests.

A pioneer in this area is the Brazilian company Semco, which has been visited by executives from all over the world.[42] In 1993, it employed 300 persons and worked closely with some 200 satellite businesses operating as subcontractors, largely set up with its help. Its product line is remarkably diverse, including commercial dishwashers, mixers, high-volume pumps, and entire biscuit factories. What has made it a focus of attention is its unusual management style, which is essentially based on trust. Establishing a climate of trust fundamentally transforms the hand–brain separation: employees manage their own work as much as possible and receive a vote on all major decisions made by the corporation. Financial information is openly discussed, and employees have been taught how to read a balance sheet and a cash-flow statement. The organizational structure is described in terms of four concentric circles with only four job titles: counsellors, who are the equivalent of a traditional vice-president and who coordinate general strategy; partners, who run the business units; coordinators, who perform the traditional roles of foreperson and supervisor; and associates, a category that includes everyone else. Ungratifying dead-end jobs have been essentially eliminated by having everyone do the mundane tasks associated with his or her particular job, including going down to meet a visitor, carrying out office chores such as word processing and photocopying, and making phone calls. Professionals can take time away from their jobs as a kind of sabbatical, to help them keep up to date or rethink their jobs. About 25 per cent of the employees set their own salaries, and leadership is evaluated by those on the receiving end through questionnaires, the results of which are publicly posted. Employees help redesign the products they make, the procedures for manufacturing them, and the marketing strategies. Top brass do not interfere with the management of individual business units. Time clocks are merely devices that help employees keep track of their time, and there are no security checks at the plant gates. If people travel for the company, they make their own arrangements as they see fit. When the owner is on holidays, he does not leave a telephone number because everyone is supposed to be as self-sufficient as possible. Even profit sharing is carried out on a democratic basis by giving employees a voice. In sum, the company believes that a capitalist society must be capitalist for everyone. It practises the values of capitalism – namely, personal freedom, individualism, and competition – within limits that control greed by the sharing of information and power and by making operations as flexible as possible. There are no management fads and no consultants, who in any case have no inside knowledge of the business and will not be around to live with the consequences of their recommendations. There is an ongoing evolution of its organization and relations with the outside world through, as much as possible, having everybody manage their own affairs, thus shortening negative-feedback loops. Everyone is treated as a responsible member of a community of stakeholders. We would expect such a company to do poorly because we have been told that employees cannot be trusted. Yet, by 1993, sales had increased sixfold, and profits 500 per cent.

This company has consistently restructured itself from top to bottom, demonstrating how mutual trust and respect and other values of modern societies can work in a business setting. Most companies have practised a more piecemeal approach, which provides us with glimpses of what some aspects of preventive approaches can accomplish even when short-lived because of an uneasy coexistence with the surrounding traditional management and organizational structure. For example, whenever companies have carried out design functions with a minimal technical division of labour, the results have been impressive. People with the requisite kinds of knowledge (both knowledge separated from experience and knowledge embedded in experience), left to themselves to decide on how best to get the work done, are generally able to do so in a much shorter time and with higher-quality results. Informal collaboration without arbitrary authority, trust and mutual respect, and no arbitrary reporting deadlines (so that the results can be documented at the end) completely transform the technical division of labour in the design of new products. Well-known examples are the way the Saturn vehicle was designed,[43] and the way IBM designed its Proprinter.[44]

AnnaLee Saxenian has convincingly demonstrated that the success of computer manufacturers in the Silicon Valley of California over the more established computer firms located along Route 128 outside Boston can be attributed to the horizontal, participative, and trusting management style in the former, which allowed them to react faster and with greater creativity to a rapidly changing marketplace, while the firms along Route 128, with their hierarchical and non-participative management style, were almost wiped out.[45]

Other success stories where work has been radically restructured to eliminate supervision and unnecessary management, and where employees truly took charge of their duties and their performance to the mutual benefit of the company and themselves, can be found, for example in the operation of the TRW's Oil Well Cable Division Plant in Lawrence, Kansas. In the words of G.T. Strippoli, this plant was designed 'to avoid all the traditional bugaboos such as authoritarian policies, rigid work rules, and adversarial management relationships that inhibit an organization's effectiveness.'[46] To this end, TRW eliminated time clocks and made all workers salaried. Job classifications were eliminated, and workers organized themselves into teams. Production scheduling, problem solving, and the hiring and training of new workers were now done with full worker participation. The work teams were made self-evaluating and also made recommendations regarding the acquisition of new equipment. In the first six years, plant productivity increased by 80 per cent. At the TRW Ramsey Piston Ring Division in St Louis, Missouri, management's role was similarly curtailed, and workers were radically empowered to control the pacing and execution of their tasks. Within weeks of implementation, Tom Bremer, the manufacturing manager, claimed that 'the general productivity effort and climate have improved significantly ... The productivity ratio of hours earned to hours worked has increased by 17%. With respect to quality, the cost of scrap per standard hour decreased by nearly 15%. In addition, the latest data on the percentage of rings rejected at final inspection reveals a gradual reduction of nearly 17%.'[47] The introduction of self-managed teams brought similar benefits to the Kemper Life Insurance Company corporate offices in Chicago and at Delco Remy, the battery manufacturer in Fitzgerald, Georgia.[48]

Edward E. Lawler, in a study of employee involvement practices by Fortune 1000 corporations, found that the most successful of these corporations had the highest rates of genuine employee involvement with self-managing work teams, job enrichment, and redesign, and employee input into all corporate decisions. The success of power-sharing programs were overwhelmingly rated as successful or very successful by a majority of the respondents to Lawler's detailed survey. No power-sharing approach was rated as unsuccessful by more than 12 per cent of the respondents, although a number of respondents did indicate that they were undecided about the level of success.[49]

The advantages of self-managing work teams and autonomous work groups, as well as job enrichment, have been extensively documented in advanced manufacturing.[50] These efforts can be extended to the design of manufacturing and service processes by incorporating into the informal teams those with the requisite knowledge. This should then be extended into the way products are actually produced.

What the above examples also illustrate is that for workplace reform to be successful it must be accompanied by a wholesale commitment and shift in 'cultural' outlook in all members of the organization. There is considerable data to suggest that piecemeal reforms or changes to the work regime that are not fully inclusive tend to fail or even to make existing working conditions worse. For instance, in cases where employees are given greater control over their work, *but* where there has been no concomitant flattening of the hierarchical structure of the organization or a lessening of the opacity of upper management's aims, attempts at reform have usually failed.[51]

## 10.8 The Scandinavian Experience

The Scandinavian countries present us with the clearest examples of what preventive approaches in the area of work are able to accomplish. The reasons are not difficult to understand. These are small, culturally homogeneous societies that have a long democratic and egalitarian tradition, learned something from the industrial experience of other nations, and avoided the bitter conflicts between labour and management. A partnership among government, industry, and labour unions emerged, giving workers and the general public a voice in technological and economic development. Democratic values and aspirations and the benefits of capitalism were extended to the workplace. To a somewhat lesser extent, this is also true for Germany and the Netherlands, to a lesser extent for other western European countries, and still less for the United States and Japan. One factor shaping these differences is that, in the Scandinavian countries, assembly-line work is done by citizens, unlike in other European countries, where it is mostly done by 'guest workers' from poor and developing nations with non-European values and cultures. In the United States, this work was traditionally done by waves of immigrants and minorities. In Japan, the sociocultural climate continues to be much more traditional and thus hierarchical, paternalistic. and authoritarian, with the result that its workers have different expectations of the workplace, although this is beginning to change as there, too, the traditional vestiges are being eroded.

As a result, labour markets function in rather different sociocultural contexts; and this has some important implications for the development of the relationships among work, human life, and society. For example, Swedish employers faced a more difficult task of attracting good workers to automotive assembly, which was regarded by many Swedes as undesirable and unhealthy employment. Under equity legislation, efforts by employers to overcome this reluctance by means of wage differentials were prohibited. In addition, employers faced the ergonomic challenge of adapting the workplace to accommodate a high proportion of female workers, which required adapting tools, machines, and procedures to suit people with different physical characteristics. For a long time, there was no pool of unemployed workers to draw from. Employers also had to reckon with the social consensus that men and women had important obligations outside of work, such as parenting responsibilities, which public policy did not regard as a private matter since children not only ensure the continuation of society, but also determine its future vitality. Government, industry, and unions formed a variety of partnerships to solve these problems and to ensure the benefits of capitalism for everyone.[52]

To brush all this aside on the grounds that this is simply a vestige of socialism that has no future in a global economy and that the real solution lies in allowing labour markets to regulate these issues is to ignore what is happening to Japan, for example. Like many other industrialized nations, Japan attracted workers to unhealthy and inhumane work by allowing labour markets to 'solve' the problem by means of wage increases. It was not long before Japanese industry recognized that it could not afford these higher wages in the long run and began to move their manufacturing operations to other Asian countries. This made Japan vulnerable to the economic conditions of such nations, which is currently contributing to a serious crisis in the Japanese economy. Almost all the industrially advanced nations have experienced the problems of 'de-industrialization,' resulting from permitting labour markets to deal with the problem they were incapable of resolving. If we can move beyond ideological clichés, it may be possible to recognize that a full and equal partnership among government, industry, and labour unions assures negative feedback in technological and economic development, since those on the receiving end (workers and the general public) have a voice in this partnership. This ensures that technological and economic development will not occur at the expense of the 'social capital' of a nation on which the long-term viability of its economy depends. Where this argument leads is that lean production, which is the main competitor to Scandinavian reflective production, is both cause and effect of sociocultural conditions detrimental to the social capital of a nation. The gross efficiency of lean production systems is indeed spectacular, but their social costs are in all likelihood going to be equally impressive, with the probable result that Scandinavian approaches to production will have a long-term future whereas lean production systems will not. In any case, the celebration of the lean-production miracle is highly premature and entirely based on a performance-value perspective best exemplified by the work of Womack and colleagues.[53]

The Scandinavian countries developed a new approach to industrial production based on what has become known as 'sociotechnical work design.' It is an approach to the design of work organizations that seeks to overcome a long-standing problem where the technical infrastructure is designed by some professionals and the social organization by others. This approach recognizes that work organizations are wholes and must be designed as such, which involves more than simply balancing the technical with the social segments. The experience of the former causes it to be enfolded into the latter, thus establishing a reciprocal relationship in which, as people change technology, technology simultaneously changes people. This has created a fundamental interdependence, because modern people would not be what they are without technology, and without these people there would not be a modern technology. Hence, any undermining of one segment by the other weakens the vitality of corporations and societies.

Sociotechnical system design was created by Eric Trist and Frederic Emery at the Tavistock Institute.[54] They were consultants to the Norwegian Industrial Democracy Project,[55] to the early Swedish sociotechnical efforts, [56] and to the Dutch sociotechnical movement.[57] However, the concept was first formed during path-breaking work in the British coal mines in 1949.[58] In this postwar reconstruction period, the institute had two new research projects. One focused on groups on all levels, including the management and labour segments of an industrial organization, and another on the diffusion of innovative work practices that had the potential to improve productivity.[59] The former focused on social systems, but the latter considered both social and technical factors, including their interdependence. Coal was of critical importance to England, but the nationalized coal mines were doing poorly in terms of improving productivity and keeping their workers. The Tavistock researchers discovered a new kind of work organization later known as 'autonomous working groups,' which did their work with a minimum of supervision because they were largely self-managing. Cooperation and personal commitment ensured that productivity was high while absenteeism and the incidence of accidents were low. Such practices used to prevail in coal mines but had almost entirely disappeared with mechanization and the introduction of a technical division of labour.[60] Autonomous working groups became the paradigm for organizational redesign. This approach did not expand the technical division of labour and the associated bureaucracy along with advancing mechanization. It represented an organizational choice on which its name, 'organizational choice theory,' was based. It represented a deliberate choice away from Fordist/Taylorist and bureaucratic approaches.

The principal features of sociotechnical organizational design include the following elements. First, a work organization was treated as a single whole whose characteristics could not be comprehended by dividing it into its constituent jobs. Second, the work group, as opposed to the individual worker, was seen as the basic building block of this organizational whole. The possibility of making these working groups as self-managing as possible, thus shortening negative-feedback loops, was seen as beneficial for the work organization. It was based on the design principal of the redundancy of functions rather than the redundancy of parts. The traditional approach was to add more people to a bureaucracy when it was faced with additional challenges. The new design principle sought to maximize the response capability of a work organization by enlarging the response repertoire of working groups through multiskilling. To derive the greatest benefits for the work organization, workers were allowed a great deal of discretion, which also meant that they could not be regarded as mere extensions to machines. In other words, the evolution of a work organization in the face of a variety of challenges reduced the technical division of labour rather than increasing it, as in the conventional approach. This meant that variety on both the individual and the group level increased, as opposed to decreasing in an evolving bureaucracy.[61]

In other words, sociotechnical design approaches did not follow the technological imperative of the traditional approach. In the case of the latter, organizations were expanded and modified according to the requirements of the technical infrastructure. Any negative results on people could be dealt with only in an end-of-pipe manner, given the dictates of the technical infrastructure. These solutions took the form of socio-economic advances and improvements in human relations, and thus failed to deal with alienation. Sociotechnical approaches, on the other hand, recognized the interdependence between the technical and the social segment of work organizations, making the very idea of a technological imperative impossible. After all, sociotechnical work organizations encompassed people and technology in a mutually dependent way. A complete research agenda was beginning to unfold at the Tavistock Institute, including a new approach called 'action research,' which was based on a joint collaboration of management, unions, and workers.

Sociotechnical systems are recognized on three levels ranging from micro to macro. Primary work systems form the subsystems of a work organization and comprise one or more groups, including specialists, supporting personnel, and management representatives as well as the necessary technical infrastructure. Organization systems are the next-larger constituent wholes, ranging from a workplace, such as a plant, to entire corporations or agencies. Macro-social systems are the context of the previous wholes and include communities, industrial sectors, or institutions of a society. All three sociotechnical systems are a result of the sociocultural past and influence the present of a society. Sociotechnical systems are limited to those entities in a society that directly depend on material inputs and a technological infrastructure to produce their outputs.[62] In other words, the approach is very similar to the one developed in chapter 4.

Instrumental in the diffusion of sociotechnical approaches was Shell, which found them so successful in its U.K. operations that they were also adopted for plants in the Netherlands, Australia, Canada (Shell Sarnia),[63] and other countries. In Norway, sociotechnical approaches spread because workers were legally entitled to jobs that conformed with Emery's six psychological principles: (1) variety; (2) learning opportunity; (3) own decision power; (4) organizational support; (5) societal recognition; and (6) a desirable future. This shaped the Industrial Democracy Project based on sociotechnical approaches. Once again, to dismiss this project as a vestige of socialism is to overlook the fact that it ensures the development of 'social capital' by corporations as opposed to depleting it. The concept of industrial democracy is widely accepted in Europe but much less so in the United States, where it is often dismissed as an element of communism, and still less in Japan. It is for these reasons that in North America the term 'quality of working life' (QWL) was adopted but, given the sociocultural context, end-of-pipe fixes absorbed a lot of its energies. In Sweden, policy development in this area involves a commitment to gender equality in work, economic self-sufficiency resulting from each citizen having a job that pays enough to earn a living, and a recognition of the importance of families for the continuation of society. In other words, children are regarded as essential for both the public and the private domain, and care for elderly parents is also recognized as such.[64] This is in sharp contrast with prevailing attitudes in Canada and the United States. Other landmarks in the development of the Swedish sociotechnical movement are described by Frieder Naschold and colleagues.[65]

In sum, sociotechnical approaches are an expression of Western values and aspirations that became submerged in the workplace during industrialization in the belief that performance values were the only ones compatible with progress. Market economists still believe that societies cannot afford these values and remain competitive. This attitude is based on the fallacious assumption that we have two choices in resolving the problems of unemployment and underemployment. The first essentially became a non-choice when the concept of the welfare state was undermined by spiralling national debts. The second was that this problem could be resolved only through strong economic growth, which labour markets could deliver. However, this latter choice conveniently ignores all the externalities of markets, resulting in ever more damage from the invisible-elbow effect on social capital. It will produce an inevitable social crisis as it has already produced an environmental crisis. Karasek[66] has pointed out the fallacy of this dilemma and advocates a rethinking of public policy related to work, which converges with the present analysis and constitutes an integral component of the preventive strategies developed here.

One of the best-known attempts to change the technical division of labour for the manufacturing of complex products is Scandinavian reflective production. The example I discuss here is that of the AV Volvo Uddevalla plant during the 1970s and 1980s.[67] Car assembly was carried out in six parallel workshops where teams assembled complete vehicles with a long cycle time and were supported by new materials-handling techniques. There were essentially two variations of this theme. The automobile under assembly was stationary, but in three workshops it was moved once, in two it was not moved, and the sixth workshop was never fully utilized. In the three workshops, teams of seven workers moved between two workstations with a cycle time of 100 minutes, while in the two workshops teams of nine persons moved between four workstations with a cycle time of 80 minutes. During full production, each shop had seven work teams. Each shop had internal buffers taking the form of unworked-on automobile bodies, unoccupied working positions, and the integration of subassemblies such as doors or engines.[68] The assembly process was supported by a specially designed information system that linked components, assembly actions, and the entire product structure.[69]

The performance of this plant, in comparison with a regular assembly plant, was remarkable. Productivity levels reached those of Volvo's Gothenburg assembly plant, after which it continued to reduce vehicle-assembly time by, on average, one hour per month. The quality of workmanship and customer satisfaction was significantly better for its vehicles than for those assembled in a regular Swedish plant. The cost of annual model changes was cut in half, and full production after model change was reached in half the time required by a mass assembly plant in Sweden. The plant could deliver customer-ordered vehicles to dealers in four weeks, and further improvements were expected, giving enormous competitive advantages.[70] From the perspective of healthy work, a substantial improvement in method and timing control ensured that this work was characterized by high demand and high control instead of high demand and low control. Unlike the case with a regular assembly line, there was an increase in problem-solving demand, but this was compensated for by increases in control. According to the model, therefore, improvements in worker health and benefits for their lives, families, and communities should be realized. It would appear, therefore, that this Volvo plant created win–win situations for all parties. This was accomplished by substantially reducing the horizontal division of labour through functional reintegration, accomplished by longer cycle time and increasing the worker's range of task competence and responsibility, which overlapped in each work team.

Fortunately and unfortunately, the Uddevalla plant was shut down in 1993. The fortunate consequence of the closure was that Volvo released the many studies it carried out on all aspects of this work organization. It is for this reason that I chose this example since it is so unusually well documented. It is without question the best-known example of the sociotechnical design of a work organization.[71]

The unfortunate part of the closing of the Volvo plant is that it marked a deterioration in the sociocultural conditions in Sweden in general, and in the economy in particular. The tight labour market had changed as a result of the closing of several shipbuilding facilities, which freed up workers; Volvo sales were dropping; Volvo faced increased competition from Saab, which had been acquired by General Motors; and Volvo entered into a partnership with Renault, which was opposed to any sociotechnical approaches and favoured lean production (a position that is understandable, given the French socio-economic context).[72] The *New York Times* criticized the Uddevalla plant as a noble failure in humanistic manufacturing.[73] It is true that, at its closing, Uddevalla performed just as well as Swedish Volvo plants based on the traditional assembly line. However, the latter plants were much less efficient than Volvo's most efficient Belgian plant. In turn, this plant lagged far behind an equivalent Japanese lean-production factory.[74] These variations cannot be entirely accounted for in terms of the different manufacturing methods used. Another significant factor is that, generally, Japanese vehicles are easier to assemble as they are designed to facilitate their manufacture. The use of design for manufacturability (DfM) could have improved the performance of the Uddevalla plant, but it remains an open question how much the gap could have been bridged. From an economic perspective, it may therefore be argued that this case demonstrates that, under free trade, sociotechnical approaches in general and plants like the Uddevalla one have no future. Several things that should be pointed out here. First, this trade is not free, except for transnational corporations. It does not permit nations to make fundamental socio-economic choices, such as not undermining their social capital. Second, it will greatly increase the invisible-elbow effect to the detriment of everyone, including transnational corporations. Until externalities are taken into account to ensure development rather than growth, short-term gains are being traded for long-term pain in the form of the erosion of social and natural capital. Free trade is a new kind of bondage that does not permit the nations to impose context values on technological and economic decision making, which, unlike performance values, are genuine human values. Free trade is nothing short of a surrender to technological and economic necessities and determinisms.

Putnam's research[75] provides food for thought. He shows that those regions of Italy that are economically most successful had their origins in the free towns of the early Italian Renaissance. These cultivated broad-based horizontal cooperation and partnerships. Decentralized networks of cooperating small businesses developed egalitarian, entrepreneurial, and trust-based socio-economic conditions through which social capital was strengthened. This was in sharp contrast with the poorer regions in the south, characterized by the opposite socio-economic conditions, which emerged from centuries of rigid feudal hierarchies that left little room for individual creativity and enterprise. This confirms the reciprocal relationship among individual life, work, and society.

The examples discussed above together begin to provide an outline of what a prevention-oriented corporation might look like and what the implications would be. Could such a work organization compete with lean production? We could find only one case-study comparing the impact on people of lean production and traditional assembly-line work organization.[76] The lean-production line had a lower timing control and an increased production pressure, relative to the conventional assembly line. There was no evidence of a change in psychological strain, but there was a reduction in job satisfaction. In terms of the demand-control model, both work organizations fell into the domain of high demand/low control, and it appears that the assembly-line reorganized according to lean-production principles did not constitute an improvement in this domain.

Can lean production be regarded as a stepping stone into sociotechnical design approaches? It is difficult to assess to what extent lean production, taken as a further technocentric step beyond the Fordist/Taylorist system, makes it more difficult for corporations to switch to anthropocentric alternatives if it turns out that lean production is not sustainable, as expected. I will briefly comment on some of the evaluations of the experience at New United Motor Manufacturing Incorporated (NUMMI), a joint venture between General Motors and Toyota.[77] It involved reopening the GM Fremont plant, which was shut down in 1982, by rehiring a significant portion of the original workforce, now running a lean-production work organization managed by Toyota but producing vehicles for GM. The critical question is: To what extent does this work organization owe its remarkable performance to exporting costs to its suppliers, workers and society? The work organization differs from that of a regular assembly line in that it has become tightly coupled, to use Perrow's term.[78] When assembly-line workers perceive a quality problem, they pull a cord, causing warning lights to go on and summoning the quality team leader and group leader to come to the rescue. If the problem cannot be resolved within sixty seconds, the line will be stopped, bringing the problem to everyone's attention. This system is made sensitive to any imperfection, which clearly increases the pressure on employees and team leaders.[79] Another way in which the system becomes tightly coupled is through the elimination of all buffers by means of the well-known just-in-time materials-flow system. The evolution of the system is governed by what the Japanese call '*kaizen*,' which translates as 'continuous improvement.' Any unit in the system is always provided with less than what are expected to be the required resources, to necessitate improvement. When work procedures have been improved, permitting the job to be done with these resources, a further reduction takes place, thus applying more pressure for ongoing improvement. The structure and evolution of this work organization are such that it constantly increases demands placed on workers and intensifies the work.

It is in this context that the meaning of terms such as 'lean production,' 'teamwork,' 'empowerment,' 'job rotation,' 'multiskilling,' 'commitment,' and 'trust' should be examined. The term 'lean production' is unfortunate in the sense that it does not leave room for any alternatives. The opposite of 'lean' is 'fat' or 'bloated,' thus creating a term no one wants to be against. One author suggests that there is a strong tendency for this kind of work organization to become anorexic, as opposed to lean.[80] Can the workers sufficiently re-create themselves each day to sustain this work intensity? Can society afford work organizations that constantly push people to their limits? Can society absorb what are bound to be higher health and social costs?

The work on each station on the lean-production assembly line is organized by means of Taylorist methods, and workers are expected to do the work according to the 'one best way.' A significant improvement has been made by teaching workers the Taylorist approach to rationalizing work and giving them stopwatches, so that they can try out any new ideas they have for improvement to see if indeed they are more efficient. If this turns out to be the case, it is discussed with other workers on the team (as well as with upstream and downstream teams) and, if all parties agree that it is indeed more efficient, everyone adopts it. This builds in short negative-feedback loops and allows for a better symbiosis between knowledge separated from experience and knowledge embedded in experience. It also 'democratizes' Taylorism,[82] which 'empowers' workers to some extent. However, the work itself is still composed of an endless repetition of minute operations with a cycle time of a few minutes at best. When the Fremont plant was being run as a Fordist/Taylorist system, experienced workers could carry out a work cycle in about 75 per cent of the allotted time, leaving 25 per cent for blowing off steam or re-creating one's resources. As a result of *kaizen*, the same figure at the NUMMI plant rose, until a work cycle took 95 per cent of the allotted time, leaving only 5 per cent, which indicated an enormous increase in work intensity. In other words, from the perspective of the demand-control model, there is no significant method or timing control, while the demands have been substantially increased. In this context, the role of the team, as compared with those in sociotechnical work organizations, is trivial, in the sense that they are not self-managing in any real sense of the word.

Interviews with workers suggest that virtually every worker preferred working in the lean work organization as compared with the one they were used to. Overall work satisfaction improved substantially. So did worker satisfaction with job security, and participation in suggestion programs grew to an average of six suggestions per worker per year.[83] There is no question that workers are more involved in the production process as human beings. However, is it possible to speak of multiskilling, empowerment, commitment, and trust when the basic operations are still the boring, monotonous, repetitive, meaningless operations of an assembly line? Are we witnessing a kind of Hawthorn effect? Is it possible that the negative effects are long-term and not immediately noticeable? There is a great deal we do not understand about lean production, but the demand-control model would suggest that we can expect serious negative effects for workers' health and well-being. For this reason it may be expected that, when the studies are done, lean production may well turn out to be humanly, socially, and eventually economically unsustainable and that we are left with reflective production as the only genuine alternative to the Fordist/ Taylorist system.

## 10.9 Conclusion

Empowering workers and making greater use of knowledge embedded in experience in a work organization appears to be exceedingly beneficial for pollution prevention and energy efficiency. Many successful pollution-prevention programs depend on the suggestions of empowered workers. Energy-efficiency programs can also greatly benefit. One of the most spectacular examples is cited by von Weizsäcker and colleagues: 'Dow is one of the world's largest and most sophisticated chemical companies – a leader in a cut-throat industry noted for penny-pinching. Yet Dow has made the astounding discovery that there are $10,000 and $100,000 bills lying all over its factory floors – and that the more of them it picks up, the more it finds.'[83]

In conclusion, it would appear that preventive approaches for redesigning work organizations along the principles of sociotechnical work design can bring enormous benefits to corporations, their employees, communities, nations, and the biosphere. When the lean production system was invented out of necessity by Toyota to compete with major industrial producers,[84] it had an enormous impact on manufacturing, ending the supremacy of the Fordist/Taylorist system. We can as yet only dream about the possibilities that may be unleashed when corporations organized preventively begin to compete with present ones. One thing is certain: The effects will be more dramatic than the impact of the lean production system since it brings substantial advantages for all stakeholders and its effects will be felt by individuals, communities, nations, and in the quality of the biosphere. It is high time that the fundamental values of Western civilization restructure the corporate world. Work-quality standards based on the findings of social epidemiology along with the appropriate economic policies can unleash the potential of preventive approaches related to work. Our common future will be the better for it.

# 11

# The Built Habitat

## 11.1 Basic Relationships

The urban habitat increasingly interposes itself between human life and the biosphere. It sustains human life (with a variety of positive and negative effects) and depends on the biosphere for all matter and energy. The proportion of humanity living in this unique habitat has been growing exponentially along with industrialization during the last two centuries, rising from less than 5 per cent in 1800 to 15 per cent in 1900 and 43 per cent in 1990. This increase may be expected to continue in the twenty-first century because, in 1990, the urban population in the industrialized world was 72 per cent of the total, as opposed to 34 per cent in the less industrialized part of the world. When one factors in the exponential growth of the global population, some 3 billion people are expected to be urban by the year 2025 (52 per cent of humanity). Some 90 per cent of this increase is likely to occur in the developing world, greatly exacerbating already serious conditions.[1] The size of cities has also been growing. In 1995, there were twenty megacities with populations of more than 10 million people. Sixty-two cities had populations exceeding 4 million. Twenty of the world's twenty-five largest cities were located in the developing countries.[2]

These developments directly and indirectly affect all three primary dimensions of sustainable development. Since cities can neither create nor destroy matter and energy, they depend on an ever-larger hinterland, thus contributing to the globalization of the networks of flows of matter and energy. The portions of these networks associated with the urban habitat involve two components: one related to human life within that habitat and the other to the evolution of that habitat itself.

All inputs and outputs must be supported by the biosphere through its role as the ultimate source and sink of all flows of matter and energy. Hence, the sustainability of the biosphere and the urban habitat are increasingly linked together and these in turn interact with human sustainability. The relationship between human beings and their life-milieu is a reciprocal one. By internalizing a life-milieu, individuals and their communities modify themselves, affecting their milieu and thus further influencing themselves according to a process earlier referred to as the cultural cycle.

I have already suggested that, during prehistory and history, the influence of the primary life-milieu permeated human consciousness and cultures. There is no reason to believe that the influence of the modern urban habitat on human life and culture will be any less significant. This opens once again the debate as to whether the modern city can be alienating in the sense that its influence on human life is greater than the influence its inhabitants exercise over it. It is in this way that the modern urban habitat interrelates the sustainability of the biosphere, the sustainability of the life-milieu (of which the urban habitat is an important component), and the sustainability of human life itself. The first two dimensions of sustainable development related to the modern urban habitat can be examined by extending the discussions begun in chapters 9 and 10. The sustainability of human life in that habitat requires its own theoretical framework, which will form the primary topic of this chapter.

With respect to human life and society, the modern urban habitat appears to have two primary functions. On the one hand, it brings together ever larger numbers of people in space and time while simultaneously separating them in space and time in other ways. The city spatially separates the rich from the poor and assigns each socio-economic stratum its own neighbourhood. Superimposed on this, there may be a separation of ethnic minorities. It also separates the activities of a way of life in space and time, creating separate locations where people work, live, and play. The personification of cities in this paragraph is not accidental. It is designed to remind us of something that was once fundamental to Western civilization – namely, that the Judaeo-Christian tradition has always regarded the city as a potential alienating force in human history, which, with the advent of the mega-city, takes on a whole new reality.[3]

This dual movement of bringing people into close proximity to one another while simultaneously separating them in space and time has profound consequences for human life and society. One of the most important is the impairment of negative-feedback loops in evolving the urban habitat. Planners, architects, engineers, managers, and administrators involved in the evolution of this habitat make their choices and decisions mostly in terms of knowledge separated from experience. Many of the consequences will fall outside of their domain of specialization and will therefore not be accessible by means of the theoretical framework, concepts, and theories that constitute their 'professional cultures.' Since the city separates these professionals in space and time from the people affected by their decisions, they have no ready access to the meaning and value of these consequences as experienced in daily life through a culture. Hence, it is very difficult for them to obtain direct negative feedback except by the kinds of approaches pioneered by people like Jane Jacobs[4] and Christopher Alexander,[5] which gave rise to movements such as the new urbanism.[6] Consequently, the evolution of the urban habitat has excessively depended on end-of-pipe solutions to its problems, a dependence that negatively affects its livability and sustainability. The destruction of negative-feedback loops in the professions that most directly influence the evolution of cities carries over into the management and administration of cities, which in turn advise elected politicians.

All the inhabitants of cities are negatively affected by this situation. The separation of the knowers, doers, managers, and the general public undermines the integrality of the social entities of cities. It creates 'us–them' perceptions that rob everyone of something essential – namely, an effective reciprocal relationship with their life-milieu on which human health and well-being fundamentally depend. Instead of allowing people to live the dialectical tension between alienation and freedom, it creates a substantial force in favour of the former.

Separating in space and time the daily activities of a way of life has another important consequence. It forces the inhabitants of cities to travel great distances that separate the places for working, living, and playing. It is extremely resource-intensive, taxing the role of the biosphere as ultimate source and sink. It robs people of valuable time and demands from the urban habitat a great deal of space that must be set aside for transportation, as well as making it much less liveable as a result of noise and pollution.

The modern urban habitat also separates its inhabitants from their communities for most of their activities. This creates an entirely different social structure from what existed in traditional cities and societies.

The changes can be most comprehensively probed through a detailed examination of the relationship between technique and culture.[7] I will therefore limit myself to a few remarks. Traditional communities derived their common unity from shared daily-life experience interpreted and valued by means of a culture.[8] Such a common unity can tolerate only a limited diversity from its own members and must restrict the influence of strangers. The former becomes all too apparent when under stress a community begins to suspect any member that is, for one reason or another, somewhat different, as being a source of its difficulties. Intolerance and phenomena such as witch-hunts have been all too common. If we recall that a healthy relationship, group, or community depends on a dialectical tension between commonality and diversity, it is clear that traditional communities have tended to value conformity rather than diversity, with both positive and negative implications for human life.

Imagine such a traditional community being engulfed by an expanding modern city. A growing number of relationships would now involve strangers who do not share its common unity. This extensive contact with other 'cultures' has frequently undermined, and even destroyed, the basis for such communities. Strangers in large numbers are no longer an added source of diversity but a fundamental threat to the community, as countless sociological and anthropological studies have shown.[9]

A modern city embodies a social structure that has evolved out of situations where most people one meets and with whom one shares daily activities are, in fact, strangers. It is what is known as a 'mass society,' in which the vestiges of the traditional common unity manifested through customs, traditions, morality, religion, and other institutions have become weak, unstable, and, more important, a matter for people's private lives as opposed to the public life of a community (i.e., they have become centred on the individual rather than the group and the community). These traditional elements have been displaced by what Jacques Ellul has called 'integration propaganda': a bath of images, symbols, and narratives mostly diffused by the mass media whose collective functions are to show the members of a community what they must look like, how they must dress, what they should eat and drink, what they should own, how they should live, what their preoccupations ought to be, and who their friends should be.[10] In other words, the social structure that has emerged with phenomena such as technique, the mass media, industrialization, and the urban habitat can be based only on diversity as opposed to a culture-based common unity. Once again, this has many positive and negative implications for human life, including the fact that much of this diversity is constrained by what technique has to offer in the many spheres of modern life. In sum, the modern city can tolerate historically unprecedented levels of diversity because cultures have largely become a matter of people's private lives, leaving the public arena to technique-based conformity and integration. From a sociological perspective, the modern city creates a stimulating diversity within the domain of technical possibilities. It has created a whole new social, moral, and spiritual setting for human life. People's private lives continue to rely on networks of more traditional relations, which may be more or less extensive and healthy, but the fact of the matter remains that loneliness, resulting from being separated from others, is a universal feature of modern life.[11] Such networks can be stimulated and made healthier through modifications to the urban habitat as in the new urbanism, for example, which seeks to re-create traditional ties between neighbours and small enclaves.[12]

In emphasizing that cities have been instrumental in transforming the social ecologies of their inhabitants, it is important to resist the temptation that comes from a secular religious commitment to make this either all good or all bad. Like all other human creations, cities have an ambivalent character that mixes good and bad features. There is a tendency in discussions about the modern city to overemphasize the one and underemphasize the other. As human creations, cities are very good for certain things, harmful to others, and simply irrelevant to still others. A realistic and self-critical examination of this human creation is invaluable for plotting a path towards more sustainable ways of life. A tentative hypothesis to point us in that direction is that the livability, vitality, and sustainability of the urban habitat can be enhanced by increasing the enfoldedness of its functions and the reciprocity of the relationships between human life and the urban habitat, on the one hand, and between the urban habitat and the biosphere, on the other. Enfoldedness in social life means that people always participate in or contribute to several things at a time, so that functions are mixed and enfolded. For example, people out shopping for groceries are, at the same time, eyes on the street contributing to safety, a social contact for a neighbour, a hand for a child that is in trouble, and part of the liveliness of streets. Modern life reflects the built habitat in that it separates out all the functions of daily life, so that people do only one thing at a time as much as possible. This is one of the reasons why everyone is running out of time and why people need to replicate separate social fabrics associated with separated activities and functions. This is very time-consuming and reifying, especially when these activities and functions are separated in space and time. It helps explain why today, although surrounded by efficient means, people are experiencing time pressures at levels that were virtually unknown in earlier societies. Modern ways of life and the urban habitat therefore reflect the results of centuries of reductionism in thinking and doing, pioneered in the sciences and later transferred into daily life, first through the phenomenon of rationality and later through technique.

## 11.2 Forces of Separation

In cities we find many forces of separation, one of which is associated with crowds. Many activities in the urban habitat are carried out within crowds encountered in places such as plants, offices, schools, hospitals, shopping malls, the subway, parks, theatres, and stadiums. This closer physical proximity of individuals within the crowd does not bring people together, however. On the contrary, it separates them from each other. This is readily understood if we recall the reciprocity of relations with the social and physical contexts. Alienation or freedom is determined by the demands these contexts place on people relative to the resources they have for meeting these demands. This can be examined on various levels, including that of the senses, individual experiences, individual lives, and the social fabric of a society. Generally speaking, research has shown that exposure to crowded conditions can produce various forms of stress as well as antisocial behaviour and attitudes. Density is positively related to high blood pressure, increases in respiration rate, and frequency of illness. People tend to have more feelings of anxiety, aggressiveness, and discomfort under high-density conditions. It tends to produce more aggressive personalities in children and a higher rate of learning disabilities. Some kinds of task performance are also negatively affected. College-dormitory residents coping with an excessive number of interactions or living in crowded dorms felt more withdrawn and less friendly, producing less cooperative and responsible social behaviour. It should be noted that density is a complex variable to define and study, and the extensive literature on this subject is often divergent and even contradictory. Nevertheless, there is more than sufficient evidence to show that crowding has important physical, psychological, and social consequences.[13]

Social scientists use the concept of environmental overload (also referred to as 'stimulus overload,' 'information overload,' and 'social overload') to describe the profound psychosocial effects of crowding and noise. As early as 1903, the sociologist Simmel[14] warned that living in cities is a source of psychological disturbances because individuals protect themselves from an excess of information by filtering out stimuli and by avoiding social contact. Simmel was concerned that this might lead to a deterioration of the social fabric in the urban habitat. Fischer summarized this view as follows: 'The aggregation of great numbers of diverse people creates both the reality and the perception of individual impotence. At the same time, the protective withdrawal this environment forces the individual into and the destruction it causes to social bonds renders man isolated from, fearful of, hostile to and manipulative of his fellow man.'[15]

By 1964, researchers recognized that environmental overload necessitates adaptive strategies,[16] and in 1976 the specific mechanisms of adaptation were described by Milgram,[17] who showed that they operate simultaneously at the cognitive level through selective attention and information sorting as well as at the social level through the avoidance of relations with others. Six adaptive strategies were identified:

1 Less time is devoted to dealing with each piece of information. Thus, social contacts among city dwellers are reduced to a minimum, while in different social settings time is usually taken to talk and listen.
2 Only immediately useful information is attended to and the rest is ignored.
3 The input load is redistributed to reduce the overload.
4 Various processes are used to block undesired inputs, including the use of answering machines or a coldness of manner to discourage callers.
5 Stimulus intensity is reduced by a series of physical and affective barriers.
6 Specific institutions are created to deal with social overload, such as emergency and social-aid telephone services.

Such defensive strategies for coping with social overload have obvious consequences. People are more likely to mistrust others, and they are less likely to help others when they are in difficulty. There is a reduction in courtesy in interpersonal relations. There are fewer apologies when people accidentally bump into others or inconvenience them. There is a greater anonymity that further isolates everyone in the crowd. People seek to protect their privacy by using only their first or last name, for example. The feeling of isolation and indifference towards others makes people more tolerant of deviant behaviour that would not be condoned in other circumstances.[18] When the size of the social units within which people carry out various activities increases, individuals may have greater difficulty making sense of unaccustomed situations, which, in turn, may lead to a sense of powerlessness and isolation.

As noted above, a characteristic response of individuals exposed to unmanageable quantities of information is a tunnelling or narrowing of attention. This affects their perception of the contexts in which activities take place. For example, in one experiment the researcher had subjects visit the shoe department of a large store during both busy and quiet times. The subjects had to describe twelve shoes on display, and were asked to draw a map of the department. The accuracies of such maps were lower when the place was crowded.[19] People driving cars in crowded downtowns tend to tune out the conversations of their passengers as well as other sounds in their environments.[20] People have always tended to select information that is immediately relevant to what they are doing, and only marginally attend to or ignore the rest. However, such a coping strategy becomes ineffective when people's attentive capacity is constantly stressed. They may then begin to avoid novel situations, withdraw from strangers or exhibit aggressive behaviour to others. Some individuals react pathologically to stimulus overload by giving up on relating effectively to their social contexts.

An alternative interpretation is that crowding may in and of itself not be stressful but the impossibility of predicting or controlling crowding is. Studies have examined the effects of feelings of helplessness and powerlessness, which lead to depression in the long term as the result of repeated exposure to uncontrollable situations, such as overcrowding. Miller and Seligman[21] found that subjects experiencing noxious stimulation or confronted with insoluble problems acquired a feeling of helplessness and eventually became incapable of learning to deal with stimuli under their control. It has also been shown that the non-controllability of noxious stimuli produces disorganization of behaviour long after the stimuli have ceased.[22] Individual differences have also been studied. Similarity between depressive symptoms and helplessness suggests that the latter could contribute to depression and be one of the causes of coronary illnesses.

> There is a behavioral style defined as typical of individuals exposed to risk of coronary illness called Type A behavior. This style typifies those individuals who are exaggeratedly concerned with how their time is used and with exactness in their work, are extremely ambitious and preoccupied with their careers ... In other words, they are people who constantly strive not to lose control of their environment. It is possible, therefore, that Type A people are more sensitive than others to situations in which they are incapable of predicting or controlling environmental stimuli when, for example, they experience traffic jams ... or population densities which hinder their plans and activities.[23]

The physiological consequences of stimulus overload and their relation to stress have been extensively studied. Under conditions of personal-space invasion or crowding, individuals can experience increased blood pressure, elevated heart rate, and more frequent illness. Significantly elevated cortisone levels and heightened levels of psychophysiological arousal have also been found. There is no evidence to suggest that physiological adaptation to crowding occurs. Both short-term laboratory studies and long-term studies of prison populations report increasing physiological symptoms of stress over time. In sum, it is clear that exposure to crowded conditions substantially increases the demands made on an individual's resources, in many cases creating unresolvable conflicts and interference that are difficult to control. This brings continual physiological arousal, leading to heightened levels of stress. In other words, the patterns are not entirely dissimilar to the ones encountered in the study of modern work. Control in situations of crowding is a key variable in determining the long-term consequences.

Of particular concern is the exposure of young children to crowding in classrooms and playgrounds. Studies show that, as with adults, this can result in less social interaction, avoidance behaviour, and withdrawal resulting in solitary play.[24] Although some studies show ambiguous results, most indicate that children become more aggressive in crowded environments. For example, four-year-olds in half-day nursery schools exhibited more deviant behaviour in crowded classrooms containing less than thirty square feet per child.[25] Competitive behaviour also increases with density levels.[26] Additional responses of children to high densities include higher levels of stress-related arousal, decreases in locomotion and motor activity, decreases in active play such as running and creative play, greater frequency of interruptions, and more fearful behaviour.[27]

Meeting the additional demands resulting from crowded conditions clearly depends on the cultural resources a community or society has evolved. Because of this cultural mediation, the consequences of crowding can vary from society to society. For example, more traditional societies such as Hong Kong and Japan experienced considerably fewer of the effects of crowding on human health than did the United States. This may well change, however, as these societies become more like modern mass societies. The relationship between crowding and pathology may, therefore, be reduced through preventive approaches that lead to improvement in the cultural resources people have for dealing with crowding, or by reconfiguring the built environment.

A second force of separation is rooted in the social structure of modern cities, which is that of a mass society. It includes but goes well beyond the effects of crowds. One of the sad ironies of the human condition in modern cities is that, even though people have more social contacts than in any other society in human history, many people suffer from endless inner loneliness. They have numerous contacts but these are often empty and unsatisfying. Accepting that human resources include a certain capacity for dealing with others, it would appear that this capacity may focus on relatively small but intense relations or may be distributed over many superficial ones. This, in turn, affects the metaconscious image people generate of their social selves that can lead to a sense of alienation. Being in crowds is part of living in a mass society. To appreciate its significance, this phenomenon must be clearly conceptualized.

In earlier societies, the individual was well integrated into the group through strong personal relationships that formed the context of most daily-life activities not separated from one another in space and time. In many cases, this led to the submergence of personal identity into that of the group. Individuals were primarily influenced by a local tradition that included a traditional morality, values, and religious institutions. Macro-level changes in society were mediated by the local group, and thus had little direct influence on the individual. For example, the propaganda of a totalitarian state had little influence on traditional peasant communities, and such regimes therefore typically attempted to break down such communities. The social changes accompanying industrialization included a relocation of the population in the industrial centres, which weakened the traditional social fabric, specifically the constraints imposed on the individual by the traditional group. These groups were gradually replaced by crowds of people associated with activities in the industrial-urban habitat. Crowds imposed new pressures on individuals who, free from the constraints of the group, were now vulnerable. This went hand in hand with a weakening of traditional cultures, leaving individuals with few reference points and little moral and religious guidance, threatening to make them the measure of all things. The culture of their society no longer effectively sustained them in the effort of making sense of and living in the world. They had to make daily-life decisions without the guidance of customs and tradition because these increasingly lost their relevance for the new industrial way of life. The resulting uncertainty, coupled with the absence of social protection from the pressure of the crowd and the lack of traditional frames of reference and values, created a social milieu that must be fed information from the outside if it was to cohere. Broad ideological influences began to shape nineteenth-century industrial societies. As the role of tradition-based cultures continued to weaken in the twentieth century, an entirely new social structure emerged, commonly referred to as a mass society. Its development is integral to those of mass production and consumption, and the mass media.

A mass society is not a disintegrating traditional society but one that integrates the individual into society by mass social currents. It is well known that crowds are easily manipulated by outside influences and that the members are inhibited in the use of their critical faculties. The result is a certain psychological unity. Beliefs and opinions are widely shared despite the fact that most people do not have direct or indirect experience or personal knowledge of what they have opinions about. This is the well-known phenomenon of public opinion. Exposure to the crowd modifies the individual's psyche, making him or her more credulous and more excitable, and thus susceptible to current or public opinion. All of this is widely accepted, but the implications are rarely drawn. In a detailed analysis of the psychological and sociological condition of the individual in a mass society, Jacques Ellul[28] demonstrates the necessity of what he calls 'integration propaganda,' which is distinguished from totalitarian propaganda. Its functions are analogous to those of custom and tradition in pre-industrial societies. The mass media envelop individuals in a bath of images, portraying the lifestyles and norms that help people belong. These are no longer provided by a tradition-based culture in individual and collective human life. A 'statistical morality'[29] makes common behavioural patterns the norm. The images of society and life within it, laden with stereotypes and myth images, are the primary means of social integration.

Effective participation in society can be achieved only when individuals are convinced of its value. Integration propaganda helps to achieve this. It is essential to understand that integration propaganda would not be effective if it were simply diffused by the media without meeting deep psychological needs and the social needs of communities. In a democracy, the actions of government are supposed to be based on popular opinion, which expresses itself in the form of public opinion in a mass society. The problem is that public opinion is not based on the experience and personal knowledge of the individual who holds it. It emerges and vanishes like the wind. It is not possible, therefore, for any democratic government to act on public opinion, which is generally unable to anticipate problems far enough in advance and incapable of sustaining government action for the length of time required to plan and implement it. Public opinion tends to extremes in times of crisis. The activities of government are based on knowledge separated from experience of many different kinds, which are difficult to integrate into a coherent picture. Hence, these activities often lack coherence, and this situation is exacerbated by the unequal distribution of power and knowledge in society. Consequently, the deep problems of the urban habitat are difficult to put on the political agenda, and even more difficult to act on by genuine democratic political activity. The rise of the phenomenon of public opinion is coupled to a growing inability of the citizen to meaningfully participate in society. Control over many activities is externalized and concentrated in large institutions. These large organizations, including government, must explain the emergence of the new issues they face and show how the actions they propose are democratic and just, and what the public would have wanted. If any resistance is encountered, techniques of public relations are drawn on to transform the decision into something that the public will accept. The citizen, or any expert, for that matter, is incapable of understanding the many complex issues facing metropolitan areas. Yet no one would readily admit that he or she is unable to form a genuine opinion about the issues that shape our lives. Thus, the citizens are ready to receive the explanations offered by the media and public-relations efforts. This must include information, preconceived positions, and value judgments. In this way, citizens feel more effective and secure in the face of complex forces and issues. It should be added that the media confront the individual with extraordinarily fragmented, discontinuous, and piecemeal images of what is happening over time. These images single out events that disturb the general order. It is difficult to live in such a world without integration propaganda. The loneliness of the crowd and the absence of strong personal relations further add to the need for finding some personal meaning in the world of images.

An integral part of the above development is the fact that it is difficult for individuals to form a strong image of their social selves due to the rarity of strong and lasting personal relations. As a result they are more vulnerable to the external influences undermining the economic and political autonomy essential in a democracy. All these developments together will deeply affect the ability of metropolitan areas to deal effectively with the challenges they face, unless these aspects are corrected through urban renewal.

The evidence would suggest that the appearance of crowds and a mass society signals a significant increase in the demands a social setting makes on people, with only modest possibilities for successfully coping through increased control. The associated forces of separation can be greatly intensified in cities by high-rise buildings. Since the effects of tall buildings are mediated through experience, they may be different for the users of the buildings, those who live or work nearby, and those who are passing through the area. It may also vary within these groups since they are not homogeneous. For example, an office tower is likely to have significantly different effects on executives, with their private quarters and elevators; those who work in the building; visitors; cleaning staff; and building-management personnel. Different kinds of high-rise buildings such as condominiums, hotels, and public housing complexes also produce different effects.

The direct effects of tall buildings on their surroundings include a loss of older buildings and familiar places, view blockage, shadow effects, and the loss of privacy in neighbouring dwellings. The indirect effects include much larger numbers of people making their way to and from the building, thus increasing traffic density and the use of nearby public transportation; parking requirements; use patterns of nearby shops, restaurants, and bars; demands on infrastructure; and so on. To fully examine the impact of tall buildings, it is necessary to conduct studies of the area before construction, during construction, and during the early and later phases of their occupation. The impacts on the neighbourhood during the period preceding construction are usually the most severe. The displacement of an existing population, the tearing down of familiar buildings, and the termination of the services they housed, all contribute to potentially significant changes in the character of the neighbourhood, which can be experienced as a serious loss by those who remain in the area. The impacts during the construction phase are also not trivial, including an increase in noise and dust levels, heavy vehicles moving through the area, and a temporary or permanent change in demand for the services rendered by local businesses as one group of people has moved out and another has not yet arrived. Even when people return, their presence may be felt only during the morning and evening rush hours and lunch time, as in the case of office buildings. Outside of working hours, nearby streets may now be abandoned, making them less safe, particularly at night. This, in turn, can influence the incidence of crime in the neighbourhood. Tall buildings dramatically increase energy consumption, which may have an impact on the local microclimate and may significantly affect wind patterns. In other words, the entire character of the neighbourhood can be profoundly affected by the presence of tall buildings.

Then there is the question of what constitutes an appropriate scale of the built habitat with respect to the human activities it is designed to support. It is obvious that a neighbourhood where the tallest buildings have no more than four storeys 'feels' very different from one mostly composed of tall buildings of more than thirty storeys. The latter reduces human life into insignificance, making the area imposing and uninviting. Erecting an occasional tall building in a low-rise neighbourhood may cause clashes of scale. The new building may be too bulky for the block in relation to other buildings, thereby dwarfing them. It may violate the patterns of spaces and volumes that give a neighbourhood its unique visual character and functions. A tall building can either destroy the gestalt of a neighbourhood or (and this is much rarer) create a landmark that complements and accentuates that gestalt. All too often a tall building decreases the livability of a low-rise neighbourhood by increasing densities and traffic patterns.

One study[30] compared the livability of blocks with and without tall buildings. It found that their effects were generally negative. This was concluded from the following observations. There were fewer instances of people engaged in various activities outdoors; there was a reduction by a factor of three of people having conversations in the streets; the number of people engaged in relaxation activities, including children at play, was cut in half; the pedestrians associated with the tall buildings outnumbered others by a ratio of five to one; and there was a sharp increase in traffic density. This dramatically changed the social character of the neighbourhood since it reduced the number of residents people were able to recognize, with a corresponding reduction in greetings and conversations between passers-by. A greater sense of anonymity and a reduction in the liveliness of streets resulted. The neighbourhood felt less lived in, less friendly, and less inviting.

Other negative effects of living in tall buildings have been confirmed by the empirical work of the architect and planner Oscar Newman[31] for New York City. He discovered startling differences in the crime rates of different housing projects, which correlated well with the design, size, and location of the projects while controlling for the background and socio-economic profile of the inhabitants. His studies indicate that, regardless of size, high-rise projects were the sites of many more crimes (per 1,000 residents) than low-rise projects. For instance, the total number of felonies per 1,000 residents rose from an average of 8.8 for three-storey buildings to an average of 20.2 for buildings of sixteen storeys or more. The rate for muggings that took place inside buildings increased even more remarkably, from 2.6 per 1,000 residents in six-storey or lower buildings to 11.5 in buildings having nineteen or more floors. Newman developed Jane Jacobs's notion of 'eyes on the street' being the best guarantee of security into his own theory of 'defensible space' – a commodity that is scarce in tall buildings where the public sidewalk is replaced by the cage of the elevator, the blind spots of the stairwell, and the anonymity of corridors. Implicit in Newman's work is the idea that tall buildings, by their very design, do not serve to empower the people in them to act in ways that make criminal activity difficult. Smaller buildings, on the other hand, make it easy for residents to survey a place and actively intervene in various situations, thus dissuading criminals from engaging in their pursuits in that area. Tall buildings create in the residents a sense that vast areas of the building are beyond their control or effective intervention. Small, surveyable projects, on the other hand, give residents a clear sense that substantial portions of the grounds belong to them and are under their control, hence obliging them to make various social interventions.

The built environment should be regarded as ecologies of buildings that more or less fit together and complement one another to help make liveable neighbourhoods. Buildings should therefore be designed to fit these ecologies. They should create a great enough diversity to make them interesting and capable of supporting a variety of human activities without destroying their synergy by stretching that diversity too far. This has become very difficult since mass production has invaded the building industry. Whether it is the replication of floor after floor in tall buildings or the almost endless replication of essentially identical homes in suburbia makes little difference. The benefits include significant reductions in construction costs, but they are realized at the expense of the liveability of the urban habitat. This trend has been further reinforced by standardizing the design of structures through computer software packages. As knowledge separated from experience became dominant in planning, architecture, and engineering, the ecological perspective was all but lost. Building on the basis of knowledge embedded in experience makes far greater use of context considerations, thus producing very different results that are obvious from any comparison. I am not saying that all modern building is inferior to that in the past. I am simply pointing out that building design, construction, and planning involved more context considerations and values in the past by virtue of the kind of knowledge bases being used. This is evident when examining the design of traditional structures and settlements in terms of their adaptation to local conditions, including climate and the culture of people living in them.[32] The extent to which human experience, culture, or knowledge separated from experience are drawn on for the evolution of the built habitat has profound consequences. Preventive approaches would seek a new synthesis between these different modes of knowing in order to balance their strengths and weaknesses to create a more liveable, healthier, and sustainable urban habitat. The patterns in the evolution of the urban habitat during the last century were closely correlated with the others we have described earlier. Generally speaking, advances are made in the domain of performance values and the problems occur in the domain of context values. The latter have led to a greater separation of people from themselves, from others, and from their built habitat.

Like crowding, noise can also be a force of separation. It increases the demands an environment makes on a person without leaving much opportunity for control. Many of the countless sounds to which individuals are exposed in the urban habitat are unwanted because they produce physiological or psychological distress by interfering with most human activities. Noise pollution is a serious problem. The effect noise has is a function of its characteristics and the context in which it is perceived. For example, other people talking may have little effect during the day but become disturbing when one is trying to sleep. Whether a sound is perceived as unwanted, making it a noise, also depends on a person's personality, attitudes, and past experiences. Hence, noise cannot be objectively defined. Nevertheless, loud noises are almost always intrusive because it is impossible to ignore them when carrying out other activities.

Research on noise has led to the following findings. It is not the physical characteristics of noise but the social and cognitive contexts in which noise occurs that play the greatest role in determining whether noise has significant effects. It is the after-effects of noise in terms of psychic costs that are the real problem. The effects of the same noise on different tasks can vary, depending on the following: whether a task requires constant alertness or is more or less routine; the duration of the task; and a person's mood, personality, age, and attitude. The predictability of the noise and whether a person can control its termination also influence its annoyance level.[33]

Behaviour following exposure to noise is most impaired when the noise is unpredictable. Once again this element of unpredictability and lack of control over what is heard is crucial. It helps to explain why the noise from your own lawn-mower is less annoying than that from your neighbour's, and why workers on a railway line are less annoyed by train noise than are passers-by. In a workshop, the noise of tools used by a neighbour is less annoying when a person can anticipate the next noisy operation because that worker is in his or her peripheral view.[34] Glass and Singer[35] report that, when noise is inescapable or unavoidable, it makes people feel that they are at the mercy of their environment; when people feel they cannot control outcomes, they are less likely to make subsequent attempts at doing so. In other words, the patterns are the same as those we already discussed with reference to healthy work and with crowding. It appears that it is not the stressful event itself that produces problems, but the implication of helplessness that comes with it. The research also shows that people exposed to high levels of noise on a constant basis are deeply affected by it. For example, when the presence of a major highway was simulated in university classrooms, students reported that they found it more difficult to hear lectures and take notes, and observers found that students showed less attentiveness and participation than students in no-noise classrooms. Students exposed to the same noise in dormitories reported problems in sleeping and studying, and increased levels of nervousness. There was also evidence of greater tensions and disagreements among students when exposed to this noise.[36]

A large number of studies have examined the effects of prolonged noise exposure on the perceptual and cognitive learning of children. Most studies conclude that the effects are deleterious. Cohen, Glass, and Singer[37] examined the long-term after-effects of exposure to noise on elementary schoolchildren living in high-rise buildings adjacent to expressways. It showed that prolonged exposure had durable effects. Children learned to adapt to noise, but at a loss of verbal and auditory capacities. It appeared that cognitive problems occur as a result of cumulative effects of noise on the learning process. Children attending noisy schools in the flight path of a Los Angeles airport were compared with children in quieter schools that matched in age, social class, and race. Children from the noisy schools had higher diastolic and systolic blood pressure levels, performed simple and difficult puzzle-solving tasks less well, and tended to give up in frustration more easily when set difficult tasks.[38] A similar study of a school near the Orly Airport in France also found that reading skills developed more slowly, with the children being more restless and having a lower tolerance for frustration than children from quieter schools.[39] Noise can also produce negative changes in social behaviour. Matthews and Canon[40] examined environmental noise levels as a determinant of helping behaviour. It appeared that, under noisier conditions, people were less likely to help others. Traffic noise had a significant influence on street life. The greater the traffic noise level, the less likely that there was active street life. Under noisy conditions people merely used the sidewalk to get where they were going. People on noisy streets reported that their street was a lonely place to live, whereas people on streets with light traffic tended to find their environment friendlier.[41] In general, it appears that noisy conditions force subjects to narrowly concentrate their attention to the neglect of other things going on, thus decreasing sensitivity to others, which is further reinforced by a negative emotional and psychic state induced by noise.

Although there may be little convincing evidence for a direct causal link between noise and physical disorders, it must be pointed out that cause–effect relations are almost impossible to establish in an environment where there are myriad complex intervening variables. Nevertheless, it is well known that noise can alter physiological processes, including the functioning of the cardiovascular, endocrine, respiratory, and digestive systems. Since such changes, when extreme, are often hazardous to health, many researchers feel that significant health effects of long noise exposure are likely. Once again the precautionary principle needs to be exercised here. In the meantime, these studies may offer a plausible explanation why so many people need pills to help them perform the most basic functions, such as sleeping, relaxing, fighting depressions or anxiety, and having regular bowel movements.

Exposure to noise in industry merits special consideration. Industrial surveys suggest that exposure to noise results in increased anxiety and emotional stress. Workers exposed to high-intensity noise (100 decibels or above) show increased incidence of nervous complaints, nausea, headaches, instability, sexual impotence, argumentativeness, changes in general mood and anxiety. In a review of recent noise literature, Welch[42] concluded that there was elevated morbidity among people who had been exposed at work to sounds of 85 decibels or greater for at least three to five years. The strongest evidence came from research on cardiovascular problems. He interpreted the data as suggesting that long-term exposure to high-intensity noise was associated with at least a 60 per cent increase in risk of cardiovascular disease. Impaired regulation of blood pressure, including hypertension, was the best documented of these effects. Cause–effect relations may never be established, but it may well be that noise causes general susceptibility to disease and thus leads to a wide variety of symptoms of physical and psychiatric disorder.[43]

Forces of separation can also be increased through pollution via its effects on human health and the ability to fully participate in daily-life activities. Social-epidemiological studies of the effects of pollution on populations have generally been limited to drastic effects such as disease or mortality. Evans and Jacobs[44] suggest that researchers should also explore discomforts, irritability, and potentially more serious mental health outcomes such as depression and anxiety. Given the well-established effects of air pollution, it is reasonable to infer that chronic discomfort may lead to more serious mental-health consequences. To make things even more complicated, there are many mediating variables. The effects of crowding, noise, and pollution, as well as those of other urban stressors, may produce negative synergistic effects, making their study extremely difficult for the scientific approach based on minimal context. There is no doubt, however, that the built habitat brings people together in space and time while simultaneously separating them in other ways, creating entirely new conditions for human life. The operation of these forces of separation must be recognized and taken into account for preventive design and decision making. Where scientific evidence is lacking, the precautionary approach will recognize that phenomena such as environmental hypersensitivity and multiple allergies are the extreme results of a massive assault on the human immune system occurring in those members of a population with the least resistance. If this is the case, it is not possible to rule out an increase in these phenomena in future generations. The same is true for the assault on human mental health through the creation of environments that are ever more demanding while restricting the ability of people to successfully cope with these demands. An ecological understanding of health would also recognize that there may well be negative synergies between these two assaults, which returns us once more to the question of human sustainability. A precautionary approach would take into account the limitations of our scientific knowledge and the harmful ignorance scientism has created. Appropriate political decisions cannot be made unless the application of specialized knowledge is restricted to its domain of validity. It will need to be supplemented by a more context-rich way of knowing based on experience and culture. It is to such approaches I will now turn.

## 11.3 Forces of Integration

Having briefly examined some of the forces of separation that operate within the built urban habitat and that thereby threaten human sustainability, we now shift our focus of attention to identifying forces of integration in support of human life. The kind of approach pioneered by Jane Jacobs in her landmark work *The Death and Life of Great American Cities*[45] is discussed here as an example of how this might be done. It is essentially a comparative approach of the fundamental constituents of cities, including streets, parks, neighbourhoods, and districts, to discover why in some human life appears to flourish, in others it merely exists, and in still others it is undermined and even threatened. It seeks to discover why some neighbourhoods degenerate into slums, and how some slums regenerate themselves into liveable neighbourhoods while others do not. It seeks to understand the processes that cause some downtowns to bring a city together, while others merely exist, and still others shift their centres. Understanding the forces of integration and disintegration is a prerequisite for making the built urban habitat a healthy one capable of enhancing human life. Such forces have their roots in the cultural cycle, in general, and in one of its components, in particular: as people affect their habitat, it simultaneously affects them. The resulting spiral of interactions over time can improve, undermine, or sustain the liveability of that habitat. It involves the daily-life experience of its components, resulting in their taking on culture-based meanings and values in relation to everything else in individual and collective human life. Hence, both knowledge embedded in and knowledge separated from experience and culture must play a role in the analysis.

People's culture-based experiences of the urban habitat must take centre stage. There can be no preconceived ideas. The first dates from the time of the Industrial Revolution and the early industrial centres, which were so riddled with problems that observers of that time saw no way out but to eliminate them. One 'solution' that became particularly influential was that of Ebenezer Howard,[46] who was so disturbed by the conditions of the poor in late-nineteenth-century London that he concluded that the city was outright evil. In 1898 he proposed to shrink London by repopulating the countryside by means of what he called 'garden cities,' each carefully planned as a small town with a cultural centre, neighbourhoods with plenty of nature, an industrial district hidden behind trees to provide work, and a green belt for agriculture. It was a grand scheme for creating a society based on these communities, planned and evolved by central authorities, that would anticipate and meet the needs of the inhabitants. The operationalization was reductionistic in its orientation, de-composing individual and collective city life into distinct and separate functions to be allocated to relatively self-contained 'parts' and based on the separation of work from the family, which began with industrialization. Geddes[47] took this scheme one step further by incorporating it into regional planning, which would rationally distribute garden cities, agricultural zones, and natural resources over large territories.

This approach separates knowledge from experience in the domain of evolving the built urban habitat. Planners and architects became the experts who worked out the details of this rational scheme, with little or no reference to the experience of those on the receiving end. The fact that this approach is paternalistic, authoritarian, and non-democratic still troubles relatively few people. The residents of a neighbourhood were by implication incapable of knowing, let alone deciding, what was good for them and their habitat. The scheme was doomed to fail because it sought to impose a rational idea of the complex economic, social, political, religious, and cultural forces that shape the rise and fall of great cities and civilizations. To legitimate this through the Judaeo-Christian tradition, which regards the city as an alienating force in human history, is to ignore the other side of the coin – namely, that Jerusalem symbolized the promise of a new creation that would be a city into which the glory of the nations would enter.[48]

According to Jane Jacobs, the second concept underlying a great deal of modern thinking about the urban habitat is that of the 'radiant city' proposed by Le Corbusier.[49] It was again a rational idea, separated from all experience, of a city of skyscrapers in a park. This was a vertical city that would achieve very high densities of 1,200 inhabitants to the acre and yet leave enough parkland around the gigantic towers. The rich and powerful would inhabit low-rise luxury housing arranged in courts. People would move around the city by means of private cars, utilizing one-way arterial roads with minimal cross-streets, while heavy vehicles were assigned to underground passageways. Pedestrians were to be restricted to parks. The result would be a Utopian society where human beings would have the maximum possible freedom within a social order that was so rational, transparent, and symbolic as to inspire many planners and architects. The two conceptions of the garden city and the radiant city were subsequently combined in more sophisticated conceptions to create the 'city beautiful' or the 'city monumental,' with tall imposing towers arranged along boulevards and dispersed with parks, and climaxing in a cultural centre, thereby creating what Jane Jacobs refers to as the 'radiant garden city beautiful.' What is so amazing about these Utopian visions is the power of their symbolism, and their complete lack of touch with the experience of city dwelling.

This approach to the building and evolution of cities stands in sharp contrast with that of the past, which was based to a much greater extent on knowledge embedded in experience. The extensive use of context created a harmony between buildings, streets, parks, the hinterland, and local ecosystems, on the one hand, and the culture and values of a society, on the other, in a manner that also reflected the wishes of an elite, but with a difference. This difference resides, as I have noted, in the capacity of culture-based thinking and doing to make full use of context through a symbolic system of meaning and values. Culture-based approaches permitted shorter negative-feedback loops and thus a different kind of evolution than one based on conformity with rational symbols and executed through knowledge separated from experience. The former is what I believe Christopher Alexander is getting at in *The Timeless Way of Building*[50] and what Donald Schön is attempting to recover in *The Reflective Practitioner*.[51]

The only prejudgments involved in the kinds of approaches pioneered by Jane Jacobs are those of a community's experiences accumulated through its culture. They provide the 'set point' for negative feedback to decide what is meaningful or meaningless and what is valuable or valueless. In other words, these comparative approaches are grounded in the experiences and cultures of the people who live in and reflect on their urban habitat. This is not to say that they cannot be sharpened by professional education, provided that the limitations of knowledge separated from experience are clearly understood. It is for these reasons that good engineering is much more than applied science, appropriate architecture much more than the building sciences, and democratic planning much more than the execution of a rational ideal. They are, first and foremost, the art of giving concrete form to the meanings, values, and aspirations of a culture in a particular domain. They make use of knowledge separated from experience but transcend its limitations to create something that excels in the domains of both performance and context values.

How the building and evolution of cities changed must be interpreted in the context of the ways of life of the first generation of industrial societies in Europe. The growing predominance of the economic over all other aspects of a way of life as well as the emergence of performance values and their dominance over context values, could not have spread to the urban habitat if it had remained a multidimensional enfolded whole in the thinking and doing of architects, planners, engineers, and administrators. This would have blocked the transformation to ways of building cities based on knowledge separated from experience that eventually contributed to the phenomenon of technique. The kinds of approaches pioneered by Jane Jacobs are at odds with these developments because they seek to evolve the urban habitat in the full context of human life.

I will briefly summarize the principal findings of Jane Jacobs. Streets and their sidewalks are the most important places in a city. If they are alive the city will be alive – they constitute a city's most vital organs. Because one meets mostly strangers on city streets, those streets are qualitatively different from those of small towns, where most people will know or recognize one another. At the very least, a person must feel safe with all these strangers, and this can be achieved only by a metaconscious law and order learned through experience. This works if there are enough people on the street that, if someone significantly deviates, negative effects can be prevented or compensated for. To ensure this population density on the street there must be a diversity of facilities providing reasons for people to use the street. On their way to and from such facilities, people provide eyes on the street, thus helping to make them safe for all to use. People themselves attract other people, since they add interest to the urban habitat. In other words, the streets are safe, and people are free to go about as they please, as long as there are enough people to develop and enforce a metaconciously developed law and order that the strangers in a neighbourhood share. When one carefully examines life on the street, it becomes evident that, as time passes, layer upon layer of users having different reasons for being there succeed one another seamlessly: shopkeepers preparing their stores early in the morning, people going to work, children walking to school, young mothers pushing baby carriages or strollers while doing some shopping, retired people getting their morning coffee or going for a stroll in the park, people moving in to have lunch at local restaurants and doing some shopping during their lunch break, people taking their picnics to a local park, and so on, in an endless texture of relations between people and the facilities on the streets that support a variety of activities. In this way, a symbiotic relationship is created between strangers who do not know one another and may not care to get to know one another but who, nevertheless, help create a neighbourhood in which they can trust others.

When this trust breaks down, people will not dare to get involved in preventing or compensating for deviant behaviour out of fear that they will not be protected from reprisals. People then walk around the streets as quickly as possible to get where they are going, avoiding others in constant fear of trouble. If there is an incident on the street, people try their utmost not to get involved and to ignore the situation. In such neighbourhoods, people take taxis home late at night so that they can be dropped off right at their front door, and automobile use in general is essential for personal security.

The difference between streets that work and those that do not is that, in the former, strangers share a great deal by participating in, and thus evolving, a metaconsciously shared law and order, and, in the latter, they share almost nothing. In the former case, there remains a clear and generally respected division between one's public role as a stranger and one's private life. People deal with one another in terms of their social roles as shopkeepers, mail persons, customers of a local facility, parents of young children, retired persons who spend a great deal of time around the neighbourhood, dog owners discussing their pets, and so on. The boundary between these social roles and people's private lives is carefully respected. People's social selves, in contrast to their social roles, evolve in the context of a network of personal relations with acquaintances, neighbours, friends, and relatives. Hence, public and private life each have their own separate footings.

Children can safely play on streets that 'work.' Conventional wisdom would have us believe that they are better off playing in the park or on the playground of a local school. However, unless such areas have the benefit of the constant presence of strangers to whom the children can turn in confidence if they encounter difficulties and who will act as a public parent if they see a child headed for trouble, such places will be much less safe. Children are quick to sense this and, if they lack a feeling of security and public support, will not be able to engage in carefree play activities. Similarly, it is on streets that work that children can safely walk to school, go to a friend's house, or run an errand for a parent or neighbour. They know that when there is trouble they can turn to other adults or find rescue in a local shop or restaurant. It is for this reason that wonderfully equipped playgrounds or other play spaces are as good as useless without the constant coming and going of adults acting as public parents. In lively and safe neighbourhoods, streets are often the most secure place where children can play. There will never be enough money to hire the adults to replace this role of street life. It is essential, therefore, that sidewalks be made wide enough to accommodate a diversity of use, not the least of which is a place for children to play. Unfortunately, city planning appears to be more concerned about making the city safe for cars.[52]

A neighbourhood is a good place to live if its streets are full of life. When this spills over into local parks, public libraries, community centres, and other facilities, they will further reinforce the diversity of facilities that, in turn, supports a constant street life. This is why the separation of the activities of daily life in space and time is such a problem. Blocks with mostly office buildings, civic or cultural centres, and neighbourhoods where most people go to work are abandoned for a large part of each day, and sometimes almost entirely on weekends. With inadequate street life, such areas can become unsafe, and may even attract antisocial activities. Of course cities evolve, and neighbourhoods that were once full of street life can turn into dangerous places, and vice versa. It is for these reasons that larger entities than streets must grow and flourish, so that more resources than any street could possibly offer can be brought to bear on problems, including the ability to bring enough pressure on city hall to make available resources to deal with particular issues. This may also involve resisting proposed changes and decisions that city administrators may make without a good knowledge of local conditions. In other words, the self-governing behaviour of street life must extend to neighbourhoods, which requires channels of communication, means of action, and a pool of volunteers supported by neighbourhood solidarity. A neighbourhood has no clear boundaries. It is a network of streets that have cooperated from time to time on particular issues of interest. Such networks frequently overlap unless there are clear boundaries. For major issues, much larger political entities may be required in order to bring the resources of city hall to the neighbourhood or to deal with large proposed projects such as an expressway. Unfortunately, it would appear that informal self-government on that scale is not easily formed and maintained, giving city halls an enormous advantage in pushing through agendas that may or may not be in the interests of neighbourhoods. This contributes greatly to a lack of negative feedback that might otherwise permit city hall to function much more effectively.

Given Jacobs's discoveries about streets, neighbourhoods, and districts, the question that imposes itself next is: how can cities create, evolve, and adjust enough diversity throughout to sustain themselves? Here the economic aspect of city life must be closely examined. Cities tend to create large numbers of small enterprises because they can draw on many resources beyond themselves, serve local niche markets, and evolve with these markets since they are an integral part of the larger fabric of relations of city life. Being dependent on local diversity, they enrich that diversity by creating new niches in a self-reinforcing process from which they themselves benefit. Because of this ability, large enterprises have a hard time competing with smaller ones firmly embedded in a city's diversity. Small businesses can serve a local diversity in a personal and direct manner, which large businesses cannot.

Jane Jacobs argues that to sustain a city's diversity four conditions are required. First, a city's constituent elements must serve as many human activities as possible and thus have multiple functions. This is essential to ensure that there will always be people on the streets for different reasons at different times. Such a diversity of activities is synergistic. For example, shoppers may also become customers for local restaurants, and vice versa. Office workers and store clerks may also frequent stores and restaurants to and from work and during lunch breaks. Residents may use different facilities at different times. Street life at night may attract people to the area to watch other people, visit a bar or see a movie. As they walk past a shop, they may be attracted to something and turn into customers. In other words, diversity thrives on diversity. It is precisely this diversity that modern planning frequently hinders by promoting homogeneity of office or apartment blocks and residential areas. Many modern downtowns have reduced their mixture of primary functions to the point that they are referred to as central business districts, with obvious consequences. A great deal of the time the streets are empty, and such streets have ceased to be the magnet that drew people to create a lively downtown. The city then loses the vital centre that draws its districts and neighbourhoods together. Reviving a downtown is a question of strengthening its diversity of primary functions and the human activities they serve, and not of erecting a few monumental buildings.

A second condition for diversity is short blocks. This is important because it brings a greater diversity within easy walking distance. If various primary functions on a neighbouring parallel street require walking down a long block before there is a side street to get to it and then to walk all the way back, the additional diversity such a street offers might as well not be there because it is not within easy reach. Chances are a person may not even know what such a street has to offer and may take the bus to a similar facility much further away. Short blocks bring the primary functions of adjacent streets within easy reach, thus complementing the diversity of each street in the neighbourhood. Short blocks cannot create that diversity by themselves. They are a necessary condition for enhancing it.

A third condition for creating and sustaining a healthy diversity is that city blocks in a neighbourhood must contain a mixture of old and new buildings. The presence of older buildings, even some that may be slightly run down, is essential because they can accommodate enterprises just getting started. If these businesses flourish, they can then move to larger quarters in a better location, possibly in a newer building, because they can now afford to pay higher rents. In other words, a diversity of buildings provides a variety of niches to suit a diversity of enterprises: some require small or large fashionable surroundings in a prime location, while others can operate only in modest quarters. A diversity requires an ecology of businesses, including start-ups, mature and successful ones, and some that are in decline. There must be a diversity of niches for each phase in their life cycle. It should be recognized that the diversity of the neighbourhood is not permanent but involves a constant flux of enterprises, somewhat like any other population.

Older and somewhat run-down buildings offer another important opportunity. Everybody has a certain metaconcious sense of their surroundings. If a neighbourhood thrives, there will be a tendency to maintain and refurbish properties in keeping with the character of the neighbourhood. It is this process, guided primarily by knowledge embedded in experience, that can over time significantly improve the attractiveness of a neighbourhood and thus make its contribution to the diversity of people it draws. If a great deal of the work is done by architects, engineers, and planners who do not live in the neighbourhood, and therefore have no metaconscious knowledge of its life, little such nurturing of the diversity of buildings within the unique unity of the neighbourhood can take place. Starting new businesses, fixing up old businesses, taking down old buildings to replace them with new ones, fixing up old buildings, widening sidewalks or converting streets into pedestrian malls, and other such changes should be guided by a shared metaconscious sense of the character of the neighbourhood, including its needs and shortcomings. The evolution of diversity requires the parallel evolution of a wide range of niches, each understood and evolved in the full context of all the others. The built habitat must provide a diversity of physical niches to suit a diversity of human activities that can sustain life on the streets, so that strangers can live safely together and be stimulated by the diversity of human experiences that causes the arts and crafts to flourish in cities. It is a mistake, therefore, to build a whole area at once, or to excessively restrict zoning so as to constrain diversity. Similarly, if most stores in an area are franchises, their ability to cater to local diversity is extremely limited.

A fourth condition for sustaining diversity is a concentration of people that must be sufficiently dense to create that criss-crossing and overlapping network of human activities that ensures constant street life. A high density is required to create a self-sustaining diversity. Downtowns are the centres of cities because they draw people from everywhere in the city to take advantage of the diversity they offer. High densities of people require a high density of habitat niches that accommodate and serve a diversity of functions and associated activities. Jane Jacobs points out that, contrary to conventional wisdom, high densities are not synonymous with slums. She suggests that the most interesting neighbourhoods have densities that fall somewhere between 80 to 140 dwellings per net acre and can go much higher.

Slums tend to have densities below this range. However, it should be remembered that density in and of itself does not create diversity, but requires the three other factors to create and sustain it. For example, high densities in low-income or subsidized regimented housing projects will not create diversity, nor will a high density on long blocks.

It is important to distinguish between a high density of dwellings and a lower density of dwellings that are overcrowded. The number of inhabitants per net acre may be the same, but the effect is very different. The census definition of overcrowding is 1.5 persons per room or more, which is independent of the number of dwellings per net acre. Jane Jacobs argues that this confusion is responsible for many of the stereotypes about high- and low-density living. For example, the garden-city concept confuses the two, assuming that overcrowded rooms and densely built-up land have the same undesirable results. High density and overcrowding are distinctly different phenomena and have a markedly different effect on a neighbourhood.

It is essential to know how many dwellings there are per net acre, including the average number of rooms such dwellings contain and how many people live there. Overcrowding of dwellings and rooms is almost always a sign of poverty or of discrimination, neither of which encourages or sustains diversity. The bottom line is that people do not live in overcrowded facilities by choice, but they do choose to live in high-density neighbourhoods with a flourishing diversity.

This begs the question as to what is an appropriate density for a neighbourhood, district, or city. For Jane Jacobs, the guiding principle is that an appropriate density is one that supports and does not inhibit diversity. When she was writing her book, suburban densities went as high as ten dwellings to the net acre. At between ten and twenty dwellings to the acre a semi-suburb is created of semi-detached homes, very small single-family residences, or more generous-sized row housing. Garden-city planning aimed at twelve dwellings to the acre. Such densities do not generate adequate public life for vital streets and neighbourhoods. It is important to note that such densities can create quite liveable neighbourhoods, provided that they are on the edges of the city so that they do not have to deal with the constant flux of strangers. In such cases, streets may still be relatively safe, but there will not, of course, be the city liveliness on the streets and this may well be preferable to those who choose to live there. Such neighbourhoods work relatively well when they are sufficiently out of the way that they do not have to contend with people that do not fit in. In any case, dwelling densities of less than twenty per net acre are not suitable for the city.

The range of twenty to seventy dwellings per net acre is too high for suburban life and insufficiently high for city life. This range creates the problem that too many people are strangers, but without the diversity that allows the neighbourhood to function. The onset of liveliness can vary significantly because it is critically dependent on other conditions. However, Jane Jacobs claims that city liveliness rarely occurs under a hundred dwellings per net acre. There is also an upper limit, which is reached when too many high-rise buildings reduce the diversity of habitat niches and thus the diversity of activities that ultimately create the liveliness of the streets. The upper limit is no longer determined by the greater risk of disease that came with increasing densities in the past, thanks to such developments as modern sanitation, epidemiology, and basic health care.

In this discussion of a few key elements of the analysis of Jane Jacobs, I am aware that, from a sociological perspective, a great deal has changed since the writing of that book. There has been a continued decline of the traditional elements in the United States and Canada. I have already mentioned that these and other modern societies have become mass societies whose social cohesion is no longer based on customs, traditions, a traditional morality and religion. The individual in such mass societies experiences a profound need for integration propaganda, to which the development of the mass media has responded by furnishing a world of myth-images that replace custom and tradition. On the psychological level, this has led to what Reisman[53] has called the 'other-directed' personality type, using the structure of experience as a kind of radar to scan what everyone else is doing, which then becomes normal and even normative. This leads to a statistical morality without any reference to a moral code. Under these conditions, there appear to be few ways in which vast numbers of strangers can live together with some kind of common unity. It would appear that implicit in the analysis of Jane Jacobs is a recognition of this social reality. Whatever our personal preferences may be in terms of the kind of neighbourhood and physical habitat we value, the fact remains that with the growing global population high-density living is here to stay, and coupled with the social realities of a mass society, what Jane Jacobs suggests continues to be a more sustainable alternative than the one held out by traditional architecture and planning.

Jacobs's methodology is particularly significant because it relies heavily on the human mind and culture. It must not be forgotten that, for as long as we can go back, human groups and societies have developed cultures to deal with a highly complex world by interpreting everything in the context of everything else through symbolic systems of names, meanings, and values. No scientific method can match this. The comparative approach seeks to grasp the complexity of city life in its full context as opposed to the limited contexts of one or more disciplines. Once fundamental interpretations are made of streets, neighbourhoods, districts, and cities in terms of their livability, specific factors can be identified as part of an explanation. This knowledge based on experience and culture can then be complemented by discipline-based knowledge separated from experience. The symbiosis of different kinds of knowing could overcome the limitations of each one.

The above approach recognizes how cities evolved in the past based on knowledge embedded in experience and culture. Everyone had a metaconscious image of their neighbourhood, which provided a kind of common background against which people made decisions about their urban habitat. This context-maximizing mode explains why traditional settlements were so well adapted to local conditions and cultures and also how this was lost once knowledge separated from experience. Jane Jacobs's approach is, first and foremost, based on knowledge embedded in experience and culture, which is able to grasp the complexity of cities on the level of the whole, and secondarily depends on knowledge separated from experience to critically examine a variety of details in a smaller context. I can conceive of no other approach capable of combining the consideration of context with the strengths of discipline-based knowing. It is essential if the urban habitat is to evolve in the direction of a greater human sustainability in the vast metropolitan areas where a growing portion of humanity will live. The interplay between the way of life of a society (with its diversity of activities, each evolved in the full context of all the others) and its interaction with the habitat niches of a particular built environment is so complex as to require the full force of all the approaches to knowledge present in our society.

If this analysis has any merit, the high expectations of how computer-based and communications technologies can 'save' the city are misplaced and will lead to end-of-pipe solutions. I am not denying that working at home can save commuting time, reduce transportation needs, and thus cut down on pollution, the need for increased road capacity, and everything else that comes with the system based on private automobiles. However, can they solve the fundamental problems of modern cities?

The city brings together large numbers of people in space and time but also separates them. The inconveniences of that separation have been reduced by technologies such as telephones and cars. However, these have also permitted an ever greater separation of people in space and time, offsetting the advantages of these technologies. Computer and communication technologies are likely to do the same thing if they distract our attention from the root problem, with all the human, social, and biosphere-related consequences. It is quite possible to create virtual offices, and even companies, by having people work at home on computers that are networked together. The problem is that more and more social interactions are being mediated through technologies of one kind or another and that these technically mediated relations are not at all the same as face-to-face ones. If these technically mediated social relations systematically filter out certain things, they will separate people in other ways. This will negatively affect the evolution of our social selves and our relations with others in a positive feedback mode. It is essential to consider the evolution of the social fabric between the inhabitants of cities in its entirety, and not to focus on certain aspects related to a specific technology in a piecemeal fashion. If working in an office or factory provides us with essential contacts, then their elimination through technological mediation may be very serious. It would be a mistake to assume that computer and communications technologies can overcome the way cities separate people in space and time, any more than the telephone, automobile, and television could. They at best modestly offset the ever greater separation of people in space and time in the ever larger metropolitan areas. So far, many technologies have been both cause and effect of this separation.

I once saw an animated short film about a Martian coming to Earth and trying to make sense of modern cities. When he had finally figured it out, it was obvious that the car was the real inhabitant of the city, since everything was built and ordered in relation to it. The only anomaly the Martian could not unravel was the significance of these funny little things going in and out of the cars. This story humorously illustrated the problem. The car was supposed to make the garden city and the radiant city livable by overcoming the separation of its inhabitants in space and time. Yet the evolution of this transportation system had the opposite effect. We continue to have difficulty recognizing the positive feedback driving the city–car interdependence.

The 'wired community' may well pose an even greater threat. If more and more relations will be technically mediated, filtering out essential elements for our social and emotional well-being, the consequences for individual and collective human life may be far greater than those of the car.[54]

The work of Jane Jacobs has inspired a different approach to the modern large city. The new urbanism[55] seeks to create habitats where many primary functions are within easy walking distance along pedestrian-friendly streets. Homes along the streets have porches instead of garages to provide 'eyes on the street' and facilitate contact with neighbours and passers-by. The idea is that people will once again learn to recognize most people on their streets, thus diminishing the problem of living with strangers and helping to create safe streets. This greatly reduces dependency on cars, particularly if public transportation can be relied on to reach those primary functions not within walking distance. In other words, the design is once again people-centred as opposed to car-centred. It is not an anti-car approach but one that simply recognizes the limitations of car-based transportation systems and designs the urban habitat accordingly. High densities are required to permit the development of public-transportation systems whose convenience can rival the car. This kind of approach can simultaneously improve the human- and biosphere-related dimensions of sustainability and deal with the reality of living in a mass society.

Jane Jacobs's influence may be seen in the work of many contemporary architects, urban designers, and city planners who are critical of conventional urban planning that is exclusively based on knowledge separated from experience, and who insist that all design must spring from an experience of place. Architect Christopher Alexander, urban designer Kevin Lynch, sociologist William Whyte, and city planners Allan Jacobs and Donald Appleyard have all acknowledged their debt to the pioneering critique of Jane Jacobs.[56]

## 11.4 Integrating Biosphere Dependence

The contribution of the discussion thus far to filling in the context matrices with respect to the urban habitat may be summed up as follows: The rows of the matrices differentiate individual and collective human life into a diversity of activities that are carried out in the corresponding diversity of physical niches in the built habitat. How the organization of these niches within the large modern cities affects sustainable development has been discussed in relation to human sustainability. The implications of the activities themselves require an extensive analysis of how the phenomenon of technique affects modern ways of life, which is the subject of another analysis.[57] The focus now shifts to the biosphere-related dimension of sustainable development.

White and Whitney[58] show that human settlements had very different interactions with their hinterlands before and after the Industrial Revolution in terms of resource and waste flows as well as energy requirements. Before the Industrial Revolution, the hinterlands or carrying-capacity regions were constituted by the immediate surroundings of settlements.[59] In these cases, there was a direct negative-feedback relationship between settlements and their carrying-capacity regions, making these relationships quasi-sustainable. If they were not, the destruction of the carrying capacity of the hinterland would follow and the settlement would disappear. The challenge for any settlement was to increase the yield of agriculture and other resources drawn from the hinterland to support the additional population. This required political, administrative, and technological adaptations with minimal environmental degradation. The size of the hinterland was limited by the capabilities of these means, as well as by the geographical characteristics of the surroundings. Transportation of resources and waste was difficult, especially in the absence of waterways. A settlement usually had a coercive relationship with its hinterland population to ensure the required resources and dispose of its wastes. For these and other reasons, the largest hinterlands were generally found in low-lying river valleys.

Industrialization substantially increased the scale of the resource requirements and waste-disposal needs of settlements that now became industrial centres. Even an entire nation was frequently inadequate as a hinterland to supply its industrial centres, and this gave further impetus to colonialism, whereby other peoples and their lands were coerced into serving as hinterlands to extend the carrying capacity of industrializing nations. Colonialism impinged on and substantially undermined the reciprocal and sustainable relations these people had established with their ecosystems, severely damaging their economies and thus their entire ways of life.

When these colonies regained their independence, it proved virtually impossible to reorient their economies away from being hinterlands for the industrialized world. This made it difficult for them either to industrialize or to return to their old ways of life, leaving them weak and vulnerable, and thus paving the way for neo-colonialism. Indirect control and appropriation of their carrying capacity continued to support the ways of life of the north. This is evident from the unequal terms of trade, the net flow of capital to the north, and their massive debts, also to the north. They were unable to appropriate the additional carrying capacity required for industrialization. For instance, the industrialization of the north relied extensively on overseas emigration to create a better balance among population, consumption, production, and the carrying capacity of the ecosystems.

What is true for the procurement of resources is equally true for the disposal of wastes. Many industrialized nations do not have an adequate sink capacity in their hinterlands, a situation made worse by the fact that their inhabitants frequently refuse to have it put in their 'backyard.' Hence, a significant export of wastes developed.[60] There exists a vast international trade in hazardous materials, in which the north pays the south a fraction of what they would normally have to pay at home to dump the most problematic wastes.[61] In this way as well, the north is appropriating some of the carrying capacity of the south. This situation is clearly unsustainable for all parties, being based on economic domination, social inequity, and injustice.[62] It is for these reasons that White and Whitney suggest that a new phase in the evolution of modern cities must occur.

The unhealthy relations between modern cities and their hinterlands are a significant factor in increasing the size of cities, further aggravating many problems. Herbert Girardet[63] has identified forces that pull and push individuals and families into cities. Population growth relative to arable land, the industrialization of agriculture requiring less farm labour, reductions in arable land (as a result of soil erosion, land degradation, desertification, and deforestation) all contribute to people being forced off the land and pushed into cities. Natural disasters, religious strife, and warfare do the same. So can the pressure of neighbouring cities on rural economies through their demands for farm produce, resulting in developments such as cash cropping by means of intensive industrialized methods. Manufactured goods undermine rural craft production by making it redundant or uncompetitive, causing still other people to move to the city. Employment opportunities, the lure of a better life, medical facilities, education, or the promise of excitement and freedom offered by urban life also encourage migration to the city. So great is the influx in many southern cities that shanty towns spring up out of necessity. As cities grow, they increase the pressures on the hinterlands, producing further growth and creating a vicious cycle that is neither healthy nor sustainable. As noted earlier, the first generation of industrial societies overcame some of these pressures on their cities by massive emigration to the colonies. For example, nearly half of the natural population increase in England between 1846 and 1932 emigrated. The south does not have this safety valve.

Many modern cities, particularly in the north, have the entire world as their hinterland, drawing on the ecological capital of many regions. They have also developed what White and Whitney call a 'waste economy' that parallels the productive economy since, as we have seen earlier, the linear throughput in the economy sooner or later turns all resources into wastes.[64] With our focus on production, we forget the first law of thermodynamics and the equally enormous effort required to deal with the direct and indirect wastes of that production. The symbiotic relationship that used to exist between cities and their hinterlands, whereby the latter was replenished by the organic wastes of the former has disappeared, representing an enormous problem in itself. Where sewage treatment occurs, human and industrial wastes are processed together, making it impossible for sewage sludge to be safe enough to be returned to the land because heavy metals and synthetic chemicals are not removed. Yet, in principle, sewage can be a viable resource of nitrogen, phosphates, potash, magnesium, and other essential ingredients for crops. New York is producing one of the largest structures in the world from its wastes. Fresh Kills, when it is entirely full sometime between 2004 and 2014, will comprise four pyramids, each roughly 450 feet high covering 2,900 acres of what was once wetlands.[65] As one of the engineers put it, 'New York City is the delta of a Nile of soil from all the farm fields of the world, flowing through the supermarkets into the landfill.'[66] This is by no means the only problem of landfills, which generate significant quantities of methane and toxic leachate (a result of the moisture and protein contents of the wastes that produce organic acids and nitrogen compounds, which pick up other components as they leach through the rubbish). It is estimated that, of the latter, Fresh Kills produces 1 million gallons a day.[67]

What is true for production is also true for transportation. The transportation system based on private cars gets people to where they are going with a consumption of resources, a production of wastes, and all the associated problems that make it clearly unsustainable. This despite impressive engineering accomplishments, including reductions in emissions of carbon monoxide by 96 per cent, hydrocarbons by

96 per cent, and nitrogen oxides by 76 per cent since the introduction of catalytic converters, unleaded gasoline, and lighter cars with better engine design, beginning in the 1960s.[68] However, these gains are completely outstripped by a spectacular explosion in the car population and increases in car usage. For example, since the 1960s the number of cars has nearly tripled in the United States.[69] The annual growth rate of automobiles in Germany exceeds that of the world population and their combined horsepower is a figure that is larger than the human global population.[70] Superimposed on these trends will be enormous expected growth rates in the car population in eastern Europe, India, and China. Meadows and colleagues[71] argue that the automobile is already on the verge of overshoot in terms of resource consumption and pollution levels, and that current growth rates are entirely unsustainable. At present, only about 8 per cent of the world's population owns a car and, if this proportion were to double or triple, the amount of land required for roads is simply not available.

The transportation system based on the private car will have to be rethought, particularly in the context of the large modern city. Jane Jacobs has shown how important diversity and human contact are for safety, security, and feelings of well-being. Separating people and activities in space and time achieves the opposite, notwithstanding modern means of transportation and communication. People have to travel longer distances between activities, requiring a greater investment of time. This reduces the opportunities for contacts with others, which might occur with other means of getting to places. Assuming that the portions of the day taken up by working, eating, sleeping, and playing have remained roughly the same, separating the places where these activities take place and minimizing the possibility of social contacts with others, have serious implications for people's social selves, their relations with others, the livability of their habitat – in sum, the sustainability of human life itself. This may well be further exacerbated by virtual offices and companies. It is essential to bring together in our minds the need to keep together daily-life activities in time and space, the need to get to places where we are going as a way of staying in touch with people and places as opposed to mere transportation, and the need to keep streets and neighbourhoods safe through life on the streets. Hagerstrand[72] recognized the importance of making contact with people and places as essential to all human activities, including transportation, but this aspect has been largely ignored in transportation planning. It is human contact that is essential for our well-being and the livability of cities. This could alleviate many problems associated with human life in the urban habitat. It would reduce the enormous throughput of matter and energy and the associated harmful effects, provide people with more time and meaningful social relations, and improve the ability of the urban habitat to support as opposed to interfere with human activities. Rethinking 'transportation' should consider the need to make the economy more cyclical in terms of its flows of matter, increase energy end-use efficiency, create meaningful and satisfying work, and re-create life on city streets through meaningful social contacts. Once again, this is not driven by anti-car sentiments but by the recognition that the transportation system based on cars, like any other human creation, has its strengths and weaknesses. We should use it only where it genuinely serves us.

Marchetti[73] has shown that the amount of time a person devotes to travel is roughly the same regardless of how fast or how far we travel. The significance of this empirical finding is that if we save time by increasing the speed of transportation, it is used to cover greater distances. People frequently spend more time getting to and from bus stations, train stations, or airports than the duration of the trip. The declining livability of cities makes growing numbers of people spend their weekends at a cottage, requiring two places of residence and a great deal of time and resources to move from one to the other. The growing deterioration of cities is causing the rich and powerful to escape to walled neighbourhoods with high levels of security, and private police now often outnumber public police in some cities. People are fleeing into recreational vehicles for greater safety or to seek some compensation for a damaged social self. Surrounded by time-saving technologies, we are all running out of time. This has been wonderfully described in Michael Ende's novel *Momo*, which is an account of a community where time thieves persuaded the residents to save time rather than 'waste' it in idle conversations, friendships that do not pay, caring for others, and almost everything else that makes human life meaningful and satisfying. The effects were dramatic; as people rushed about saving time, they ran out of time for themselves and others. People became restless, irritable, and dissatisfied.[74] We, too, are being robbed of our space and time because the built environment demands a great deal more from us than we demand from it in terms of a convivial and supportive habitat for our activities. It has become a major alienating force in modern life.

Why do we lack a technological imagination to find our way out of the labyrinth of the social and environmental effects of technology? To do this, we have to conceptually put together what belongs together. For example, when it comes to transportation, the need to save time in getting from one place to another should be satisfied so as to sustain rather than undermine human life. How successful is our present solution when highways cost upward of $15 million per mile, and when people spend a significant portion of their incomes on cars that in the cities move them at average speeds of, at best, twelve miles an hour? In many places, bicycles do almost as well as cars in terms of speed, allow for more direct contact between people and places, give people much-needed exercise, and could provide other benefits if only cars did not make streets so dangerous and polluted. Some people do their daily workout on an exercise bike while, for transportation, they use an expensive and polluting vehicle that, in many cities, cannot move much faster than a bike. This would be funny if it were not so tragic. What will we do when a growing number of people cannot, for health or financial reasons, drive a car? It is time to go back to the drawing board and rethink the kinds of contradictions Illich[75] and others struggled with. Discipline-based knowing and technical specialties using minimal context have forced us into a corner from which we can exit only if we recognize their limitations. There are no 'pure' human activities of working, eating, playing, or moving about even though we try to 'decontaminate' them from all others by separating them in space and time. Life remains enfolded, albeit in a much weaker way than in the past, and this enfoldedness is essential for what it is to be human. Living a life is not a matter of collecting essentially separate experiences of acquaintances, friends, families, and communities.

The transportation system based on cars which, in addition to moving people about in air-conditioned comfort while being entertained by music, also has a major impact on the health and mortality of the inhabitants in cities, on the ability of children to enjoy independent travel and the use of streets, on the ability of those who cannot drive cars to live normal lives in cities, on the shape of cities and the way they separate human lives in space and time, and on the unsustainable demands made by these cities on the biosphere. All these must be considered as 'outputs' of this system, making its net efficiency, productivity, and profitability extremely low and its incompatibility with human life, society, and the biosphere unsustainable. There is no question that the transportation system based on the car makes a substantial contribution to the 'invisible elbow' effect of modern ways of life.[76]

Some authors have gone much further in their claims. Haughton and Hunter state that

> American capitalism from the 1920's ... required the investment opportunities created by urban expansion, in part as it fled the constraints of the inner cities with their more organized and more costly labor forces. From hidden government subsidy of road building programs to private-sector manipulation of markets, suburbanisation was not a simple 'natural' law of urbanisation in motion, but a shift fuelled by the search for commercial gain, from property developers, road builders, oil companies and automobile manufacturers. Perhaps most disturbing is the case of the General Motors subsidiary, National City Lines, which during the 1930's and 1940's bought and then closed down networks of electric streetcars and trolley buses in 45 cities in 16 states.[77]

As suggested previously, since cities can neither create nor destroy matter and energy, they depend on the biosphere, much like any organism. Urban metabolism, input–output analysis, and urban ecology are interdisciplinary concepts to examine how cities depend on a variety of inputs performing their functions, including growth and maintenance of the infrastructure and its building stock, as well as the activities that occur within its physical niches. In so doing, these inputs are transformed, usually into degraded matter and energy which become the outputs returned by cities to their hinterlands. This metabolism can range from those that are highly linear to those that are much more circular, with fundamental implications for the health of cities and their sustainability with respect to the biosphere. Such concepts are powerful tools for design and decision making related to the urban habitat and should be further complemented by others showing the reciprocal relation between the physical niches provided by that habitat and the activities of people. Incorporated in these considerations would have to be positive- and negative-feedback loops that drive various developments. We are still very far from such models, but important advances are being made in that direction. As White puts it in his important book on the subject,

> this book [*Urban Environmental Management*] began as an attempt to operationalize the holistic approach to urban environmental management, rather than to propose solutions to the general problem of poverty, either in the North or in the South. To meet this objective, the discussion will

> focus on physical variables like litres of water and tons of carbon dioxide. However, at no time do I wish to suggest that we can simply plan our way out of the present crisis. The causes of human marginalization are social and political and no number of garden festivals or enterprise zones can change that. What this book is designed to do is to identify the physical instruments that are available to implement a politically-conscious social policy with respect to resource use and residuals management.[78]

Large cities have the possibility of essentially creating their own microclimate. The virtual elimination of all natural growth in downtown cores, suburban shopping centres, and industrial parks creates what is known as a 'heat-island effect.'[79] The built environment uses denser materials than local ecosystems which are able to absorb heat by day and release it by night. In ecosystems ground cover protects denser soils, and evaporation absorbs some of the incoming heat. Once again, heat-island effects set up a positive-feedback loop, as they encourage the greater use of air conditioners, which intensify the problem they were supposed to solve. There are also many negative effects resulting from the manufacture and disposal of these air conditioners, including the dispersal of CFCs.

Heat islands create so-called dust-dome effects. As relatively warmer air above the city rises, carrying its pollutants upwards, a low-pressure system is created that draws in air from the countryside, which is replaced by the air descending from the city as it cools at higher altitudes. The result is a cloud of dust and other airborne effluents from the city hanging over it, trapping pollutants near the ground surface. It can often be seen as a cloud of smog hanging over the city as one approaches. The problem is particularly serious in third-world nations, where some of the particles are dried faeces. Dust domes are broken up under the influence of winds, which then carry the pollutants to downwind areas. Since airborne particles provide a surface upon which water vapour can condense, higher-than-average precipitation can result in cities and their hinterlands.[80] In some instances, the geography may stabilize microclimates and dust domes, as is the case in Los Angeles, Denver, and Mexico City, because of nearby mountains or the combination of mountains and large bodies of water. In other cases, such as Athens and Bombay, the built environment mitigates the horizontal winds, exacerbating the heat-island effect and dust domes.[81]

So many processes within the urban habitat are inherently dissipative that air, water, and land pollution present serious problems. The operation of factories, refineries, power stations, cars, trucks, diesel locomotives, buses, incinerators, lawn mowers, and barbecues; the application of fertilizers, pesticides, and herbicides; salting roads in winter; tarring roofs; sandblasting stone; construction and demolition, and a great deal else dissipate various materials. The result is an explosion of synergistic effects, the permutations and combinations of which become impossible to calculate. To pretend that more research can help us understand all the possible effects and the cause–effect relationships is to ignore the limitations of the sciences. Recall what I have said earlier about what we currently know about the short- and long-term effects of all the chemicals used in modern ways of life, to which we add a significant number each year. Nothing short of a precautionary principle will assist us in dealing with a situation that cannot be fully known scientifically and that cannot be dealt with by applying discipline-based (limited context) knowledge. This is unlikely, however, until society recognizes that its ability to develop and apply knowledge of our present world is incommensurate with its ability to transform that world.

Modern cities have an enormous impact on local and regional water cycles. Fresh-water use by many cities in many parts of the world is unsustainable. It is about to become a serious bottleneck to urban and economic growth. It is likely to become a source of conflict, and possibly war, between regions and nations, especially where rivers and bodies of water are shared.[82] Currently there are unresolved disputes over water resources between many African states, between India and Pakistan, Turkey and Iraq, Israel and Jordan, and Mexico and the United States. Cities have to procure water from ever-greater distances. For example, New York and Los Angeles depend on water reservoirs and rivers hundreds of miles away for over 80 per cent of their water. Tapping aquifers in addition to surface waters is not sustainable in many parts of the world. Ground subsidence is becoming a serious problem for many cities, including Houston and Beijing, where expensive engineering projects are required to prevent or repair damage to buildings, roads, pipelines, and other structures. In other cases, the water table has become so low that aquifers become contaminated with polluted or saline water, and in many more cases the rate of replenishment is lower than the rate at which water is pumped out. The former problem is particularly serious in coastal cities, including U.S. cities along the Atlantic seaboard and in the coastal cities in the Near and Middle East.[83] In sum, in many places we are mining water, as opposed to using a renewable resource.[84]

A substantial portion of the networks of flows of matter and energy associated with modern cities is devoted to building, maintaining, and demolishing activities specifically related to the urban habitat. These activities may account for as much as one-tenth of the global economy.[85] The throughput of matter and energy is enormous. It is estimated, for example, that buildings consume between one-sixth to one-half of global physical resources.[86] Modern building methods have greatly extended the ecological footprints of buildings. Indoor air pollution continues to be a major problem. Yet buildings continue to be designed with little reference to the context in which they function, and as if they will never need to be taken apart and the materials to be disposed of. It will become increasingly important to design buildings for their entire life cycle. A few months of building construction (including the production and transportation of the materials to the construction site) can consume more resources and cause more pollution than a decade of building operation.[87] This becomes obvious when it is recognized that nearly half of the copper used in the United States ends up in buildings, primarily for pipes and electrical wiring, of which only one-fifth is recycled materials.[88] Many building materials (such as PVCs) are extremely difficult and uneconomic to recycle.[89] In 1992 built structures accounted for about one-third of world energy consumption, including 26 per cent of fossil fuels, 45 per cent of hydro power, and 50 per cent of nuclear power.[90] This does not include the embodied energy. Adding this into the figures, the building sector's share of energy use in the United States is 45 per cent, making it the largest consumer of energy.[91] It is estimated that the erection of a typical 150-ton home in the United States sends 7 tons of refuse to the local dump, and for every six houses or apartment buildings constructed, one is demolished and the materials landfilled.[92] Many modern building materials help produce 'sick buildings,' with concentrations of pollutants in their air sometimes hundreds of times higher than outside air. The medical and productivity costs are estimated in the tens of billions of dollars per year.[93] The preventive approaches being developed in the building sector are having only a modest effect, given strong competition and a failure to make decisions based on life-cycle costs. In 1937, Le Corbusier made a claim to the effect that he designed one single building for all nations and climates, and this continues to characterize many modern buildings and cities. The level of incompatibility between buildings and their human, social, and natural contexts has been compensated for by a variety of technologies driven by end-of-pipe solutions and positive feedback.[94]

In sum, the modern urban habitat is the outcome of methods and approaches making minimal use of context. It is a monument to context incompatibility and a threat to the sustainability of human life, society, and the biosphere. There is no reason why this has to be the case. Preventive approaches, including the ones that were discussed in earlier chapters, can fundamentally change the metabolism of cities. The use of the kinds of methods and approaches pioneered by Jane Jacobs and worked out in the new urbanism could make the city much more compatible with individual and collective life. There are few forces today that have a greater influence in shaping human life than the built habitat. Its alienating power is possibly matched only by that of technique. Yet a great deal of modern architecture, building engineering, planning, and administration continue to be based on an acceptance of the two-stage approach described at the beginning of this book. We owe it to ourselves and our children to do much better, and to give the development and implementation of preventive approaches for transforming our built habitat a much greater priority relative to other activities such as the development of the Internet or space exploration. It is time we wake up to the symptoms of distress and make it impossible for the rich and powerful to escape to their own walled neighbourhoods to minimize the experience of the consequences of their decisions.

One of the big problems that a political consciousness will have to overcome is why, despite considerable evidence and policies to utilize negative feedback to reduce serious problems or unsustainable trends (as in the case of leaded gasoline, or the effects of private cars and the need for public transportation), there is so much inertia in many 'systems' that even when steering corrections can be made it takes decades for significant results to become evident. Couple this to the preoccupation of politicians to get re-elected and the resulting pressure for short-term, quick-payoff political platforms, and the dilemma of modern urban management comes clearly into focus. Current democratic political systems in the face of these issues fail miserably and will have to be rethought as an integral part of developing policies for sustainable development. Do we really want to wait until the problems become so severe that a new Big Brother, fascism, or other extreme movement feels fully justified to take absolute power into its hands in the name of 'saving' humanity? Are we going to tinker around until some new extreme secular political religion imposes itself? Conspiracy-theory explanations are not very helpful here. There is no question that large corporations and powerful interest groups lobby and distort the political process, but is this possible because there is a vacuum in political imagination in today's technological age?[95] We should dig deeper to understand the profound constraints that lie buried within the individual and collective metaconscious as a symbolization of the fundamental reality of a way of life. To a far greater extent than was ever the case in human history, humanity, because of globalization, shares one common future from which even transnationals will not be able to escape.

It has now become apparent that there is a nexus between the findings of the three previous chapters and the present one. Achieving a more circular economy in terms of its networks of flows of matter will make cities more sustainable. It can be extended by including urban farming and the return of nutrients to the hinterlands of cities.[96] Other synergies may also be achieved, such as the one in Calcutta, where sewage treatment is combined with fish-farming.[97] Making the urban habitat more supportive of human activities will, at the same time, improve the practicality of the intensive use of public transportation, ushering in a transformation of the transportation system based on cars so that it will correspond increasingly to what it can do best and curtail those uses that make no sense. A greater mixing of primary functions will reduce the separation of human activities in space and time, with a host of benefits for human life and the sustainability of cities. People will be able to walk to more places, get more exercise while carrying out daily-life activities, and save a great deal of time. Children will be able to go to school, visit their friends, or go to a public library without needing to be escorted by adults in more neighbourhoods than is currently the case. This will also reduce some of the pressures on single-parent families. Some of the most liveable cities in the United States, such as Portland, Oregon,[98] and San Diego, California,[99] owe a great deal to a 'transportation policy' that is very much in keeping with the thrust of this chapter. In the past, transportation policy, like energy policy, was entirely focused on supply. In these cases, the focus on the supply of ever more roads and expressways shifted when people became aware of the problems of this positive feedback–based approach. It began to dawn on them that what needed to be dealt with was the demand side for transportation. The design of the whole urban habitat thus had to be taken into account, requiring an integration of considerations that until then had been dealt with separately.

Alternative ecological-design approaches for the urban habitat are beginning to spring up that recognize that the urban habitat is an ecology of niches facilitating a diversity of human activities that make up the way of life of a city. The design of buildings, streets, neighbourhoods, and districts are not separate activities in these ecological approaches because each is regarded as a unit of a larger whole, ultimately enfolded in a society, local ecosystems, and the biosphere.[100] Ecological approaches make a much greater use of context considerations and are based on the premise that cities are there to serve human beings and not the other way around. Creating healthy cities is another important component in choosing the direction of our common future. Do we want to move in the direction of a society that once again embraces fundamental Western values, or do we want to return to the kinds of societies and civilizations that predated the West, where, in most cases, the dialectical tension between the individual, on the one hand, and groups, organizations, institutions, and society itself, on the other, favoured the latter? The choice is ours if we have the courage to exercise a technological and economic imagination, and dare to be iconoclastic with respect to the conventional wisdoms and dogmatisms of our present economic world-view.

# Postscript

I, too, have a dream. In it, I find myself teaching in a new kind of engineering school. It all began when the plan of the former premier's Council of Ontario, for a round table on professional education and the long-term viability of the province, was resurrected. It quickly became evident that the existing situation did not serve anyone's interests. The intellectual division of labour in the professions was clearly causing a great deal of needless and preventable harm to the economy, human life, communities, and ecosystems. It necessitated an equally senseless patchwork of end-of-pipe approaches, which rarely got to the root of any problem. Despite considerable private and public expenditures, the situation had gradually grown worse, including a steady drop in the standard of living as the creation of gross wealth was undermined by the growing costs of expanding and maintaining the patchwork of end-of-pipe approaches. When a consensus was reached on this diagnosis, the prescription was obvious. To begin with, a reform of higher education was implemented, which, in turn, led to a great many other changes.

The reforms to universities happened relatively rapidly since they were able to build on a great deal of prior work, such as that of the Bonneau–Corry Commission[1] and the Boyer report.[2] A distinction was made between different kinds of knowledge-gathering strategies, ranging from frontier research to reflective research. The former is discipline-based and advances through endless specialization; the latter enquires into the significance of the highly specialized findings in relation to one another, and to human life, society, and the biosphere. Creating a dialogue between these research strategies greatly reduced knowledge externalities. The gap between the 'two cultures' was also diminished. For example, reflective research made steady progress in showing how our modern world was unthinkable without science and technology, and this, in turn, brought substantial changes to the theories and approaches of disciplines such as economics, sociology, political science, psychology, philosophy, and religious studies. It became acceptable for some researchers in each discipline to enquire into the significance of what they knew for other disciplines, and vice versa. Such researchers were cross-appointed to a central interfaculty, and their accomplishments were measured and rewarded by means of tenure and promotion criteria based on a set of standards distinct from those for discipline-based research. In return, departmental budgets were significantly affected by the extent to which collaboration with this central interfaculty affected the evolution of their curricula and research endeavours. In this way, universities created vital symbiotic relationships between their faculties and departments via the central interfaculty. At long last, universities established 'conversations' between their specialties.

All this developed much more rapidly than anyone expected. It so happened that each department had one or more 'philosophers' who had interests well beyond their disciplines and who were motivated to create the vital link between their departments and the central interfaculty, thus contributing to the teaching and research efforts of both. An intellectual home was made for faculty members who pursued their disciplines in the context of many others to create a broader understanding so typical of thinkers who still have a significant influence on many disciplines today, including Hegel, Marx, Weber, Durkheim, Freud, Darwin, Heidegger, and Toynbee. Until these developments were set in motion, it was difficult to imagine how such thinkers could have found an intellectual home in a department of a modern university. Yet their existence was and continues to be indispensable if humanity is to have an intellectual future. Only reflective research and teaching can prevent *homo oeconomicus* from being succeeded by *homo informaticus*. The seminal works referred to in the last four chapters in particular as key entry points into certain bodies of literature are but some examples of how the reflective-research tradition had remained alive in spite of the modern university and discipline-based refereeing and publishing. All these kinds of problems were now rapidly becoming a thing of the past.

The new symbiotic relationship between central interfaculty and discipline-based departments had a number of important effects. The central interfaculty explored the public interest through themes such as sustainability, justice, unemployment, homelessness, competitiveness, sustainable free trade, and health by means of 'conversations' between the 'philosophers' from relevant disciplines. Schools were established around these themes, offering graduate seminars and research thesis topics. Departments began to accept responsibility for the knowledge they developed by exploring the uses society made of it. For example, the mathematics department began to research the use society made of mathematics and the implications this had for modern ways of life. Engineering departments began to make the kinds of changes suggested in chapter 3. The consequences of particular engineering methods and approaches were explored, and this understanding was used in a negative-feedback mode to make adjustments to these methods and approaches to ensure the greatest possible compatibility between technological development and its contexts. Medical schools were beginning to examine how changes in society and the biosphere were transforming the demands placed on the cultural, psychological, and physiological resources of individuals. For example, the findings of socio-epidemiology as to how human health is affected by workplaces cannot be resolved through drugs or operations. Neither can the growing load on the human immune system from the contamination of food, water, and air be effectively dealt with by end-of-pipe approaches. It became quite evident why a growing number of people had been turning to alternative medicine: being based on knowledge embedded in experience, it was able to deal with a variety of conditions that knowledge separated from experience could not. A new symbiosis between the two medical approaches was rapidly emerging, with enormous benefits to all parties. Similarly, business schools were exploring the context implications of various methods and approaches to business administration and using this understanding to make adjustments, thereby improving the capabilities of businesses to earn healthy profits without undermining their support systems. Law faculties were beginning to examine the kinds of changes in society that were driving ongoing legal reforms, including their implications. The aim was to use this understanding to develop a more context-sensitive legal framework that saw litigation for what it is: a breakdown of less adversarial methods for dispute resolution. All these developments made the university a much more interesting, stimulating, and relevant institution. The new changes constituted an advantageous alternative to forcing linkages with industry and government that tied the research agenda to limited scope and time spans, creating explosive conflicts of interest and ethical dilemmas. The university was now better able to contribute to genuine solutions to the benefit of all parties.

In my dream, as an engineering professor I find myself engaged in creating a new generation of engineering schools, graduating engineers with a difference. With the support of a central interfaculty, it is now possible to further advance the kinds of developments my Centre for Technology and Social Development has been putting into place for well over a decade. All first-year engineering students are introduced to the concept of preventive approaches based on negative feedback. This provides them with an understanding of how the complementary-studies and technical-studies components of the curriculum contribute to engineering design and decision making. In a first-year course, students learn the principles of preventive approaches, and the extent to which these are being developed in the best-managed corporations and public policies. The process of industrialization is examined to understand how, in the course of two centuries, society has changed technology and how, in turn, this has changed society. It provides students with a map of how modern technology interacts with and depends on its contexts. Two follow-up courses complete this theoretical framework and its preventive applications. By simultaneously increasing the context included in engineering courses, the two components of the curriculum gradually blend into each other in a symbiotic relationship that greatly facilitates the teaching of design, communications, and applied ethics. All this is greatly reinforced by the decision to begin to practise what we are preaching. Hence, in addition to the ongoing modifications to the curriculum, the findings of socio-epidemiology are used to redesign the work of faculty members, staff, and students. This is further reinforced by recognizing the distinctive contributions of knowledge embedded in experience and knowledge separated from experience. Whenever a new building is required, faculty, students, and staff are involved in its preliminary design specifications to ensure sustainability in all three dimensions. A similar approach is taken to the retrofitting of existing buildings. In other words, my faculty has gradually become a showcase of what is being 'professed' there, and a steady stream of visitors helps transfer preventive approaches to other sectors of society.

These developments are gradually helping to resolve a fundamental problem in the relationship between the technology-related professions and society. Generally speaking, society had granted these professions the right to regulate their own affairs, including the setting of standards of professional practice and licensing in return for protecting the public interest. This had been a difficult bargain to keep because the consequences of professional practice almost always fall outside the domain of expertise of its practitioners, with the result that they are left to others. There is no mechanism in the governance of most professions to deal with this problem; hence, there is no negative feedback to guide adaptation to a turbulent context. The result has been a loss of prestige and status for some professions and their members, and their clients have not been served as well as they might have been. It is, therefore, in everyone's interest to modify the regulation of professions to build in negative feedback. The universities are helping to resolve this situation by graduating professionals with a difference, able to help their professions work through these problems. Using the kinds of research instruments reported in chapter 3, the Canadian Engineering Accreditation Board and its U.S. counterpart have developed new long-term targets and the ability to measure progress.

At first, many people were highly sceptical as to whether the reforms noted above would change the engineering, business, and accounting practices of corporations. Would the new generation of engineers find a receptive environment for what they had learned? The situation was certainly difficult, but far from hopeless. It is true that the divisions and departments of corporations make many decisions whose consequences fall outside their jurisdictions. They can therefore be dealt with by others only in an end-of-pipe manner. How can the corporate organization chart be modified to keep together what belongs together – namely, the actions and their consequences? For example, the responsibilities of an environment department should be incorporated into other units whose actions have significant environmental implications. This would oblige the environment department to, as much as possible, work itself out of a job, which is contrary to normal organizational and bureaucratic tendencies. These could be changed if the environment department were considered as an in-house resource whose effectiveness is assessed in terms of the ability of its client units to deal with their environmental problems in a preventive manner. Ideally, such environment departments would be incorporated in a business-context unit cooperating with all others to ensure sustainability in all three dimensions. Governments could then adjust tax rates according to the social and environmental performance of the corporation, measured in terms of the costs it passes on to society. Balance sheets are now accompanied by statements of how operations affect the social and natural capital of a society.

A small beginning has been made in transforming the structure and evolution of transnational corporations. Once social consensus began to build around the recognition that the production of consumer goods and technical commodities fundamentally shapes the services these render, the human activities involved in these services, and the social fabric constituted by them, it gradually became evident that current perceptions of the corporation were entirely outmoded. Corporations should be seen as communities of stakeholders because they do far more than simply produce goods and services for a profit. They shape the relationships that involve us all, and this must be recognized in the way this institution is framed. This is not an anti-business agenda but an attempt to create viable enterprises that make healthy profits without undermining the viability of society and the biosphere and, thus, of themselves. The transnationals present a particular challenge because they are out of the context of a nation and, therefore, out of the context of any human community. By designing goods and services for global markets, they are necessarily out of context with local conditions and only in context with global technical systems. These institutions are the principal challenge. However, as the discipline of economics is busy testing its fundamental assumptions in the light of what the other social sciences know about the consequences of economic growth, it is becoming evident that what it has regarded as externalities now shape much of human life, society, and the biosphere. This also modifies the understanding of the principal economic institution – namely, the transnational corporation. Once its complex and diverse role beyond the economy is more clearly understood, more effective legislative frameworks can be developed to ensure the public interest.

As expected, the reactions of corporations to the potential of preventive approaches vary considerably. Some of the best-managed corporations that have already been experimenting with some aspects of preventive approaches welcome the new developments as providing them with the resources to move more rapidly and comprehensively down this road, which is becoming profitable. However, other corporations are persisting in the traditional ways of lobbying the government against any changes and are installing end-of-pipe abatement or paying for social or health services only when legally required to. The differences in behaviour of the two groups of corporations are becoming obvious to everyone. Lobbyists now face the increasingly impossible task of convincing politicians, the general public, and the media that prevention-oriented policies and regulations jeopardize the economy by making corporations less competitive with dire consequences for communities and nations. The new-found capability of universities to mobilize intellectual resources in the direction of prevention, coupled with government policies and incentives, is convincing a growing number of corporations to make the switch to a prevention-oriented corporate culture and business strategy. All this is considerably helped by an aware public interested in green products and green investment. In fact, the very concept of 'green' has become more comprehensive to include all three dimensions of sustainability.

A few governments are beginning to respond to all these changes by rethinking their own internal operations, and by creating a central ministry or department charged with ensuring that the collective effort of all ministries or departments will move society towards greater sustainability in all three dimensions. Such a ministry relies extensively on the resources of several central interfaculties of interested universities to develop an integrated policy that incorporates components directed towards industrial ecology, integrated resource planning and energy efficiency, healthy work, and sustainable habitat. This work makes extensive use of round tables of the kind described in the opening chapter of this work. It also depends on an equivalent of citizens' jury duty to resolve complex technological and economic issues on the basis of knowledge embedded in experience and culture. These developments make decision processes somewhat less vulnerable to lobbying and distortion by powerful interest groups. Governments are beginning to use new and more comprehensive indicators of *net* economic performance and the state of human communities and ecosystems. Three kinds of books are being kept: the usual fiscal books, and the other two related to the 'social' and 'natural' capital of a nation. These efforts, in turn, provide central interfaculties with excellent research opportunities.

Some governments are beginning to design appropriate frameworks for markets that can function in such a way as to allow the invisible hand to operate with minimal effects of the invisible elbow. Pension-fund legislation has been drafted to ensure that investments will not damage the long-term interests of the community and nation. By relatively straightforward tax measures, the use of markets as gambling casinos has been reduced, and green investment is being encouraged.

The control of the pension funds has been shifted from the employer to independent trustees capable of ensuring that investments do not damage the long-term viability of the societies in which people will retire. Once again, comprehensive investment strategies are being developed with the assistance of central interfaculties to prevent negative impacts on the long-term viability of communities and ecosystems.

Suddenly, I wake up with a scream as this new and better world comes crashing down. In a flash I see that, while all this was going on, technique continued to undermine human consciousness and cultures. Their ability to provide individual and collective human life with set points that reflect genuine human aspirations instead of performance ratios masquerading as human values had progressively diminished. I suddenly realize that the emphasis on biosphere-related sustainability had overshadowed the other two dimensions, eventually bringing down the entire enterprise. My dream world was missing one essential ingredient – namely, the ability to create a civilization that includes modern science and technique but whose cultures are not dominated by them, so that they can be regulated by means of negative feedback on human terms. The one ingredient that was missing in my dream world was the essential iconoclasm necessary to bring down the economic world-view and its associated hierarchy of values. Our imagination has to have a strong iconoclastic dimension for, without it, we will not be able to pull down our 'secular golden calves,' and our well-intentioned efforts will quickly be assimilated by the 'system.' The ultimate challenge will be to overcome the profound sociocultural influence technological and economic growth are having on human consciousness and cultures. Without this, technology and the economy will remain judge and jury of themselves.

In the last four chapters, I reported the findings of a study comparing state-of-the-art methods and approaches for dealing with social and environmental issues related to modern technology with their conventional counterparts. I must insist on one essential point: It is not that we are lacking in preventive alternatives to conventional approaches. It is not that such alternatives are being developed by fringe groups. It is not that such alternatives have been found wanting by governments or corporations that have put them to the test. It is not that there is evidence suggesting that preventive approaches cannot make companies and economies more competitive while reducing the burdens imposed on communities and ecosystems. It is certainly not a question of the engineering profession not having been informed of the findings of this research. This begs the question why preventive approaches are not rapidly displacing their conventional counterparts.

What appears to be lacking is a critical awareness of how much the world has changed in the last half of the twentieth century. This becomes obvious when we examine the explicit and implicit assumptions underlying most disciplines and professions. These may have been reasonable and defensible at the time they were made, but few of them stand up to present critical scrutiny. The methods and approaches that may have been good enough for our intellectual and professional fathers and mothers are not good enough for us. To claim that they are is to surrender to an intellectual or professional fundamentalism deeply rooted in secular religious beliefs and values that help to constitute the cultural ground on which we stand. Once again, it is not that we are lacking the scholars who have critiqued these assumptions, given the way the world has changed, but that they are banished to the margins of their disciplines as intellectual heretics. It would appear that intellectual fundamentalisms of various kinds are both cause and effect of professional fundamentalisms.

Is it reasonable to suppose that capital remains the prime factor of production, only modestly influenced by science and technology? Has their influence not qualitatively changed since the end of the Second World War? Is it reasonable to continue to suppose that communities and societies can generate the moral boundaries to constrain the self-interested behaviour of producers and consumers, given the nature of modern mass societies and their lack of traditional values and reference points? Is it reasonable to assume that the invisible elbow is of little consequence relative to the invisible hand? Is it reasonable to continue to assume that free trade cannot lead to substantial unemployment and underemployment as if the international mobility of capital were low, so that capital freed up in one area of an economy would be reinvested in another? Is it reasonable to continue to assume that the enormous increases in advertising expenditures over the last fifty years have had no influence on consumers and economic democracy? Is it reasonable to assume that the biosphere can indefinitely support economies by acting as the ultimate source and sink of all flows of matter and energy, as life support and habitat, given the considerable change in the relative size of the global economy?

Is it reasonable for contemporary sociology, despite references to post-industrialism or post-capitalism, to fail to develop an adequate sociology of technology, including its implications for the entire social fabric and institutional framework? Is it reasonable to describe modern democracies with little consideration given to how technologies and techniques have transformed the workings of governments, corporations, and political parties, the formation of public political opinions, the role of political advertising, and a great deal else? Is it possible to discuss ethics in general and professional ethics in particular without a clear understanding of how modern cultures no longer have any equivalent of traditional values and morality? Is it reasonable for modern medicine to cling to conceptions of health and disease that ignore how science and technology are changing the air we breathe, the food we eat, the water we drink, the way we work, how and where we live?

Why should we bother with the 'grand theories' of yesterday when the computer permits us to carry out large surveys? Yet without such theories how can the relevance of the survey questions be assessed? All these issues and much more have been eloquently argued by intellectual and professional heretics, but their voices have had little effect on the reigning fundamentalisms. The autonomy of disciplines and professions does not readily admit that the assumptions they inevitably have to make about their 'boundary conditions' at the interfaces with other disciplines and professions do not reflect the fact that reality is seamless, and certainly not compartmentalized to suit the structure of the modern university. We are desperately trying to make modern universities more relevant by creating overall mission statements to which faculties, departments, and individual courses must contribute in accordance with their own mission statements in a way that is *measurable*. We are also trying to align these mission statements with the interests of governments and industry. However, if the diagnosis presented in this work has any merit, we are dealing with the symptoms and not the real problems.

There can, therefore, be no conclusion to this work in the usual sense of that term. There is no point in providing a long list of concrete and specific recommendations of what we ought to do. Colleagues will not get involved in them unless they become convinced of the need for an iconoclastic attitude. Such attitudes are no longer born of modern education. They may be born of earlier religious, political, or humanitarian traditions in so far as these have not been assimilated into the mainstream. Perhaps these can still provide some equivalent of a set point by means of which we can learn to say no. No, this senseless destruction of life will not do. We are needlessly and senselessly sacrificing millions of fellow human beings to our modern secular idols. The intellectual and professional fundamentalists would have us believe that there is no other way, admonishing us to be realistic. However, the intellectual and professional heretics on the fringes of many disciplines and professions claim that things don't have to be this way. It is not a question of taking sides but of recognizing the need for a critical examination of our assumptions, given the way the world has changed in the recent past and continues to change in the present. Chances are that this will lead to intellectual and professional journeys similar to the one reported in this work. In undertaking the extensive research for this book, I discovered that many have walked this path before me, that others are walking it with me, and that it is to be hoped that more will walk it after us. These voices are of critical importance to prevent our civilization from doing irreparable harm to itself, to all life, and to our planet. This work hopefully adds another voice to the chorus.

There is no evidence to suggest that our civilization is about to heed the voices of our scientific and professional heretics. Hope is sought somewhere else. Many cling to the possibility of a political solution: just wait until the politicians we favour take control of the state. If it were as simple as that we should have reached a political consensus on a more preventive approach for technological and economic growth a long time ago, because there is something in it for everyone. Amory Lovins put it well, although somewhat optimistically, when he wrote:

> a soft path simultaneously offers jobs for the unemployed, capital for businesspeople, environmental protection for conservationists, enhanced national security for the military, opportunities for small business to innovate and for big business to recycle itself, exciting technologies for the secular, a rebirth of spiritual values for the religious, traditional virtues for the old, radical reforms for the young, world order and equity for globalists, energy independence for isolationists, civil rights for liberals, states' rights for conservatives.[3]

It is often argued that genuine political solutions are being blocked by powerful vested interests. Such conspiratorial explanations fall short of the mark. This is not at all a denial of the existence of the powerful interests that lobby against any progressive legislation. However, people can take to the streets, and they have done so in the past when the stakes were high. Not even dictatorial and oppressive regimes have been able to withstand that kind of moral pressure in the long term. There is, therefore, another part to this story, which must be explored in some detail. It has to do with how living in the labyrinth of technology has affected the consciousness of modern people and the cultures of their societies. I have already mentioned one of the primary symptoms of this influence – namely, the ability of performance ratios to masquerade as human values. The result is that modern technology is regulated on its own terms, with disastrous results. It is only by means of an iconoclasm towards the deep cultural values and beliefs of our age that a mutation towards a humane and common future can occur. An iconoclastic attitude cannot be legislated or induced by psychological, social, and pedagogical techniques. Yet only by means of such an attitude can a genuine human 'set point' be established by which technological and economic growth can be regulated. Many readers may feel that this conclusion is at best naïve, but it is the only one I have been able to reach after examining in detail the influence that living in the labyrinth of technology has on human consciousness and cultures.

# Notes

## 1: Preventive Approaches as a New Technology and Economic Strategy

1 I have developed the political ramifications of this in 'Political Imagination in a Technical Age,' in *Democratic Theory and Technological Society,* ed. Richard B. Day, Ronald Beiner, and Joseph Masciulli, 3–35 (New York: M.E. Sharpe, 1988).
2 Juliet B. Shor, *The Overworked American: The Unexpected Decline of Leisure* (New York: Basic Books, 1991); R.T. Golembiewski, Robert Boudreau, Robert Munzenrider, and Huaping Luo, *Global Burnout: A Worldwide Pandemic Explored by the Phase Model* (Greenwich, CT: JAI Press, 1996).
3 Margaret A. Shotton, 'Should Computer Dependency Be Considered a Serious Problem?,' in *Social Issues in Computing: Putting Computing in Its Place,* ed. Chuck Huff, and Thomas Finholt, 673–89 (New York: McGraw-Hill, 1994).
4 Barry Commoner, 'The Environmental Cost of Economic Growth,' in *Energy, Economic Growth and the Environment,* ed. S.H. Schurr, 83–98 (Baltimore, MD: Johns Hopkins University Press, 1972).
5 Craig Brod, *Technostress: The Human Cost of the Computer Revolution* (Reading, MA: Addison-Wesley, 1984).
6 Lester Brown, *Tough Choices: Facing the Challenge of Food Scarcity* (New York: Norton, 1996).
7 John Whitelegg, *Transportation for a Sustainable Future: The Case for Europe* (London: Bellhaven, 1993).
8 See, for instance, M.J. Eden and J.T. Parry, eds., *Land Degradation in the Tropics: Environmental and Policy Issues* (New York: Pinter, 1996); M.S. Gamser, H. Appleton, and N. Carter, *Tinker, Tiller, Technical Change* (New York: Bootstrap, 1990); UNESCO, *Science and Technology in Developing Countries: Strategies for the 90s* (Paris: UNESCO, 1992).

9 The present work builds on my earlier study of values and cultures: *The Growth of Minds and Cultures: A Unified Theory of the Structure of Human Experience* (Toronto: University of Toronto Press, 1985) and on Jacques Ellul, *The Technological Society,* trans. by John Wilkinson (New York: Vintage, 1964).

10 *Our Common Future: World Commission on Environment and Development* (Oxford: Oxford University Press, 1987).

11 Ellul, *The Technological Society.*

12 W.H. Vanderburg and N. Khan, 'How Well Is Engineering Education Incorporating Societal Issues?.' *Journal of Engineering Education* 83 (1994): 357–61. The curriculum continues to be monitored for deviations in the patterns described in this reference. Thus far, no significant deviations have been observed; thus, the findings of this study are as relevant today as when it was carried out.

13 A. Jürgensen, N. Khan, and W.H. Vanderburg, 'Industrial Ecology for Greater Sustainability' (Scarecrow Press, forthcoming); N. Khan, and W.H. Vanderburg, 'Energy for Greater Sustainability' (Scarecrow Press, forthcoming); K. Jalowica, N. Khan, and W.H. Vanderburg, 'Toward Healthier Work' (unpublished manuscript); N. Khan and W.H. Vanderburg 'Healthy, Sustainable Cities' (Scarecrow Press, forthcoming).

14 W.H. Vanderburg, 'Rethinking End-of-Pipe Engineering and Business Ethics,' *Bulletin of Science, Technology and Society* 17/2–3 (1997): 141–53.

15 W.H. Vanderburg, 'Preventive Engineering: Strategy for Dealing with the Negative Social and Environmental Implications of Technology,' *Journal of Professional Issues in Engineering Education and Practice* 121/3 (July 1995): 155–60.

16 José Goldemberg, Thomas Johansson, Amulya Reddy, and Robert Williams, *Energy for a Sustainable World* (Washington, DC: World Resources Institute, 1987).

17 Clifford F. Cobb, Ted Halstead, and Jonathan Rowe, *The Genuine Progress Indicator: Summary of Data and Methodology* (San Francisco: Redefining Progress, 1995). For a good summary, see C. Cobb, Ted Halstead, and Jonathan Rowe, 'If the GDP Is Up, Why Is America Down?,' *The Atlantic Monthly* 276 (October 1995), pp. 58–78. See also H.E. Daly and J.B. Cobb, Jr, *For the Common Good: Redirecting the Economy Toward Community, The Environment, and a Sustainable Future* (Boston: Beacon, 1989).

18 Ed Andrew, *Closing the Iron Cage: The Scientific Management of Work and Leisure* (Montreal: Black Rose, 1981). See also Gunilla E. Bradley and Hal W. Hendrick, eds., *Fourth International Symposium on Human Factors in Organizational Design and Management* (Stockholm: Elsevier, 1994).

19 See for example Peter Brodner and Waldema Karwowski, eds., *Ergonomics of Hybrid Automated Systems*, vol. 3 (Stockholm: Elsevier, 1994).
20 W.H. Vanderburg, 'Living in Technique' (unpublished manuscript).
21 Peter Brodner, *The Shape of Future Technology: The Anthropocentric Alternative* (New York: Springer-Verlag, 1990); David F. Noble, 'Present-Tense Technology,' *Democracy* 3/1, 2, 3 (1983): 8–24, 70–82, 71–92, and *Forces of Production: A Social History of Industrial Automation* (New York: Oxford University Press, 1986).
22 R. Rudolph and S. Ridley, *Power Struggle: The Hundred Year War over Electricity* (New York: Harper and Row, 1986).
23 Donella Meadows, Dennis Meadows, and Jorgen Randers, *The Limits to Growth: A Report for the Club of Rome Project on the Predicament of Mankind* (New York: New American Library, 1972).
24 Walter C. Patterson, *The Energy Alternative: Changing the Way the World Works* (London: Channel 4 Books, 1990).
25 T.W. Zosel, 'How 3M Makes Pollution Pay Big Dividends,' *Pollution Prevention Review*, Winter 1990–1, pp. 67–72.
26 *Responsible Care: A Total Commitment* (Ottawa: Canadian Chemical Producers Association, Oct. 1990).
27 W.A. Irwin, 'The Legal Basis for Waste Avoidance and Waste Utilization in the Federal Republic of Germany,' in *The Environmental Challenge of the 1990's: Proceedings of the International Conference on Pollution Prevention*, EPA/600/9-90/039 (Washington, DC: EPA, 1990).
28 B. Nussbaum and J. Templeman, 'Built to Last – Until It's Time to Take It Apart,' *Business Week*, 17 September 1990, pp. 18–21.
29 M.E. Porter, 'America's Green Strategy,' *Scientific American*, April 1991, p. 168.
30 R. Karasek and T. Theorell, *Healthy Work: Stress, Productivity, and the Reconstruction of Working Life* (New York: Basic, 1990).
31 K. Matsushita, quoted in Peter J. Denning, 'Beyond Formalism,' *American Scientist* 79 (January–February 1991): 19.
32 The widely reported phenomenon of *karoshi*, a word coined by the Japanese medical establishment to denote death by overwork, is but one indicator of these problems.
33 AnnaLee Saxenian, *Regional Advantage: Culture and Competition in Silicon Valley and Route 128* (Cambridge, MA: Harvard University Press, 1994).
34 *Ontario beyond Tomorrow: Ideas for Building a Sustainable Society (Summary Version)* (Toronto: The Premier's Council and the Ontario Roundtable on Environment and Economy, 24 March 1995).
35 Gretchen C. Daily, ed., *Nature's Services: Societal Dependence on Natural*

*Ecosystems* (Washington, DC: Island Press, 1997).; Michael Jacobs, *The Green Economy: Environment, Sustainable Development and the Politics of the Future* (Vancouver: University of British Columbia Press, 1993).

36 Donald A. Schon, *The Reflective Practitioner: How Professionals Think in America* (New York: Basic Books/Harper Colophon, 1983).

37 Vanderburg, 'Living in Technique.'

**2: Individual Prerequisites for Preventive Approaches**

1 Garret Hardin, 'The Tragedy of the Commons,' *Science* 162 (April 1968): 243–8.

2 W.H. Vanderburg, 'Living in Technique' (unpublished manuscript).

3 James P. Womack, Daniel T. Jones, and Daniel Roos, *The Machine That Changed the World: Based on the Massachusetts Institute of Technology's 5-Million Dollar, 5-Year Study on the Future of the Automobile* (New York: Rawson Associates, 1990).

4 J.M. Ham, P.A. Lapp, and I.W. Thompson, 'Careers of Engineering Graduates 1920–70, University of Toronto,' in Engineering Alumni Association, *Studies in Engineering Education* (Toronto: University of Toronto, March 1973), 1–71.

5 L. Bailyn and E. Shein, *Where Are They Now and How Are They Doing?* (Cambridge, MA: Alumni Association of MIT, 1972).

6 L. J. Brooks, 'Corporate Codes of Ethics,' *Journal of Business Ethics* 8/2 & 3 (February–March 1989): 117–29.

7 See for example Canadian Standards Association, with the Canadian Environmental Auditing Association, *Guidelines for Environmental Auditing: Principles and General Practices*, and Canadian Standards Association, *A Guideline for a Voluntary Environmental Management System* (Ottawa: Canadian Standards Association, February 1993).

8 W.H. Vanderburg, *The Growth of Minds and Cultures: A Unified Theory of the Structure of Human Experience* (Toronto: University of Toronto Press, 1985).

9 Andrew Bard Schmookler, *The Parable of the Tribes: The Problem of Power in Social Evolution* (Berkeley: University of California Press, 1984).

10 Vanderburg, 'Living in Technique.'

11 See for example T.E. Graedel and B.R. Allenby, *Industrial Ecology and the Automobile* (Upper Saddle River, NJ: Prentice-Hall, 1998).

12 C. Brod, *Technostress: The Human Cost of the Computer Revolution* (Reading: Addison-Wesley, 1984); S. Turkle, *The Second Self* (New York: Simon and Schuster, 1984); Gunilla Bradley, *Computers and the Psychosocial Work Environment* (London: Taylor and Francis, 1988).

13 Barbara Garson, *The Electronic Sweatshop: How Computers Are Transforming the Office of the Future into the Factory of the Past* (New York: Simon and Schuster, 1988).
14 Chuck Huff and Thomas Finholt, eds., *Social Issues in Computing: Putting Computing in Its Place* (New York: McGraw-Hill, 1994); Vincent Mosco and Janet Wasko, eds., *The Political Economy of Information* (Madison: University of Wisconsin Press, 1988), especially Andrew Clement, 'Office Automation and the Technical Control of Information Workers,' 53–63; Judith A. Perrolle, *Computers and Social Change: Information, Property and Power* (Belmont, CA: Wadsworth, 1987).
15 David H. Flaherty, *Protecting Privacy in Surveillance Societies: The Federal Republic of Germany, Sweden, France, Canada and the United States* (Chapel Hill: University of North Carolina Press, 1989); Warren Freedman, *The Right of Privacy in the Computer Age* (New York: Quorom, 1987); David Lyon and Elia Zureik, eds., *Computers, Surveillance and Privacy* (Minneapolis: University of Minnesota Press, 1996). See Mosco and Wasko, eds., *The Political Economy of Information*. Also Jeffrey Rothfelder, *Privacy for Sale: How Computerization Has Made Everyone's Private Life an Open Secret* (New York: Simon and Schuster, 1992).
16 David Icove, Karl Seger, and William von Storch, *Computer Crime: A Crimefighter's Handbook* (Sebastopol, CA: O' Reilly and Associates, 1995); Deborah G. Johnson and Helen Nissenbaum, eds., *Computers, Ethics and Social Values* (Englewood Cliffs, NJ: Prentice-Hall, 1995); Martin Wasik, *Crime and the Computer* (Oxford: Clarendon Press, 1991).
17 L. Winner, *The Whale and the Reactor: A Search for Limits in an Age of High Technology* (Chicago: University of Chicago Press, 1986), ch. 3. Interestingly enough, 'high reliability' organizational theory invariably turns to the military for its own examples. For a discussion of 'normal accident' versus 'high reliability' theory, see Scott D. Sagan, *The Limits of Safety: Organizations, Accidents and Nuclear Weapons* (Princeton, NJ: Princeton University Press, 1993).
18 C. Perrow, *Normal Accidents: Living with High Risk Techniques* (New York: Basic, 1984).
19 See Winner, *The Whale and the Reactor.*
20 C. Pateman, *Participation and Democratic Theory* (Cambridge: Cambridge University Press, 1984).
21 Vandana Shiva, *Monocultures of the Mind: Perspectives on Biodiversity and Biotechnology* (London: Zed, 1993).
22 A. Lovins, *Soft Energy Paths: Toward a Durable Peace* (Cambridge, MA: Ballinger, 1977).

23 S. Giedion, *Mechanization Takes Command* (New York: Norton, 1969).
24 See Winner, *The Whale and the Reactor.*
25 D. Noble, *Forces of Production* (New York: Knopf, 1986).
26 A. Maslow, *Psychology of Science: A Reconnaissance* (New York: Harper and Row, 1966).
27 Vanderburg, *The Growth of Minds and Cultures*, especially ch. 7.

**3: Collective Prerequisites for Preventive Approaches**

1 W.H. Vanderburg, 'Living in Technique' (unpublished manuscript).
2 W.H. Vanderburg, *The Growth of Minds and Cultures: A Unified Theory of the Structure of Human Experience* (Toronto: University of Toronto Press, 1985). For a more complete discussion, see Vanderburg, 'Living in Technique.'
3 W.H. Vanderburg and N. Khan, 'How Well Is Engineering Education Incorporating Societal Issues?,' *Journal of Engineering Education* 83/4 (October 1994): 357–61.
4 Canadian Engineering Accreditation Board, 'Policy Statement,' Toronto, 1992.
5 Exceptions were courses to which the research instruments were not applicable such as language electives, public speaking, and report writing, and fourth- year thesis courses. In the engineering science program, the options, with a significant number of courses from outside the faculty, were not scored; only the aerospace, chemical, and electrical engineering options are included.
6 We are not including the so-called hidden curriculum in this study. See Benson R. Snyder, *The Hidden Curriculum* (New York: Knopf, 1971).
7 We are well aware of the objections to any rating system such as the one used in the Gourman report. Nevertheless, all evidence indicates the Faculty of Applied Science and Engineering ranks among the best North American schools.
8 W.H. Vanderburg, 'Measuring Professional Education for Sustainable Development,' in *ENTREE 96 Proceedings*, Conference on Measuring Progress Towards Sustainable Development, 349–59 (Sunderland, UK: November 1996); W.H. Vanderburg, 'On the Measurement and Integration of Sustainability in Engineering Education,' *Journal of Engineering Education* 88/2 (April 1999): 1–5.
9 This began with the work of T.S. Kuhn, *The Structure of Scientific Revolutions*, 2d ed. (Chicago: University of Chicago Press, 1970).
10 Snyder, *The Hidden Curriculum.*
11 *Waterloo Engineering Alumni Letter*, February 1993, p. 8.

12 U.F. Franklin, 'The Real World of Technology,' CBC Massey Lectures, 1990.

13 B. Snyder, 'Literacy and Numeracy: Two Ways of Knowing,' *Daedalus* 119 (1990): 233–56.

14 S. Tobias, *They're Not Dumb, They're Different: Stalking the Second Tier* (Tucson, AZ: Research Corporation Foundation for the Advancement of Science, 1990).

15 See, for instance, the *Bulletin of Science, Technology & Society*, Sage Publications.

16 This example is adapted from Louis L. Bucciarelli and Sarah Kuhn, 'Engineering Education and Engineering Practice: Improving the Fit,' in *Between Craft and Science: Technical Work in U.S. Settings*, ed. Stephen R. Barley and Julian E. Orr, 92–122 (Ithaca, NY: Cornell University Press, 1997).

17 C.W. Smith, 'The Qualitative Significance of Quantitative Representation,' in *The Qualitative-Quantitative Distinction in the Social Sciences*, ed. B. Glassner and J.D. Moreno, 29–42 (Boston: Kluwer Academic, 1978).

18 C. Perrow, *Normal Accidents: Living with High Risk Technologies* (New York: Basic, 1984).

**4: Tools for Map-Making**

1 E. Skjei and M.D. Whorton, *Of Mice and Molecules* (New York: Dial, 1983).

2 E.O. Wilson, 'Is Humanity Suicidal?,' *New York Times Magazine*, 30 May 1993, pp. 24–9.

3 F. Capra, *The Turning Point: Science, Society and the Rising Culture* (New York: Bantam, 1988), 271.

4 For a good description of the difference between living and mechanistic wholes, see Vandana Shiva, *Biopiracy: The Plunder of Nature and Knowledge* (Boston: South End Press, 1997).

5 H.L. Dreyfus, *What Computers Still Can't Do: A Critique of Artificial Reason* (Cambridge, MA: MIT Press, 1992).

6 Neil Evernden, *The Social Creation of Nature* (Baltimore: Johns Hopkins University Press, 1992).

7 W.H. Vanderburg, *The Growth of Minds and Cultures: A Unified Theory of the Structure of Human Experience* (Toronto: University of Toronto Press, 1985).

8 E. Devereux, *From Anxiety to Method in the Behavioral Sciences* (Paris: Mouton, 1967). See also W. Labov, *Language in the Inner City* (Pittsburgh: University of Pennsylvania Press, 1972).

9 J.H. Van Den Berg, *The Changing Nature of Man* (New York: Delta, 1961).

10 Neil Evernden, *The Natural Alien* (Toronto: University of Toronto Press, 1985).

11 Van Den Berg, *The Changing Nature of Man.*
12 James Lovelock, *The Age of Gaia: A Biography of Our Living Earth,* rev. ed. (New York: Norton, 1995).
13 H.H. Gerth and C.W. Mills, *From Max Weber* (New York: Oxford University Press, 1958); Jacques Ellul, *The Technological Society,* trans. by John Wilkinson (New York: Vintage, 1964).
14 For an insightful account of this, see Harvey Cox, 'The Market as God,' *Atlantic Monthly,* March 1999, pp. 18–23.
15 This mode is similar to Hobbes's famous resolutive-compositive method which he applied mechanistically to society.
16 Vanderburg, *The Growth of Minds and Cultures,* especially ch. 7.
17 Kuhn, *The Structure of Scientific Revolutions.*
18 Devereux, *From Anxiety to Method in the Behavioral Sciences.*
19 D. Bohm, *Wholeness and the Implicate Order* (London: Routledge, 1980).
20 Gregory A. Keoleian, Krishnendu Kar, Michelle M. Manion, and Jonathan W. Bulkley, *Industrial Ecology of the Automobile: A Life Cycle Perspective* (Warrendale, PA: Society of Automotive Engineers, 1997).
21 Ruth Schwartz Cowan, *More Work for Mother: The Ironies of Household Technologies from the Open Hearth to the Microwave* (New York: Harper, 1982). See also Susan Strasser, *Never Done: A History of American Housework* (New York: Pantheon, 1982).
22 R. Stivers, *The Culture of Cynicism: American Morality in Decline* (Oxford: Blackwell, 1994).
23 L.B. Lave, Elisa Cobas-Flores, Chris Hendrikson, and Francis McMichael, 'Using Input–Output Analysis to Estimate Economy-wide Discharges,' *Environmental Science & Technology* 29/9 (1995): 420A–6A.
24 Kuhn, *The Structure of Scientific Revolutions.*
25 M. Polanyi, *Personal Knowledge* (Chicago: University of Chicago Press, 1962).
26 B. Wynne, 'Uncertainty and Environmental Learning,' in *Clean Production Strategies: Developing Preventive Environmental Management in the Industrial Economy,* ed. Tim Jackson, 63–83 (Boca Raton, FL: Lewis, 1993).

**5: Context Values for Map-Making**

1 The fact that these are essentially contested concepts only points to their importance to our culture. William Connolly has given us a classic account of these in his essay 'Essentially Contested Concepts,' in *The Terms of Political Discourse,* 10–44 (Princeton, NJ: Princeton University Press, 1984). A good entry point into the literature is Radoslav Selucky, *Marxism, Socialism,*

*Freedom: Towards a General Democratic Theory of Labour-Managed Systems* (New York: St Martin's Press, 1979). See also Sir Isaiah Berlin, *Concepts and Categories: Philosophical Essays*, ed. by Henry Hardy (New York: Viking, 1979), and C.B. Macpherson, *Democratic Theory: Essays in Retrieval* (Oxford: Clarendon Press, 1973). For Marxist interpretations of alienation, see Bertell Ollman, *Alienation: Marx's Conception of Man in Capitalist Society* (Cambridge: Cambridge University Press, 1971) and István Mészáros, *Marx's Theory of Alienation* (London: Merlin, 1970). Richard Schacht, *Alienation* (Garden City, NY: Doubleday, 1970), is a good introductory text that explores other existentialist and sociological meanings of the term. For the Judaeo-Christian notion of alienation, see Jacques Ellul, *Ethics of Freedom*, trans. by Geoffrey W. Bromiley (Grand Rapids: Eerdmans, 1976), 1–70. The notion of integrality is implicit in the works of Plato and Aristotle, specifically in their concepts of justice being dependent on a harmony between reason and appetite in a person. See for example Aristotle, *The Politics*, trans. by Lord Carnes (Chicago: University of Chicago Press, 1985) and Plato, *Plato's Republic*, trans. by G.M.A. Grube (Indianapolis: Hackett, 1974).

2 W.H. Vanderburg, *The Growth of Minds and Cultures: A Unified Theory of the Structure of Human Experience* (Toronto: University of Toronto Press, 1985).

3 W.H. Vanderburg, 'Living in Technique' (unpublished manuscript).

4 Vanderburg, *The Growth of Minds and Cultures*.

5 There is a vast literature on the physiocrats. A good entry point is provided by Elizabeth Fox-Genovese, *The Origins of Physiocracy: Economic Revolution and Social Order in Eighteenth-Century France* (Ithaca, NY: Cornell University Press, 1976).

6 This is surprisingly similar to the young Marx's view of money. See Karl Marx, *Economic and Philosophic Manuscripts of 1844*, trans. by Martin Milligan (New York: International, 1964). For an opposing view, see Ayn Rand, *Capitalism: The Unknown Ideal* (New York: New American Library, 1967). For the biblical interpretation of the city and money, see Jacques Ellul, *The Meaning of the City*, trans. by Dennis Pardee (Grand Rapids, MI: Eerdmans, 1970), and *Money and Power* (Downers Grove, IL: Inter-Varsity Press, 1984).

7 See note 1, above.

8 David Riesman, *The Lonely Crowd*, rev. ed. (New Haven, CT: Yale University Press, 1961); Jacques Ellul, *Propaganda*, trans. by Konrad Kellen (New York: Knopf, 1968), chs. 2–4.

9 Walter Lippmann, *Public Opinion* (New York: Harcourt Brace, 1922).

10 For an interesting discussion of how 'hive' versus 'market' societies can so dominate people in opposite ways, leading to their individuality or their

social self, respectively, becoming 'negatively hallucinated,' see Asher Horowitz, and Gad Horowitz, *'Everywhere They Are in Chains': Political Theory from Rousseau to Marx* (Scarborough, ON: Nelson Canada, 1988). By 'negative hallucination,' the Horowitzes mean the tendency to not see something that is there.

11 Nowhere is this more emblematic than in Hobbes's work. See Thomas Hobbes, *Leviathan* (New York: Everyman's Library, 1973).

12 T. Hancock and L. Duhl, *Promoting Health in the Urban Context*, WHO Healthy Cities Project, WHO Healthy Cities Papers No. 1, FADL 54 (Copenhagen: WHO, 1988).

13 For a good critique of the genome project in general, and genetic research in particular, see Ruth Hubbard and Elija Wald, *Exploding the Gene Myth: How Genetic Information Is Produced and Manipulated by Scientists, Physicians, Employers, Insurance Companies, Educators, and Law Enforcers* (Boston: Beacon, 1997) and Ruth Hubbard, 'Eugenics: New Tools, Old Ideas,' in *Embryos, Ethics and Women's Rights: Exploring the New Reproductive Technologies*, ed. Hoffman Baruch, Amadeo F. D'Adamo, Jr, and Joni Seager, 22–36 (New York: Hawthorn, 1988).

14 For a widely debated, but nonetheless fascinating, account of the origin of various forms of institutional segregation, see Michel Foucault, *Madness and Civilization: A History of Insanity in the Age of Reason*, trans. by Richard Howard (New York: New American Library, 1965); *The Birth of the Clinic: An Archaeology of Medical Perception*, trans. by A.M. Sheridan Smith (New York: Vintage, 1982); and *Discipline and Punish: The Birth of the Prison*, trans. by Alan Sheridan (New York: Pantheon, 1977).

15 Ivan Illich, *Disabling Professions* (Don Mills, ON: Burns and MacEachern, 1977).

16 Priscilla Ellis, 'Environmentally Contaminated Families: Therapeutic Considerations,' *American Journal of Orthopsychiatry* 62/1 (January 1992): 44–54; Donald G. Unger, 'Living Near a Hazardous Waste Facility: Coping with Individual and Family Distress,' *American Journal of Orthopsychiatry* 62/1 (January 1992): 55–70.

17 Norman Myers, *Ultimate Security: The Environmental Basis of Political Stability* (New York: Norton, 1993); Thomas F. Homer-Dixon, Jeffrey H. Boutwell, and George W. Rathjens, 'Environmental Change and Violent Conflict: Growing Scarcities of Renewable Resources Can Contribute to Social Instability and Civil Strife,' *Scientific American* 268/2 (1993): 38–46; Sean M. Lynn-Jones and Steven E. Miller, eds., *Global Dangers: Changing Dimensions of International Security* (Cambridge, MA: MIT Press, 1995).

18 For a good discussion of this from the perspective of healthy cities, see

Herbert Girardet, *The Gaia Atlas of Cities: New Directions for Sustainable Urban Living* (New York: Anchor, 1992).

19 Hancock and Duhl, *Promoting Health in the Urban Context*, 7.

20 The World Commission on Environment and Development, *Our Common Future* (Oxford: Oxford University Press, 1987), 8–9.

21 W.H. Vanderburg, 'Political Imagination in a Technical Age,' in *Democratic Theory and Technological Society*, ed. R. Day, Ronald Beiner, and Joseph Masciulli, 3–35 (New York: M.E. Sharpe, 1988).

22 Herman E. Daly and John B. Cobb, Jr, *For the Common Good: Redirecting the Economy toward Community, the Environment, and a Sustainable Future*, with contributions by Clifford W. Cobb (Boston: Beacon, 1989). See also C. Cobb, Ted Halstead, and Jonathan Rowe, 'If the GDP Is Up, Why Is America Down?,' *The Atlantic Monthly* 276 (October 1995), pp. 58–78.

23 Aristotle, *The Politics*.

24 Francis Fukuyama, 'The Great Disruption: Human Nature and the Reconstitution of Social Order,' *Atlantic Monthly* 283 (May 1999), pp. 55–80.

25 Charles Perrow, *Normal Accidents* (New York: Basic, 1984); Scott D. Sagan, *The Limits of Safety: Organizations, Accidents and Nuclear Weapons* (Princeton, NJ: Princeton University Press, 1993).

26 Langdon Winner, *The Whale and the Reactor: A Search for Limits in an Age of High Technology* (Chicago: University of Chicago Press, 1986), especially Part 1, ch. 2.

27 Robert Jay Lifton, *The Broken Connection: On Death and the Continuity of Life* (New York: Simon and Schuster, 1979); Christopher Lasch, *The Minimal Self: Psychic Survival in Troubled Times* (New York: Norton, 1984).

28 J. Edwards Clayton and Henry A. Regier, *An Ecosystem Approach to the Integrity of the Great Lakes in Turbulent Times: Proceedings of a 1988 Workshop Supported by the Great Lakes Fishery Commission and the Science Advisory Board of the International Joint Commission*, Great Lakes Fishery Commission Special Publication 90-4 (Ann Arbor, MI: Great Lakes Fishery Commission, July 1990).

## 6: Differentiating Sustainable Development

1 Peter Chapman and F. Roberts, *Metal Resources and Energy* (London: Butterworths, 1983), 99–100; Peter Chapman, *Fuel's Paradise: Energy Options for Britain* (London: Cox and Wyman, 1975); Gerald Leach and International Institute for Environment and Development, *A Low Energy Strategy for the United Kingdom* (London: International Institute for Environment and Development, 1979).

2 This discussion is largely based on P.A. Colinvaux, *Why Big Fierce Animals Are Rare: An Ecologist's Perspective* (Princeton, NJ: Princeton University Press, 1978). See also Eugene P. Odum, 'Energy Flow in Ecosystems: A Historical Review,' *American Zoologist* 8 (February 1968): 11–18, and R.H. Whittaker, *Communities and Ecosystems*, 2d ed. (New York: Macmillan, 1975).

3 The discussion in this section is largely based on Michael Jacobs, *The Green Economy: The Environment, Sustainable Development and the Politics of the Future* (Vancouver: University of British Columbia Press, 1993). See also William Ashworth, *The Economy of Nature: Rethinking the Connections between Ecology and Economics* (Boston: Houghton Mifflin, 1995); Robert Costanza, ed., *Ecological Economics: The Science and Management of Sustainability* (New York: Columbia University Press, 1991); Robert Costanza, Bryan G. Norton, and Benjamin D. Haskell, eds., *Ecosystem Health: New Goals for Environmental Management* (Washington, DC: Island Press, 1992); Robert Costanza, Olman Segura, and Juan Martinez-Alier, eds., *Getting Down to Earth: Practical Applications of Ecological Economics* (Washington, DC: Island Press, 1996); Robert Costanza, Olman Segura, and Juan Martinez-Alier, *An Introduction to Ecological Economics* (Boca Raton, FlL: St Lucie, 1995); Herman E., Daly, ed., *Economics, Ecology, Ethics: Essays toward a Steady-State Economy* (San Francisco: W.H. Freeman, 1980), and *Beyond Growth: The Economics of Sustainable Development* (Boston: Beacon, 1996); Gretchen C. Daily, ed., *Nature's Services: Societal Dependence on Natural Ecosystems* (Washington, DC: Island Press, 1997); Faye Duchin and Glenn-Marie Lange, with Knut Thonstad and Annemarth Idenburg, *The Future of the Environment: Ecological Economics and Technological Change* (New York: Oxford University Press, 1994); Thomas Prugh, *Natural Capital and Human Economic Survival*, with Robert Costanza, Olman Segura, and Juan Martinez-Alier (Solomons, MD: International Society for Ecological Economics [ISEE] Press/White River Junction, VT: Chelsea Green, 1995) and David Malin Roodman, *The Natural Wealth of Nations: Harnessing the Market for the Environment* (New York: Norton, 1998).

4 B. Commoner, 'The Environmental Cost of Economic Growth,' in *Energy, Economic Growth and the Environment: Papers Presented at a Forum Conducted by Resources for the Future Inc.*, ed. Sam H. Schurr, 30–65 (Baltimore, MD: Johns Hopkins University Press, 1972).

5 Herman E. Daly and John B. Cobb, Jr, *For the Common Good: Redirecting the Economy toward Community, the Environment, and a Sustainable Future* (Boston: Beacon, 1989).

6 A. Toynbee, *A Study of History*, abridged version ed. by D.C. Somervell

(New York: Dell, 1978). See also Pitirim Sorokin, *Social and Cultural Dynamics* (Boston: Porter Sargent, 1957). Toynbee's theories were incorporated into the general theory of culture in Vanderburg, *The Growth of Minds and Cultures: A Unified Theory of the Structure of Human Experience* (Toronto: University of Toronto Press, 1985).

7 Clive Ponting, *A Green History of the World: The Environment and the Collapse of Great Civilizations* (New York: Penguin, 1991).

8 Donella H. Meadows, Dennis L. Meadows, and Jørgen Randers, *Beyond the Limits: Confronting Global Collapse, Envisioning a Sustainable Future* (Post Mills, VT: Chelsea Green, 1992).

9 Commoner, 'The Environmental Cost of Economic Growth.'

10 Mathis Wackernagel and William Rees, *Our Ecological Footprint: Reducing Human Impact on the Earth* (Gabriola Island, BC: New Society, 1996).

11 World Commission on Environment and Development, *Our Common Future* (Oxford: Oxford University Press, 1988).

12 While this is explained in the next chapter, a fuller explanation is developed in W.H. Vanderburg, 'Living in Technique' (unpublished manuscript).

13 Full details of society as a cultural whole and the transformation of societies by the phenomenon of technique is dealt with in a separate work. However, the principal elements are summarized in a later chapter.

**7: Differentiating a Way of Life**

1 W.H. Vanderburg, 'Living in Technique' (unpublished manuscript).

2 B. Commoner, 'The Environmental Cost of Economic Growth,' in *Energy, Economic Growth and the Environment*, ed. Sam H. Schurr, 30–65 (Baltimore, MD: Johns Hopkins University Press, 1972). For a similar use of formulas, see also Paul Harrison, *The Third Revolution: Population, Environment and a Sustainable World* (New York: Penguin, 1993); Norman Myers, 'Population, Environment, and Development,' *Environmental Conservation* 20/3 (1993): 205–16; and Thomas E. Graedel and Braden R. Allenby, *Industrial Ecology and the Automobile* (Upper Saddle River, NJ: Prentice-Hall, 1998).

3 Vanderburg, 'Living in Technique.'

4 B. Commoner, *Making Peace with the Planet* (New York: New Press, 1992), and 'The Environmental Cost of Economic Growth.'

5 B. Commoner, 'The Failure of the Environmental Effort,' in *Environment in Peril*, ed. Anthony B. Wolbarst, 38–63 (Washington, DC: Smithsonian Institution Press, 1991).

6 Harrison, *The Third Revolution.*

7 Vanderburg, 'Living in Technique.'

8 John Passmore, *Man's Responsibility for Nature* (London: Duckworth, 1994), 177.

9 Mathis Wackernagel and William Rees, *Our Ecological Footprint: Reducing Human Impact on the Earth* (Gabriola Island, BC: New Society, 1996).

10 I have slightly modified the definition Ellul gives in 'Note to the Reader,' from *The Technological Society*, trans. by John Wilkinson (New York: Vintage, 1964), which reads: The term *technique*, as I use it, does not mean machines, technology, or this or that procedure for attaining an end. In our technological society, *technique is the totality of methods rationally arrived at and having absolute efficiency* (for a given stage of development) in *every* field of human activity. Its characteristics are new; the technique of the present has no common measure with that of the past.

This definition is not a theoretical construct. It is arrived at by examining each activity and observing the facts of what modern man calls technique in general, as well as by investigating the different areas in which specialists declare they have a technique.

In the course of this work, the word *technique* will be used with varying emphasis on one or another aspect of this definition. At one point, the emphasis may be on rationality, at another on efficiency or procedure, but the over-all definition will remain the same. (pp. xxv–xxvi)

11 Wassily Leontief, Letter to *Science* 217 (9 July 1982): 104–5.

12 H.E. Daly and J.B. Cobb, Jr, *For the Common Good: Redirecting the Economy toward Community, the Environment, and a Sustainable Future* (Boston: Beacon, 1989).

13 For some of the notable works in this area, see the references supplied in note 3, chapter 6. See also F. Archibugi and P. Nijkamp, eds., *Economy and Ecology: Towards Sustainable Development* (Boston: Kluwer Academic, 1989); F. Herbert Borman and Stephen R. Kellert, eds., *Ecology, Economics, Ethics: The Broken Circle* (Binghampton, NY: Vail-Ballou, 1991); Carl Folke and Tomas Kåberger, eds., *Linking the Natural Environment and the Economy: Essays from the Eco-Eco Group* (Boston: Kluwer Academic, 1991); Robert Gale and Stephan Barg, with Alexander Gillies, eds., *Green Budget Reform: An International Casebook of Leading Practices* (London: Earthscan, 1995); Matthias Ruth, *Integrating Economics, Ecology and Thermodynamics* (Boston: Kluwer Academic, 1993); Herman E. Daly and Kenneth N. Townsend, eds., *Valuing the Earth: Economics, Ecology, Ethics* (Cambridge, MA: MIT Press, 1993); Douglas E. Booth, *The Environmental Consequences of Growth: Steady-state Economics as an Alternative to Ecological Decline* (New York: Routledge, 1998); Juan Martinez-Alier, with Klaus Schlüpmann, *Ecological Economics: Energy, Environment and Society* (New York: Blackwell, 1987).

14 Costanza and colleagues have estimated that the contribution of the biosphere to the global economy exceeds the value of the latter: see Robert Costanza, Herman E. Daly, and Joy A. Bartholomew, 'Goals, Agenda, and Policy Recommendations for Ecological Economics,' in *Ecological Economics: The Science and Management of Sustainability,* ed. Robert Costanza, 1–21 (New York: Columbia University Press, 1991).
15 Adam Smith, *The Wealth of Nations* (New York: Modern Library, 1937).
16 Karl Polanyi, *The Great Transformation* (New York: Octagon, 1975).
17 Ibid.
18 Michael Jacobs, *The Green Economy: The Environment, Sustainable Development and the Politics of the Future* (Vancouver: UBC Press, 1993).
19 J.K. Galbraith, *The New Industrial State* (Scarborough, ON: New American Library, 1978).
20 Vanderburg, 'Living in Technique.'
21 See Paul Hawken, *The Ecology of Commerce: A Declaration of Sustainability* (New York: Harper Business, 1993), and Ian Christie, *Cleaner Production in Industry* (London: Policy Studies Institute, 1995).
22 W.H. Vanderburg, ed., *Perspectives on Our Age: Jacques Ellul Speaks on His Life and Work* (Toronto: CBC Enterprises/Seabury Press, 1986).

**8: Materials and Production**

1 Tim O'Riordan and James Cameron, eds., *Interpreting the Precautionary Principle* (London: Earthscan, 1994).
2 James Lovelock, *The Age of Gaia: A Biography of Our Living Earth,* rev. ed. (New York: Norton, 1995).
3 Charles Perrow, *Normal Accidents* (New York: Basic, 1984). Michael Crichton's *Jurassic Park* (New York: Ballantine, 1990) is a fictional account of a 'normal accident.' Examples of other possible 'normal accidents' may be found in R. Vacca's *The Coming Dark Age* (New York: Doubleday, 1973). See also Scott D. Sagan, *The Limits of Safety: Organizations, Accidents and Nuclear Weapons* (Princeton, NJ: Princeton University Press, 1993).
4 Tim Jackson, ed., *Clean Production Strategies: Developing Preventive Environmental Management in the Industrial Economy* (Boca Raton, FL: Lewis, 1993), particularly chs. 2 and 3.
5 Organisation for Economic Co-operation and Development, *Existing Chemicals: An Investigation, Priority Setting and Chemical Review* (Paris: OECD, 1986); Tim Jackson, *Material Concerns: Pollution, Profit and Quality of Life* (New York: Routledge, 1996), p. 30.
6 Steering Committee on Identification of Toxic and Potentially Toxic Chemi-

cals for Consideration by the National Toxicology Program, Board on Toxicology and Environmental Health Hazards, Commission on Life Sciences, National Research Council, *Toxicity Testing: Strategies to Determine Needs and Priorities* (Washington, DC: National Academy Press, 1984).

7 William M. Sloan, 'Basel Convention on the Control of Transboundary Movements of Hazardous Wastes,' in *Site Selection for New Hazardous Waste Management Facilities*, Annex 1 (Copenhagen: World Health Organization, Regional Office for Europe, 1993).

8 Organisation for Economic Co-operation and Development, *The State of the Environment* (Paris: OECD, 1991).

9 U.S. Office of Technology Assessment, *Serious Reduction of Hazardous Wastes: For Pollution Prevention and Industrial Efficiency* (Washington, DC: Congress of the U.S. Office of Technology Assessment, 1986).

10 For the relationship between the growing population and environmental pollution, see especially Paul Harrison, *The Third Revolution: Population, Environment and a Sustainable World* (New York: Penguin, 1993); Norman Myers, 'Population, Environment, and Development,' *Environmental Conservation* 20/3 (1993): 205–16; George D. Moffett, *Critical Masses: The Global Population Challenge* (New York: Penguin, 1994); Constance Mungall and Digby J. McLaren, eds., *Planet under Stress: The Challenge of Global Change* (Toronto: Oxford University Press, 1990); Lourdes Arizpe, M. Priscilla Stone, and David C. Major, eds., *Population and Environment: Rethinking the Debate* (Boulder, CO: Westview, 1994) and Gayl D. Ness, William D. Drake, and Steven R. Brechin, eds., *Population–Environment Dynamics: Ideas and Observations* (Ann Arbor: University of Michigan Press, 1993). For the deteriorating relationship between population and human health, see Eric Chivran, Michael McCally, Howard Hu, and Andrew Haines, eds., *Critical Condition: Human Health and the Environment: A Report by Physicians for Social Responsibility* (Cambridge, MA: MIT Press, 1993). For the production of wastes, see Jackson, *Clean Production Strategies*, chs. 1 and 4.

11 Amory B. Lovins and John H. Price, *Non-Nuclear Futures: The Case for an Ethical Energy Strategy* (Cambridge, MA: Ballinger, 1975), especially Part 2, 'Dynamic Energy Analysis and Nuclear Power.'

12 Ernst von Weizsäcker, Amory B. Lovins, and L. Hunter Lovins, *Factor Four: Doubling Wealth – Halving Resource Use: The New Report to the Club of Rome* (London: Earthscan, 1997), 52.

13 Jackson, ed., *Clean Production Strategies*, ch. 1.

14 Jackson, *Material Concerns.* See also Gretchen C. Daily, ed., *Nature's Services: Societal Dependence on Natural Ecosystems* (Washington, DC: Island Press, 1997) and Jackson, *Clean Production Strategies*, ch. 1.

15 P. Vitousek, P. Ehrlich, and P. Matson, 'Human Appropriation of the Products of Photosynthesis,' *Bioscience* 36/6 (1986): 368.
16 Jackson, *Clean Production Strategies*, ch. 1.
17 von Weizsäcker, Lovins, and Lovins, *Factor Four: Doubling Wealth – Halving Resource Use*; Joel E. Cohen, *How Many People Can the Earth Support?* (New York: Norton, 1995); Kenneth Blaxter, *People, Food and Resources* (New York: Cambridge University Press, 1986); Kingsley Davis and Mikhail S. Bernstam, eds., *Resources, Environment, and Population: Present Knowledge, Future Options* (New York: Oxford University Press, 1991).
18 For some of these ratios, see T.E. Graedel and B.R. Allenby, *Industrial Ecology* (Englewood Cliffs, NJ: Prentice-Hall, 1995).
19 V. Dethlefsen, T. Jackson, and P. Taylor, 'The Precautionary Principle – Towards Anticipatory Environmental Management,' in *Clean Production Strategies*, ed. Jackson, 55.
20 This analysis is the subject of W.H. Vanderburg, 'Living in Technique' (unpublished manuscript).
21 See, for instance, C. Moore and A. Miller, *Green Gold: Japan, Germany, the United States and the Race for Environmental Technology* (Boston: Beacon, 1994), ch. 5; N. Gertler and J.R. Ehrenfeld, 'A Down to Earth Approach to Clean Production,' *Technology Review* 99/2 (February/March 1996): 48–54; H. von Lersner, 'Outline for an Ecological Economy: Countries Can Indeed Prosper While Protecting Their Environment,' *Scientific American*, September 1995, p. 188; Peter Wiederkehr, *Control of Hazardous Air Pollutants in OECD Countries* (Paris: OECD, 1995); L.W. Baas and H. Dieleman, 'Cleaner Technologies and the River Rhine,' *International Journal for Clean Technology* 1/1 (1989): 54.
22 Robert Socolow, cited in Thomas E. Graedel and Braden R. Allenby, *Industrial Ecology and the Automobile* (Upper Saddle River, NJ: Prentice-Hall, 1998), 8.
23 von Weizsäcker, Lovins, and Lovins, *Factor Four: Doubling Wealth – Halving Resource Use.*
24 Faye Duchin and Glenn-Marie Lange, with Knut Thonstad and Annemarth Idenburg, *The Future of the Environment: Ecological Economics and Technological Change* (New York: Oxford University Press, 1994).
25 Organisation for Economic Co-operation and Development, *Existing Chemicals: Systematic Investigation, Priority Setting and Chemicals Reviews* (Paris: OECD, 1986).
26 Steering Committee on the Identification of Toxic or Potentially Toxic Chemicals in the National Toxicology Program, *Toxicity Testing: Strategies to Determine Needs and Priorities.*

27 Arpad Horvath, Chris Hendrikson, Lester Lave, Francis McMichael, and Tse-Sung Wu, 'Toxic Emissions Indices for Green Design and Inventory,' *Environmental Science and Technology* 29/1 (1995): 86–90.
28 Ontario Ministry of Environment and Energy, *Candidate Substances for Bans, Phase-outs or Reductions* (Toronto: Ontario Ministry of Environment and Energy, 1993); Sloan, *Site Selection for New Hazardous Waste Management Facilities.*
29 E.B. Petersen, *Cumulative Effects Assessment in Canada: An Agenda for Action and Research* (Hull, PQ: Canadian Environmental Assessment Research Council, 1987).
30 Organisation for Economic Development and Co-operation, *Environmental Indicators: A Preliminary Set* (Paris: OECD, 1991).
31 Mathis Wackernagel and William Rees, *Our Ecological Footprint* (Gabriola Island, BC: New Society, 1996).
32 Norman Myers, *Ultimate Security: The Environmental Basis of Political Stability* (New York: Norton, 1994); Thomas F. Homer-Dixon, 'On the Threshold: Environmental Changes as Causes of Acute Conflict,' *International Security* 16/2 (Fall 1991): 76–116; Jessica T. Mathews, 'Redefining Security,' *Foreign Affairs* 68/2 (Spring, 1989): 162–77; Neville Brown, 'Climate, Ecology and International Security,' *Survival* 31/6 (November/December 1989): 519–32; Sean M. Lynn-Jones and Steven E. Miller, eds., *Global Dangers: Changing Dimensions of International Security* (Cambridge, MA: MIT Press, 1995).
33 Society of Environmental Toxicology and Chemistry, *A Conceptual Framework for Life-Cycle Impact Assessment* (Washington, DC: SETAC, 1993), xxv.
34 T.E. Graedel and B.R. Allenby, *Industrial Ecology* (Englewood Cliffs, NJ: Prentice-Hall, 1995), 108.
35 Lester B. Lave, Elisa Cobas-Flores, Chris Hendrikson, and Francis McMichael, 'Using Input–Output Analysis to Estimate Economy-wide Discharges,' *Environmental Science and Technology* 29/9 (1995): 420A–6A. See also David Hawden and Peter Pearson, 'Input–Output Simulations of Energy, Environment, Economy Interactions in the U.K.,' *Energy Economics* 17/1 (1995): 73–86; Faye Duchin, 'Input–Output Analysis and Industrial Ecology,' in *Greening of Industrial Ecosystems*, ed. Braden R. Allenby and Deanna J. Richards, 61–8 (Washington, DC: National Academy Press, 1994); and Ronald Miller and Peter D. Blair, *Input–Output Analysis: Foundations and Extensions* (Englewood Cliffs, NJ: Prentice-Hall, 1985).
36 Paul Hawken, *The Ecology of Commerce: A Declaration of Sustainability* (New York: Harper Business, 1993), 37.
37 von Weizsäcker, Lovins, and Lovins, *Factor Four: Doubling Wealth – Halving Resource Use*, xx.

38 Graedel and Allenby, *Industrial Ecology*, 109.
39 A.B. Lovins, Michael Brylawski, and David Cramer, *Hypercars: Materials, Manufacturing and Policy Implications* (Aspen, CO: Rocky Mountain Institute, The Hypercar Centre, 1996).
40 Gertler and Ehrenfeld, 'A Down to Earth Approach to Clean Production'; Ernest Lowe, 'Industrial Ecology – An Organizing Framework for Environmental Management,' *Total Quality Environmental Management*, Autumn 1993, pp. 73–85; Karen Schmidt, 'The Zero Option,' *New Scientist*, 1 June 1996, pp. 32–7.
41 T.E. Graedel and B.R. Allenby, *Design for Environment* (Upper Saddle River, NJ: Prentice-Hall, 1996).
42 Jackson, *Clean Production Strategies*, ch. 1; see also Herman E. Daly, 'On Economics as a Life Science,' *Journal of Political Economy* 73/3 (1965): 392–406, and Nicholas Georgescu-Roegen, *The Entropy Law and Economic Processes* (Cambridge, MA: Harvard University Press, 1971).
43 Graedel and Allenby, *Design for Environment*; see also Joseph Fiksel, *Design for Environment: Creating Eco-Efficient Products and Processes* (New York: McGraw-Hill, 1996); Janine Sekutowski, 'Design for Environment,' *MRS Bulletin*, June 1991, p. 3, and K.B. Misra, *Clean Production: Environmental and Economic Perspectives* (New York: Springer-Verlag, 1996), especially Section 2, 'Clean Production.'
44 Graedel and Allenby, *Industrial Ecology*; Graedel and Allenby, *Design for Environment.*
45 Graedel and Allenby, *Industrial Ecology*,163. See also B. Steen and S. Ryding, *The EPS Enviro-Accounting Method: An Application of Environmental Accounting Principles for Evaluation and Valuation of Environmental Impact in Product Design* (Stockholm: Swedish Environmental Research Institute [IVL], 1992).
46 Annotated bibliographies are being prepared in the four areas included in this book, as well as some others.
47 The following analysis converges with the argument suggesting that energy should be regarded as a commodity, a social resource, an ecological resource, and a strategic resource: see Paul Stern and Elliot Aronson, *Energy Use: The Human Dimension* (New York: Freeman, 1984) ch. 2. See also Jesse S. Tatum, *Energy Possibilities: Rethinking Alternatives and the Choice-Making Process* (Albany: State University of New York Press, 1995), ch. 1.
48 Jackson, *Materials Concerns*. See also Paul Hawken, *The Ecology of Commerce: A Declaration of Sustainability* (New York: Harper Business, 1993).
49 Hawken, *The Ecology of Commerce.*
50 John Stuart Mill, *Autobiography*, cited in Jackson, *Material Concerns*, 182.

51 See for instance 'The Catalog of Happiness in Nations,' in Ruut Veenhoven, *The World Database of Happiness* (Rotterdam: Erasmus University, 1999).
52 Robert Goodland and Herman Daly,' Why Northern Income Growth Is Not the Solution to Southern Poverty,' Environment Department Divisional Working Paper No. 1993-43 (Washington, DC: World Bank, 1993).
53 F. Schmidt-Bleek, 'MIPS – A Universal Ecological Measure?,' *Fresenuis Environmental Bulletin* 2 (January 1996): 306–11.
54 J.K. Galbraith, *The New Industrial State* (Scarborough, ON: New American Library, 1978).
55 Jacques Ellul, *The Political Illusion*, trans. by Konrad Kellen (New York: Knopf, 1967).
56 W.H. Vanderburg, ed., *Perspectives on Our Age: Jacques Ellul Speaks on His Life and Work* (Toronto: CBC Enterprises/Seabury Press, 1997).
57 Jackson, *Material Concerns*, ch. 10.
58 Cited in ibid., 192.
59 von Weizsäcker, Lovins, and Lovins, *Factor Four: Doubling Wealth – Halving Resource Use*.

**9: Energy**

1 With some notable exceptions; see for instance Henry R. Linden, 'Energy and Industrial Ecology,' in *The Greening of Industrial Ecosystems*, ed. Braden R. Allenby and Deanna J. Richards, 83–92 (Washington, DC: National Academy Press, 1994); David Hawden and Peter Pearson, 'Input–Output Simulations of Energy, Environment, Economy Interactions in the U.K.,' *Energy Economics* 17/1 (1995): 73–86; K.B. Misra, ed., *Clean Production: Environmental and Economic Perspectives* (New York: Springer-Verlag, 1996), Section 4; Mark Ross, 'Efficient Energy Use in Manufacturing,' *Proceedings of the National Academy of Sciences* 89 (February 1992): 827–31.
2 This was certainly the case for the province of Ontario in Canada. See for instance Neil B. Freeman, *The Politics of Power: Ontario Hydro and Its Government* (Toronto: University of Toronto Press, 1996).
3 Charles Perrow, *Normal Accidents: Living with High-Risk Technologies* (New York: Basic, 1984).
4 E.F. Schumacher, *Schumacher on Energy: Speeches and Writings of E. F. Schumacher* (London: Cape, 1982).
5 M. King Hubbert, 'The Energy Resources of the Earth,' *Scientific American* 224 (September 1971), p. 70.
6 Workshop on Alternative Energy Strategies, *Energy: Global Prospects, 1985–2000* (New York: McGraw-Hill, 1977).

7 Conservation Commission of the World Energy Conference, *World Energy: Looking Ahead to 2020* (Guildford: IPC Science and Technology Press, 1978); Conservation Commission of the World Energy Conference, *Energy 2000–2020: World Prospects and Regional Stresses* (London: Graham and Trotman, 1983).
8 International Institute for Applied Systems Analysis, *Energy in a Finite World: Report by the Energy Systems Program Group* (Cambridge, MA: Ballinger, 1981).
9 Walter C. Patterson, *The Energy Alternative: Changing the Way the World Works* (London: Boxtree, 1990).
10 Ford Foundation Energy Policy Project, *A Time To Choose: America's Energy Future. Final Report* (Cambridge, MA: Ballinger, 1974).
11 Robert Hill, Phil O'Keefe, and Colin Snape, *The Future of Energy Use* (London: Earthscan, 1995); Gerald Leach, 'The Energy Transition,' *Energy Policy*, February 1992, pp. 116–23.
12 Amory B. Lovins, *World Energy Strategies* (Boulder, CO: Earth Resources Research, 1973); rev. ed. published as *World Energy Strategies* (Cambridge, MA: Ballinger, 1975); Amory B. Lovins, 'Energy Strategy: The Road Not Taken?,' *Foreign Affairs* 55 (1976): 65–96; Amory B. Lovins, *Soft Energy Paths: Toward a Durable Peace* (Cambridge, MA: Ballinger, 1975).
13 US Atomic Energy Commission, *Nuclear Power Growth, 1974–2000*, WASH-1139 (74) (Washington, DC: Government Printing Office, 1974).
14 Peter Chapman, *Fuel's Paradise: Energy Options for Britain* (Harmondsworth: Penguin, 1975).
15 Gerald Leach, *Energy and Food Production* (Guildford: IPC Science and Technology Press, 1976).
16 Gerald Leach, *Energy and Growth*, quoted in Walter C. Patterson, *The Energy Alternative* (London: Boxtree, 1990), 81.
17 Gerald Leach, Christopher Lewis, Frederic Romig, Ariana Van Buren, and Gerald Foley, *A Low Energy Strategy for the United Kingdom* (London: International Institute for Environment and Development, 1979).
18 José Goldemberg, Thomas Johansson, Amulya Reddy, and Robert Williams, *Energy For a Sustainable World* (New York: Wiley, 1988). See also José Goldemberg, *Energy, Environment and Development* (London: Earthscan, 1996).
19 Goldemberg et al., *Energy For a Sustainable World.*
20 This is implicit in all of Lovins's work. See for instance Amory B. Lovins, *World Energy Strategy* (London: Ballinger, 1975).
21 Amory B. Lovins, 'The Negawatt Revolution: New Developments in Energy Efficiency,' in *Energy and the Environment*, ed. B. Abeles, A.J. Jacobson, and P. Sheng, 2–34 (New Brunswick, NJ: World Scientific, 1992).

22 Ernst von Weizsacker, Amory B. Lovins, and L. Hunter Lovins, *Factor Four: Doubling Wealth – Halving Resource Use: The New Report to the Club of Rome* (London: Earthscan, 1997), 161.
23 Ibid.
24 Ibid., 162.
25 Howard S. Geller, 'Implementing Electricity Conservation Programs: Progress towards Least-Cost Energy Services among US Utilities,' in *Electricity: Efficient End-Use and New Generation Technologies, and Their Planning Implications*, ed. Thomas B. Johansson, Birgit Bodlund, and Robert H. Williams, 741–58 (Lund: Lund University Press, 1989).
26 von Weizsäcker, Lovins, and Lovins, *Factor Four: Doubling Wealth – Halving Resource Use*, 166ff.
27 Quoted in F. Kreith, 'Integrated Resource Planning,' *Journal of Energy Resources Technology* 115 (June 1993): 80.
28 M. Schweitzer, E. Hirst, and E. Yourstone, 'The Process of Integrated Resource Planning for Electric Utilities,' in *ACEEE Summer Study on Energy Efficiency in Buildings*, 219–26 (Pacific Grove, CA: ACEEE, 1990).
29 von Weizsäcker, Lovins, and Lovins, *Factor Four: Doubling Wealth – Halving Resource Use*, 162.
30 Lovins, *Soft Energy Paths.*
31 Amory B. Lovins, and L. Hunter Lovins, 'Reinventing the Wheels,' *Atlantic Monthly* 271(January 1995), pp. 75–81.
32 von Weizsäcker, Lovins, and Lovins, *Factor Four: Doubling Wealth – Halving Resource Use*, 11-12.
33 Ibid., ch. 1.
34 Information on the Toronto Healthy House can be found on the CMHC Web site at http://www.cmhc-chl.gc.ca/HealthyHousing/Toronto/ index.html.
35 Goldemberg et al., *Energy For a Sustainable World.*
36 Steven Vogel, *Cat's Paws and Catapults: Mechanical Worlds of Nature and People* (New York: Norton, 1998).

**10: Work**

1 Theo Colborn, Dianne Dumonowski, and John Peterson Myers, *Our Stolen Future: Are We Threatening Our Fertility, Intelligence and Survival? A Scientific Detective Story* (New York: Dutton, 1996). See also Deborah Cadbury, *The Feminization of Nature: Our Future at Risk* (London: Hamish Hamilton, 1997).
2 Cadbury, *The Feminization of Nature.*
3 Robert Karasek, Requirements for an Alternative Economic Future,' *Scandi-*

*navian Journal of Work, Environment and Health*, vol. 23, Supplement 4 (1997), p. 60.

4 Adam Smith, *The Wealth of Nations*, quoted in Karasek, 'Labor Participation and Work Quality Policy,' 64.

5 Adam Smith, *The Wealth of Nations* (New York: Modern Library, 1937).

6 Siegfried Giedion, *Mechanization Takes Command: A Contribution to Anonymous History* (New York: Oxford University Press, 1948). See especially the section on the Gilbreths. Frederick W. Taylor, *The Principles of Scientific Management* (New York: Norton, 1967).

7 Max Weber, *Economy and Society*, ed. by Guenther Toth and Claus Wittich (Berkeley: University of California Press, 1978). See also Ed Andrew, *Closing the Iron Cage: The Scientific Management of Work and Leisure* (Montreal: Black Rose, 1981).

8 Robert Karasek and Tores Theorell, *Healthy Work: Stress, Productivity and the Reconstruction of Working Life* (New York: Basic, 1990).

9 Ibid., ch. 2.

10 F.E. Emery, E.L. Hilgendorf, and B.L. Irving, *The Psychological Dynamics of Smoking* (London: Tobacco Research Council, 1969); R.L. Ackoff and F.E. Emery, *On Purposeful Systems* (Chicago: Aldine-Atherton, 1972).

11 For this and the following discussion, I am indebted to the literature review carried out by my student Peter Benda in 'Extending the Preventive Engineering Paradigm to the Analysis and Design of Discrete Product Manufacturing Systems: The Role of Functional Integration in the Production of Social and Psychological Outputs,' MASc thesis, Department of Mechanical and Industrial Engineering, University of Toronto, 1997.

12 Micheil Kompier, 'Job Design and Well-Being,' in *Handbook of Work and Health Psychology*, ed. M.J. Schabracq, J.A.M. Winnubst, and C.L. Cooper, 352–3 (Chichester: Wiley and Sons, 1996); Barbara A. Israel, 'Occupational Stress, Safety, and Health: Conceptual Framework and Principles for Effective Prevention Interventions,' *Journal of Occupational Health Psychology* 1/3 (1996): 261–86.

13 Johannes Siegrist, 'Adverse Health Effects of High-Effort/Low Reward Conditions,' *Journal of Occupational Health Psychology* 1/1 (1996): 261–86.

14 Ibid., 27.

15 Toby D. Wall, Paul Jackson, Sean Mullarkey, and Sharon Parker, 'The Demands-Control Model of Job Strain: A More Specific Test,' *Journal of Occupational and Organizational Psychology* 69 (1996): 155; Carles Muntaner and Patricia J. O'Campo, 'A Critical Appraisal of the Demand/Control Model of the Psychosocial Work Environment: Epistemological, Behavioral and Class Considerations,' *Social Science and Medicine* 36/11 (1993): 1511.

16 These dimensions of demand and control are taken from Paul R. Jackson, Toby D. Wall, Robin Martin, and Keith Davids, 'New Measures of Job Control, Cognitive Demand, and Production Responsibility,' *Journal of Applied Psychology* 78/5 (1993): 753–62, and Toby D. Wall, Paul R. Jackson, and Sean Mullarkey, 'Further Evidence on Some New Measures of Job Control, Cognitive Demand and Production Responsibility,' *Journal of Organizational Behavior* 16/5 (1995): 431–55.

17 Wall et al., 'The Demand-Control Model of Job Strain'; Muntaner and O'Campo, 'A Critical Appraisal of the Demand/Control Model of the Psychosocial Work Environment.'

18 Muntaner and O'Campo, 'A Critical Appraisal of the Demand/Control Model of the Psychosocial Work Environment.'

19 Norman Halpern, 'Sociotechnical Systems Design: The Shell Sarnia Experienc,' in *Quality of Working Life: Contemporary Cases*, ed. J. Cunningham and T.H. White, 31–75 (Ottawa: Labour Canada, 1984). Also see Karasek and Theorell, *Healthy Work*, 234–5.

20 Gillian Symon, 'Human-Centred Computer Integrated Manufacturing,' *Computer Integrated Manufacturing Systems* 3/4 (1990): 224; Peter Brödner, *The Shape of Future Technology: The Anthropocentric Alternative* (New York: Springer-Verlag, 1990).

21 See for example Philip J. Kirby, *A Socio-Technical Approach to Reducing Manufacturing Cycle-Time: A Case Study*, Technical Report MM85-721 (Dearborn, MI: Society of Manufacturing Engineers, 1985); Karasek and Theorell, *Healthy Work*, 253–9; Mark van Bijsterveld and Fred Huijgen, 'Modern Sociotechnology: Exploring the Frontiers,' in *The Symbiosis of Work and Technology*, ed. J. Benders, J. de Haan, and D. Bennet, 25–40 (New York: Wiley and Sons, 1994); Agnes T. Haak, *Dutch Sociotechnical Design in Practice: An Empirical Study of the Whole Task Group* (Assen: Van Gorcum, 1994); and William A. Pasmore and John J. Sherwood, eds., *Sociotechnical Systems: A Sourcebook* (La Jolla, CA: University Associates, 1978).

22 John E. Gibson, 'Taylorism and Professional Education,' in *Manufacturing Systems: Foundations of World-Class Practice*, ed. Joseph H. Heim and W. Dale Compton, 149 (Washington, DC: National Academy Press, 1992).

23 Karasek and Theorell, *Healthy Work*, 24.

24 Benda, 'Extending the Preventive Engineering Paradigm,' 19–20.

25 W.H. Vanderburg, *The Growth of Minds and Cultures: A Unified Theory of the Structure of Human Experience* (Toronto: University of Toronto Press, 1985).

26 Pravin Varaiya, 'Productivity in Manufacturing and the Division of Mental Labor,' in *Knowledge and Industrial Organization*, No. 2, ed. Ake E. Anders-

son, David F. Batten, and Charlie Karlsson, 21 (New York: Springer-Verlag, 1989).

27 Shoshana Zuboff, *In the Age of the Smart Machine: The Future of Work and Power* (New York: Basic, 1988).

28 Hubert L. Dreyfus and Stuart E. Dreyfus, *Mind Over Machine: The Power of Human Intuition and Expertise in the Era of the Computer* (New York: The Free Press, 1986).

29 Vanderburg, *The Growth of Minds and Cultures.*

30 Hubert L. Dreyfus, *What Computers Still Can't Do: A Critique of Artificial Reason* (Cambridge, MA: MIT Press, 1992); Roger Penrose, *The Emperor's New Mind: Concerning Computers, Minds and the Laws of Physics* (London: Vintage, 1990).

31 Dreyfus and Dreyfus, *Mind Over Machine.*

32 Hubert L. Dreyfus and Stuart E. Dreyfus, 'What Is Morality? A Phenomenological Account of the Development of Ethical Expertise,' in *Universalism vs. Communitarianism,* ed. B. Rasmussen, 237–64 (Cambridge, MA: MIT Press, 1990).

33 Dreyfus and Dreyfus, *Mind Over Machine,* 32.

34 Thomas S. Kuhn, *The Structure of Scientific Revolutions,* 2d ed. (Chicago: University of Chicago Press, 1970). See 'Postscript,' 182, for a discussion of Kuhn's idea of the disciplinary matrix.

35 Michael Polanyi, *Personal Knowledge: Towards a Post-Critical Philosophy* (Chicago: University of Chicago Press, 1962).

36 Silvano Arietti, *Creativity: The Magic Synthesis* (New York: Basic Books/ Harper Colophon, 1976), especially 268–70 for a discussion of Poincaré.

37 W.H. Vanderburg, 'Living in Technique' (unpublished manuscript).

38 Brödner, *The Shape of Future Technology.*

39 John Kenneth Galbraith, in *The New Industrial State* (Boston: Houghton-Mifflin, 1978), is perhaps the only economist who has adequately recognized this division. His technostructure (collective brain) is separated from the collective hand, fundamentally changing the structure of the corporation and the way it interacts with its contexts. However, he failed to recognize the full depth of this phenomenon, a point I discuss in another work: Vanderburg, 'Living in Technique.'

40 Brödner, *The Shape of Future Technology,* 82–3.

41 Ibid., 83.

42 Ricardo Semler, *Maverick: The Success Story behind the World's Most Unusual Workplace* (New York: Warner, 1993).

43 Saul Rubenstein, Michael Bennett, and Thomas Kochan, 'The Saturn Partnership: Co-Management and the Reinvention of the Local Union,' in

*Employee Representation: Alternatives and Future Directions*, ed. Bruce E. Kaufman and Morris M. Kleiner, 181–92 (Madison, WI: Industrial Relations Research Association, 1993).

44 Varaiya, 'Productivity in Manufacturing and the Division of Mental Labor,' 21–2.

45 AnnaLee Saxenian, *Regional Advantage: Culture and Competition in Silicon Valley and Along Route 128* (Cambridge, MA: Harvard University Press, 1994).

46 Joseph H. Boyett, and Henry P. Conn, *Workplace 2000: The Revolution Reshaping American Business* (New York: Dutton, 1991), 183–4.

47 Ibid., 197–8.

48 Ibid., 82–3.

49 Edward E. Lawler III, Susan Albers Mohrman, and Gerald E. Ledford, Jr, *Employee Involvement and Total Quality Management: Practices and Results in Fortune 1000 Companies* (San Francisco: Jossey-Bass, 1992), 57.

50 Gavriel Salvendy and Waldemar Karwowski, *Design of Work and Development of Personnel in Advanced Manufacturing* (New York: Wiley and Sons, 1994).

51 Lawler III, Albers Mohrman, and Ledford, Jr, *Employee Involvement and Total Quality Management;* James R. Lincoln, *Culture, Control and Commitment: A Study of Work Organization and Work Attitudes in the United States and Japan* (Cambridge: Cambridge University Press, 1990).

52 Lotte Bailyn, 'Changing the Conditions of Work: Responding to Increasing Work Force Diversity and New Family Patterns,' in *Transforming Organizations*, ed. Thomas A. Kochan and Michael Useem, 189 (New York: Oxford University Press, 1992); Christian Berggren, *The Volvo Experience: Alternatives to Lean Production in the Swedish Auto Industry* (Houndsmills, Basingstoke: Macmillan, 1993). See especially Mark Van Bijsterveld and Fred Huijgen, 'Modern Sociotechnology: Exploring the Frontiers,' in *The Symbiosis of Work and Technology*, ed. Jos Benders, Job de Haan, and David Bennett, 138 (London: Taylor and Francis, 1995); Hans Van Beinum, 'The Kaleidoscope of Workplace Reform,' in *Constructing the New Industrial Society*, ed. Frieder Naschold (Assen: Van Gorcum, Stockholm: Swedish Center for Working Life, 1993), 169f.

53 James P. Womack, Daniel T. Jones, and Daniel Roos, *The Machine That Changed the World* (New York: Macmillan, 1990).

54 The following discussion owes much to Eric Trist and Hugh Murray, 'Historical Overview: The Foundation and Development of the Tavistock Institute,' and Eric Trist, 'Introduction to Volume II,' in *The Social Engagement of Social Science: A Tavistock Anthology*, ed. Eric Trist and Hugh Murray, vol. 2:

*The Socio-Technical Perspective*, 1–60 (Philadelphia: University of Pennsylvania Press, 1993).

55 Fred Emery and Einar Thorsrud, 'The Norskhydro Fertilizer Plant,' in *The Social Engagement of Social Science: A Tavistock Anthology*, ed. Trist and Murray, vol. 2: *The Socio-Technical Perspective*, 492–507; Fred Emery and Einar Thorsrud, *Form and Content in Industrial Democracy: Some Experiences from Norway and Other European Countries* (London: Tavistock, 1969).

56 Ulf Karlsson, 'The Swedish Sociotechnical Approach: Strengths and Weaknesses,' in *The Symbiosis of Work and Technology*, ed. Benders, de Haan, and Bennett, 47–58.

57 Emery and Thorsrud, *Form and Content in Industrial Democracy*; Haak, *Dutch Sociotechnical Design in Practice*.

58 E.L. Trist, G.W. Higgin, H. Murray, and A.B. Pollock, *Organizational Choice: Capabilities of Groups at the Coal Face under Changing Technologies: The Loss, Rediscovery and Transformation of a Work Tradition* (London: Tavistock, 1963).

59 Ibid.

60 Ibid.

61 Fred E. Emery, 'Characteristics of Socio-Technical Systems,' in *The Social Engagement of Social Science*, ed. Trist and Murray, vol. 2: *The Socio-Technical Perspective*, 157–86.

62 Ibid.

63 Halpern, 'Sociotechnical Systems Design.'

64 Bailyn, Changing the Conditions of Work,' 189.

65 Van Beinum, 'The Kaleidoscope of Workplace Reform.'

66 Karasek and Theorell, *Healthy Work*.

67 Tomas Engstrom, Jan A. Johansson, Dan Jonsson, and Lars Medbo, 'Empirical Evaluation of The Reformed Assembly Work at the Volvo Uddevalla Plant: Psychosocial Effects and Production Responsibility,' *International Journal of Industrial Ergonomics* 16/5 (1995): 293–308.

68 Ibid., 296.

69 Tomas Engstrom and Lars Medbo, 'Production Systems Design – A Brief Summary of Some Swedish Design Efforts,' in *Enriching Production: Perspectives on Volvo's Uddevalla Plant as an Alternative to Lean Production*, ed. Åke Sandberg, 65 (Aldershot: Avebury, 1995); Tomas Engstrom, Lars Medbo, and Dan Jonsson, 'An Assembly-Oriented Product Description as a Precondition for Efficient Manufacturing with Long Cycle Time Work,' in *Productivity and Quality Management Frontiers – IV*, ed. David J. Sumanth, Johnson A. Edosomwan, Robert Poupart, and D. Scott Sink, 460–1 (Norcross, GA: Industrial Engineering and Management Press, 1993).

70 Christian Berggren, 'The Fate of Branch-Plants – Performance Versus

Power,' in *Enriching Production: Perspectives on Volvo's Uddevalla Plant as an Alternative to Lean Production*, ed. Sandberg, 107–8.

71 Karlsson, 'The Swedish Sociotechnical Approach: Strengths and Weaknesses,' in *Technological Change and Co-determination in Sweden*, ed. Ake Sandberg, 32–46 (Philadelphia: Temple University Press, 1992).

72 Benders, de Haan, and Bennett, 'Symbiotic Approaches: Contents and Issues,' 4.

73 'Edges Fray on Volvo's Brave New Humanistic World,' *New York Times*, 7 July 1991.

74 Berggren, *The Volvo Experience*, 164–5.

75 Robert D. Putnam, *Making Democracy at Work* (Princeton, NJ: Princeton University Press, 1993).

76 Paul R. Jackson and Robin Martin, 'Impact of Just-In-Time on Job Content, Employee Attitudes and Well-Being: A Longitudinal Study,' *Ergonomics* 39/1 (1996): 2; Mike Parker and Jane Slaughter, 'Unions and Management by Stress,' in *Lean Work: Empowerment and Exploitation in the Global Auto Industry*, ed. Steve Babson, 41–53 (Detroit: Wayne State University Press, 1995).

77 Paul Adler, '"Democratic Taylorism": The Toyota Production System at NUMMI,' in *Lean Work: Empowerment and Exploitation in the Global Auto Industry*, ed. Babson, 207–19.

78 Charles Perrow, *Normal Accidents: Living with High-Risk Technologies* (New York: Basic, 1984).

79 For example, the 'killer software' developed by Cypress Semiconductor Corporation shuts down computer systems in the event of minor problems leading to delays (cited in Parker and Slaughter, 'Unions and Management by Stress,' 45).

80 Ibid., 44.

81 Adler, '"Democratic Taylorism".'

82 Ibid., 214.

83 Ernst U. von Weizsäcker, Amory Lovins, and L. Hunter Lovins, *Factor Four: Doubling Wealth – Halving Resource Use: The New Report to the Club of Rome* (London: Earthscan, 1997), 65.

84 Womack and Roos, *The Machine That Changed the World.*

**11: The Built Habitat**

1 Herbert Girardet, *The Gaia Atlas of Sustainable Cities: New Directions for Sustainable Urban Living* (New York: Anchor, 1993); George D. Moffet, *Critical Masses* (New York: Viking, 1994), 8–15.

2 Girardet, *The Gaia Atlas of Sustainable Cities*, and Moffet, *Critical Masses.*

3 Jacques Ellul, *The Meaning of the City*, trans. by Dennis Pardee (Grand Rapids, MI: Eerdmans, 1970).
4 Jane Jacobs, *The Death and Life of Great American Cities* (London: Pelican, 1961).
5 Christopher Alexander, *The Timeless Way of Building* (New York: Oxford University Press, 1979).
6 Nan Ellin, *Postmodern Urbanism* (Cambridge, MA: Blackwell, 1996).
7 W.H. Vanderburg, 'Living in Technique' (unpublished manuscript).
8 W.H. Vanderburg, *The Growth of Minds and Cultures: A Unified Theory of the Structure of Human Experience* (Toronto: University of Toronto Press, 1985).
9 David Reisman, *The Lonely Crowd* (New Haven, CT: Yale University Press, 1969); Robert N. Bellah, *Habits of the Heart: Individualism and the Commitment in American Life* (New York: Harper and Row, 1986).
10 In my estimation, the best description of the existential condition of the individual in a mass society is found in Jacques Ellul, *Propaganda: The Formation of Men's Attitudes*, trans. by Konrad Kellen (New York : Knopf, 1969), particularly chs. 2, 3, and 4.
11 Reisman, *The Lonely Crowd;* Bellah et al., *Habits of the Heart.*
12 See for example Peter Calthorpe, 'The Post-Suburban Metropolis,' *Whole Earth Review* 73 (Winter 1991), 44–51.
13 For this and following discussions I am indebted to my research assistant Ken Jalowica, who prepared an overview of this research. He suggests the following entry points into this literature: A. Baum and Y. Epstein, eds., *Human Response to Crowding* (Hillsdale, NJ: Lawrence Erlbaum Associates, 1978); A. Booth, *Urban Crowding and Its Consequences* (New York: Praeger, 1976); M. Baldassare, *Residential Crowding in Urban America* (Berkeley: University of California Press, 1979); G.W. Evans, ed., *Environmental Stress* (New York: Cambridge University Press, 1982); D. Stokols, ed., *Perspectives on Environment and Behavior* (New York: Plenum, 1977).
14 G. Simmel, 'The Metropolis and Mental Life,' in *Cities and Society*, ed. P.K. Hatt and A.J. Reiss, Jr, 32–57 (New York: The Free Press, 1957).
15 C. Fischer, 'On Urban Alienations and Anomie: Powerlessness and Social Isolation,' *American Sociological Review* 38 (1973): 311.
16 J.G. Miller, 'Adjusting to Overloads of Information,' in *Disorders of Communication*, ed. D. McKrioch and E.A. Weinstein, 92–109 (Baltimore: Williams and Wilkins, 1964).
17 S. Milgram, 'The Experience of Living in Cities,' in *Environmental Psychology*, ed. H.M. Proshansky, William H. Ittelson, and Leanne G. Rivlin, 494–508 (New York: Holt, Rinehart and Winston, 1976).
18 C. Levy-Leboyer, *Psychology and Environment*, trans. by David Canter (Beverly Hills: Sage, 1982), 115.

19 S. Saegert, 'Stress-Inducing and Reducing Quality of Environments,' in *Environmental Psychology*, ed. Proshansky et al., 218–23.
20 J.D. Brown and E.C. Poulton, 'Measuring the Spare "Mental Capacity" of Car Drivers by a Subsidiary Task,' *Ergonomics* 4 (1961): 35–40.
21 W.R. Miller and M.E. Seligman, 'Depression and Learned Helplessness in Man,' *Journal of Abnormal Social Psychology* 84 (1973): 228–38.
22 D.C Glass and J.E. Singer, 'Behavioral and Physiological Effects of Uncontrollable Environmental Events,' in *Perspectives in Environment and Behavior*, ed. D. Stokol (New York: Plenum, 1977).
23 Levy-Leboyer, *Psychology and Environment*, 120.
24 John R. Aiello, Donna E. Thompson, and Andrew Baum, 'Children, Crowding and Control: Effects of Environmental Stress on Social Behavior,' in *Habitats for Children: The Impacts of Density*, ed. J. Wohlwill, 97–124 (Hillsdale, NJ: Erlbaum, 1985).
25 Ibid., 100.
26 W. Rohe and A.H. Patterson, 'The Effects of Varied Levels of Resources and Density on Behavior in a Day Care Centre,' in *Man-Environment Interaction: Evaluation and Application*, D.H. Carlson, Part 3, vol. 3, Part 3: 161–71 (Stroudsberg, PA: Dowden, Hutchinson and Ross, 1974); John R. Aiello, G. Forosia, and Donna E. Thompson, 'Physiological, Social and Behavioral Consequences of Crowding on Children and Adolescents,' *Child Development* 50 (1979): 195–202.
27 Aiello, Thompson, and Baum, 'Children, Crowding and Control,' 103.
28 Jacques Ellul, *Propaganda: The Formation of Man's Attitudes*, trans. by Konrad Kellen (New York: Knopf, 1965), especially chs. 2, 3, and 4.
29 Daniel J. Boorstin, *The Americans: The Democratic Experience* (New York: Random House, 1973).
30 Pat Gelb, 'High-Rise Impact on City and Neighbourhood Livability,' in *Human Responses to Tall Buildings*, ed. Donald Conway, 131 (Stroudsberg, PA: Dowden, Hutchinson and Ross, 1977).
31 Newman's work is discussed in A. Baum and S. Valins, *Architecture and Social Behaviour* (New York: Wiley and Sons, 1977).
32 David Pearson, *Earth to Spirit: In Search of Natural Architecture* (San Francisco: Chronicle, 1995).
33 For entry points into this research, see Levy-Leboyer, *Psychology and Environment*; N. Heimstra and L. McFarling, *Environmental Psychology* (Monterey, CA: Brooks/Cole, 1978); Karl D. Kryter, *The Handbook of Hearing and the Effects of Noise* (San Diego: Academic, 1994), especially ch. 8–10.
34 Levy-Leboyer, *Psychology and Environment*, 109.

35 David C. Glass and Jerome E. Singer, *Urban Stress: Experiments on Noise and Social Stressors* (New York: Academic, 1972).
36 Lawrence Ward and Peter Swedfeld, 'Human Responses to Highway Noise,' *Environmental Research* 6 (1976): 306–26.
37 S. Cohen, David C. Glass, and Jerome E. Singer, 'Apartment Noise, Auditory Discrimination and Reading Ability in Children,' *Journal of Experimental Social Psychology* 9 (1973): 407–22.
38 S. Cohen, Physiological, Motivational and Cognitive Effects of Aircraft Noise on Children: Moving from the Laboratory to the Field,' *American Psychologist* 35 (1980): 231–43; S. Cohen, 'Aircraft Noise and Children: Longitudinal Cross-Sectional Evidence on Adaptation to Noise and the Effectiveness of Noise Abatement,' *Journal of Personality and Social Psychology* 40 (1981): 331–45.
39 A. Sibony, 'Le bruit à l'école' (unpublished manuscript, 1979).
40 K.E. Mathews, Jr, and L.K. Canon, 'Environmental Noise Level as a Determinant of Helping Behavior,' *Journal of Personality and Social Psychology* 32 (1975): 571–7.
41 D. Appleyard and M. Lintell, 'The Environmental Quality of City Streets: The Residents' Viewpoint,' *Journal of the American Institute of Planners* 38 (1972): 84–101.
42 B.L. Welch, 'Extra-Auditory Health Effects of Industrial Noise in Survey of Foreign Literature,' Wright-Patterson Aerospace Medical Research Laboratory Air Force Systems Command, June 1979.
43 Sheldon Cohen and Neil Weinstein, 'Non-Auditory Effects of Noise on Behavior and Health,' in *Environmental Stress*, ed. Gary W. Evans, 64 (Cambridge: Cambridge University Press, 1982).
44 Gary W. Evans and Stephen V. Jacobs, 'Air Pollution and Human Behavior,' in *Environmental Stress*, ed. Evans, 108.
45 Jacobs, *The Death and Life of Great American Cities.*
46 Sir Ebenezer Howard, *Garden Cities of To-morrow* (London: Faber and Faber, 1946).
47 Sir Patrick Geddes, *Cities in Evolution: An Introduction to the Town Planning Movement and to the Study of Civics* (London: Benn, 1968).
48 Ellul, *The Meaning of the City.*
49 Le Corbusier, *The City of Tomorrow and Its Planning*, 3d ed., trans. by Frederick Etchells (London: The Architectural Press, 1971).
50 Alexander, *The Timeless Way of Building.*
51 Donald Schön, *The Reflective Practitioner: How Professionals Think in Action* (New York: Basic, 1983).

52 Mayer Hillman, John Adams, and John Whitelegg, *One False Move: A Study of Children's Independent Mobility* (London: Policy Studies Institute, 1990).

53 Reisman, *The Lonely Crowd.*

54 Stephen L. Talbott, *The Future Does Not Compute: Transcending the Machines in Our Midst* (Sebastopol, CA: O'Reilly and Associates, 1995).

55 See for example Calthorpe, 'The Post-Suburban Metropolis'; Peter Calthorpe, *The Next American Metropolis: Ecology, Community and the American Dream* (New York: Princeton Architectural Press, 1993); Alex Krieger and William Lennertz, eds., *Andres Duany and Elizabeth Plater-Zyberk: Towns and Town-Making Principles* (New York: Rizzoli, 1991); and Philip Langdon, 'A Good Place to Live,' *Atlantic Monthly* 261 (March 1988): 39–60.

56 Christopher Alexander, 'The City as a Mechanism for Sustaining Human Contact,' in *People and Buildings*, ed. R. Gutman, 406–34 (New York: Basic, 1972); Christopher Alexander, Murray Silverstein, Shlomo Angel, Sara Ishikawa, and Deny Abrams, *The Oregon Experiment* (New York: Oxford University Press, 1975); Christopher Alexander, *A Pattern Language: Towns, Buildings, Construction* (New York: Oxford University Press, 1979); Christopher Alexander, *The Timeless Way of Building* (New York: Oxford University Press, 1979); Donald Alexander and Ray Tonalty, *Urban Policy for Sustainable Development: Taking a Wide-Angle View* (Winnipeg: Institute of Urban Studies, University of Winnipeg, 1994); Donald Appleyard, *Livable Streets* (Berkeley: University of California Press, 1981); Dolores Hayden, *Redesigning the American Dream: The Future of Housing, Work and Family Life* (New York: Norton, 1984); Allan B. Jacobs and Donald Appleyard, 'Toward an Urban Design Manifesto,' in *The City Reader*, ed. Richard T. LeGates and Frederic Stout, 164–75 (New York: Routledge, 1996); Allan B. Jacobs, *Great Streets* (Cambridge, MA: MIT Press, 1993); James Howard Kunstler, *Home from Nowhere* (New York: Touchstone, 1966); James Howard Kunstler, *The Geography of Nowhere* (New York: Touchstone, 1993); Kevin Lynch, *Growing Up in Cities: Studies of the Spatial Environment of Adolescence* (Cambridge, MA: MIT Press and UNESCO, 1977); Kevin Lynch, *A Theory of Good City Form* (Cambridge, MA: MIT Press, 1981); Kevin Lynch, *Wasting Away* (San Francisco: Sierra, 1990); William Whyte, *City: Rediscovering the Center* (New York: Doubleday, 1988).

57 Vanderburg, 'Living in Technique.'

58 Rodney White and Joseph Whitney, 'Cities and the Environment: An Overview,' in *Sustainable Cities: Urbanization and the Environment in International Perspective*, ed. Richard Stren, Rodney White, and Joseph Whitney, 8–12 (Boulder, CO: Westview, 1992).

59 For the discussion in this section, I am indebted to my research associate Namir Khan for the preliminary report he drafted on healthy cities.
60 White and Whitney, 'Cities and the Environment,' 11.
61 Ibid., 13.
62 Ibid., 15.
63 Girardet, *The Gaia Atlas of Sustainable Cities*, 16.
64 White and Whitney, 'Cities and the Environment,' 24.
65 Thomas Outerbridge, 'The Big Backyard: Composting Strategies in New York City,' *The Ecologist* 24/3 (May/June, 1994): 106–9. In fact, Fresh Kills was closed in 1996, almost a decade before its planned closing.
66 Karl Hammer, *Personal Communication*, quoted in Outerbridge, 'The Big Backyard,'108.
67 Outerbridge, 'The Big Backyard, 108.
68 Wilfred Owen, 'Transportation and the Environment: Lessons from the Global Laboratory,' in *The City as a Human Environment*, ed. D.G. Levine and A.C. Upton, 73–81 (Westport, CT: Praeger, 1994); L.B. Lave, and E. Seskin, *Air Pollution and Human Health* (Baltimore, MD: Johns Hopkins University Press, 1977); Steve J. Nadis and James J. MacKenzie, *Car Trouble* (Boston: Beacon, 1993).
69 Ibid.
70 H. Holzapfel, 'Violence and the Car' (unpublished manuscript), University of Kassel, cited in John Whitelegg, *Transport for a Sustainable Future: The Case for Europe* (New York: Halsted, 1993), 4.
71 Donella H. Meadows, Dennis L. Meadows, and Jørgen Randers, *Limits to Growth*, cited in Whitelegg, *Transport for a Sustainable Future*, 4.
72 T. Hagerstrand, 'Space, Time and the Human Condition,' in *Dynamic Allocation of Urban Space*, ed. A. Karlquist, L. Lundqvist, and F. Snickars, 3–14 (Lexington, MA: Saxon House, 1975).
73 C. Marchetti, *Building Bridges and Tunnels: The Effect on the Evolution of Traffic*, Publication SR-88-01 (Vienna: International Institute for Applied Systems Analysis, 1988).
74 Michael Ende, *Momo* (Harmondsworth: Penguin, 1984), cited in Whitelegg, *Transport for a Sustainable Future*, 80–3.
75 Ivan Illich, *Energy and Equity* (London: Boyers, 1974).
76 For the effects of the car on human health, children, the environment, and buildings, see A. Watson, *Air Pollution, the Automobile and Public Health* (Washington, DC: National Academy Press, 1988); F. Godlee, 'Transport: A Public Health Issue,' *British Medical Journal* 304 (1992): 1539–43; Mayer Hillman, John Adams, John Whitelegg, *One False Move: A Study of Children's Independent Mobility* (London: Policy Studies Institute, 1991); R. Davis,

*Death on the Streets: Cars and the Mythology of Road Safety* (New York: Leading Edge, 1993); C.S. Weinstein and T.G. David, eds., *Spaces for Children: The Built Environment and Child Development* (New York: Plenum, 1987); Sanford Gaster, 'Urban Children's Access to Their Neighbourhoods, Changes Over Three Generations,' *Environment and Behavior* 23/1 (1991): 70–85; P. Nieuwenhuis and P. Wells, *Motor Vehicles in the Environment: Principles and Practice* (Chichester: Wiley and Sons, 1994), ch. 1; Wolfgang Zuckerman, *The End of the World: The World Car Crisis and How We Can Solve It* (Post Mills, VT: Chelsea Green, 1991); P. Newman and J. Kenworthy, *Cities and Automobile Dependence* (Aldershot: Gower, 1989); Whitelegg, *Transport for a Sustainable Future*; Nadis and MacKenzie, *Car Trouble* (Boston: Beacon Press, 1993).

77 Graham Haughton and Colin Hunter, *Sustainable Cities* (Bristol, PA: Jessica Kingsley, 1994), 47. See also R. Walker, 'A Theory of Suburbanization: Capitalism and the Construction of Urban Space in the US,' in *Urbanization and Urban Planning in Capitalist Society*, ed. M. Dear and A.J. Scott, 131–52 (London: Metheun, 1991).

78 Rodney R. White, *Urban Environmental Management: Environmental Change and Urban Design* (Chichester: Wiley and Sons, 1994), 9.

79 Girardet, *The Gaia Atlas of Sustainable Cities*, 28.

80 Ibid.

81 Ibid.

82 Norman Myers, *Ultimate Security: The Environmental Basis of Political Stability* (New York: Norton, 1993), 18ff.

83 Sandra Postel, 'Forging a Sustainable Water Strategy,' in *State of the World 1996*, ed. L.R. Brown, 41–3 (New York: Norton, 1996).

84 Ibid.

85 D.M. Roodman and N. Lennssen, 'Our Buildings, Ourselves,' *Worldwatch* 7/6 (1994): 22.

86 Ibid.

87 Ibid., 23.

88 Ibid., 23–4.

89 Ibid., 25.

90 Ibid. Also see J. Byrne and D. Rich, eds., *Energy and Cities* (New Brunswick, NJ: Transaction, 1985), ch. 3.

91 Roodman and Lennssen, 'Our Buildings, Ourselves,' 24. See also M.C. Baechler, D.L. Hadley, T.J. Marseille, R.D. Stenner, M.R. Peterson, D.F. Naugle, and M.A. Berry, *Sick Building Syndrome: Sources, Health Effects, Mitigation* (Parkridge, NJ: Noyes, 1991).

92 Roodman and Lennssen, 'Our Buildings, Ourselves,' 24.

93 Ibid., 27.
94 Ibid.
95 W.H. Vanderburg, 'Political Imagination in a Technological Age,' in *Democratic Theory and Technological Society*, ed. Richard B. Day, Ronald Biener, and Joseph Masciulli, 3–35 (New York: M.E. Sharpe, 1988).
96 Toni Nelson, 'Urban Agriculture: Closing the Nutrient Loop,' *Worldwatch* 9/6 (1996): 10–17.
97 Ibid., 11.
98 Hank Dittmar, 'A Broader Context for Transportation Planning,' *APA Journal*, Winter 1995, pp. 7–13; Robert T. Dunphy, 'Transportation-Oriented Development: Making a Difference?,' *Urban Land* 54 (July 1995): 32–6.
99 Ibid.
100 For entry points into the literature on ecological design, see Sim Van Der Ryn and Stuart Cowan, *Ecological Design* (Washington, DC: Island Press, 1996); John Tillman Lyle, *Regenerative Design for Sustainable Development* (New York: Wiley and Sons, 1994); Ian McHarg, *Design with Nature* (New York: Natural History Press, 1969); David Wann, *Deep Design: Pathways to a Liveable Future* (Washington, DC: Island Press, 1996); Doug Aberley, ed., *Futures by Design: The Practice of Ecological Planning* (Gabriola Island, BC: New Society, 1994); and Nancy Jack Todd and John Todd, *From Eco-Cities to Living Machines: Principles of Ecological Design* (Berkeley, CA: North Atlantic Books, 1994).

**Postscript**

1 L.P. Bonneau and J.A. Corry, *Quest for the Optimum: Research in the Universities of Canada* (Toronto: Association of Colleges and Universities of Canada [AUCC], Fall 1972).
2 Ernest L. Boyer, *Scholarship Reconsidered: Priorities of the Professionate* (Princeton, NJ: The Carnegie Foundation for the Advancement of Teaching, Princeton University Press, 1990).
3 Amory B. Lovins, *Soft Energy Paths: Toward a Durable Peace* (New York: Harper Colophon, 1979), 23.

# Index